T0401731

Urban Biodiversity and Ecological Design for Sustainable Cities

Keitaro Ito
Editor

Urban Biodiversity and Ecological Design for Sustainable Cities

Editor
Keitaro Ito
Laboratory of Environmental Design,
Faculty of Civil Engineering
Kyushu Institute of Technology
Kitakyushu-city, Fukuoka, Japan

ISBN 978-4-431-56895-7 ISBN 978-4-431-56856-8 (eBook)
https://doi.org/10.1007/978-4-431-56856-8

This Springer imprint is published by the registered company Springer Japan KK, part of Springer Nature.
The registered company address is: Shiroyama Trust Tower, 4-3-1 Toranomon, Minato-ku, Tokyo 105-6005, Japan

Preface

When you know the fourfoil in all its seasons root and leaf and flower, by sight and scent and seed, then you may learn its true name, knowing its being: which is more than its use. What, after all, is the use of you? Or of myself? Is Gont Mountain useful, or the Open Sea? Ogion went on a half mile or so, and said at last, "To hear, one must be silent."
- A Wizard of Earthsea, Ursula K. Le Guin

What is urban biodiversity? Previous studies have discussed the concept from various points of view, for example, flora and fauna conservation and management in cities, sometimes in social and cultural contexts. Urban biodiversity became one of the important issues for city planning and management of the built environment after CBD (the Conventions on Biological Diversity) in Rio de Janeiro in 1992. The aims of the CBD are as follows (UN 1992; Müller and Werner 2009):

1. The conservation of biological diversity; maintaining the earth's life support systems and future options for human development.
2. The sustainable use of its components, that means livelihoods to people without jeopardizing future options.
3. The fair and equitable sharing of the benefit arising from the use of genetic resources.

In 2008, the conference "Urban Biodiversity and Design" was held in Erfurt, Germany. It was a very interesting and important conference to think about biodiversity issues and share ideas among ecologists, city planners, architects, and landscape architects. The approaches were:

- Investigation and evaluation of biodiversity in urban areas
- Cultural aspects of urban biodiversity
- Social aspects of biodiversity
- Urban biodiversity and climate change
- Design and future of urban biodiversity

Since I have learned essential issues from Ian McHarg's "Design with Nature (1969)", urban biodiversity has become an important component of sustainability in cities. These days, "green infrastructure", such as coastal forests, urban forests, rivers, and urban parks, not only can improve the biological integrity of an area but such infrastructure can improve social, cultural, and economic concerns. In Japan, we have been designing nature restoration projects for the past 20 years. As a rapid decrease has been observed in the amount of open or natural space over the last century, especially in urban areas, we believe that preserving these areas as wildlife habitats and spaces which the residents can use for their daily lives will be necessary for present and future city planning. In particular, urban green infrastructure can serve as a place for children to play outdoors. "Children's Play" is an important experience in learning about the structure of nature, whilst "Environmental Education" has been afforded much greater importance in primary and secondary school education in Japan since 2002. Therefore, it is an important issue of how to manage and design open space and semi-natural space in cities for urban biodiversity.

In this book, we focus on urban biodiversity and landscape design, not only through academic discussion but also through practical approaches by landscape designers and activists. We also discuss planning and management for urban green spaces and their functions, describing the fascinating interaction between European and Asian urban biodiversity researchers, planners, and landscape architects. Thus, the collaborative design and wise use of green infrastructure will be really essential for sustainable cities.

Each contributor provides a unique perspective on the relationship between nature and people in urban areas, and the ecosystem and biodiversity in urban areas and how to manage them. The structure of the book is:

[Part I] Landscape Design and Urban Biodiversity: Practical landscape design examples given by landscape designers and activists.

[Part II] Landscape Management for Biodiversity in Urban areas: More wide-ranging discussion about landscape planning and management

[Part III] Towards Ecological Landscape ecology and planning for future cities: Future visions for landscape ecology and planning in urban areas are presented and discussed.

All chapters explore and consider the relationship between humans and nature in cities, a subject which is taking on increasing importance as new cities are conceptualized and planned. These discussions and examples are useful for urban ecology researchers, biologists, city planners, government staff working in city planning, architects, landscape architects, and university instructors. Various planning, design, and management examples should also be useful for undergraduates and postgraduates in civil engineering, architecture, city planning, and landscape design courses.

As *Le Guin* (1968) told us the relationship between nature and humans in her book, now the time for rethinking biodiversity for sustainability. We should listen to the sound of nature, learn from the basic structure of them, and design our sustainable future.

Kitakyushu-city, Fukuoka, Japan
December 2020

Keitaro Ito

Acknowledgments

I would like to express my thanks to all of the authors who contributed to this book, especially thanks to Dr. Mahito Kamada and JALE (Japanese Association for Landscape Ecology) members for discussion about landscape ecology and vernacular landscape planning. Also special thanks to Prof. emeritus Ingunn Fjørtoft for 20 years of pedagogical and ecological learning approach in our landscape design and research in Norway and Japan. I deeply appreciate Dr. Ingo Kowarik, Dr. Nobert Müller, Dipl.-Biologist Peter Werner, Dr. Stefan Zerbe, and Dr. Andreas Langer for providing many inspirations for urban ecology in Berlin and Japan. And joy to work with Dr. Mark Hostetler, who generously shared wonderful experiences and good discussions from an ecological point of view in Florida. Also thanks to Dr. David Maddox (The Nature of Cities) for thinking about urban nature in the world. Concerning local history and cultural studies, thanks to Prof. emeritus Ian Ruxton for discussion from a different point of view. In the process of editing this book, it has been supported by Dr. Kazuhito Ishimatsu, Dr. Tomomi Sudo, Hayato Hasegawa, and students in the lab. of Kyushu Institute of technology and Dr. Tohru Manabe of Kitakyushu Museum of Natural History and Human. I thank Kyushu Institute of Technology for supporting my research, Department of Civil Engineering and Architecture, International Affairs Division, and all staff at our university. At Springer, I am grateful to Mei Han Lee, Momoko Asawa, and Yuko Matsumoto for proposing and supporting this project. I appreciate PREC Institute staff and Dr. Tamio Suzaki, Dr. Masami Sugimoto, Dr. Morio Imada, Dr. Akira Saito, Dr. Koichiro Gyokusen, Dr. Noriko Sato, Dr. Satoshi Ito, Dr. Nobuya Mizoue, Dr. Kotaro Sakuta, Keisuke Harada, and friends in Kyushu University for supporting my earlier ecological research, planning and studying together. I also thank my family and all the friends for their support. This project was supported by Fulbright Japan and IIE, Fulbright scholar program in 2020. Also supported by Japanese Science Research Grant "Kakenhi" (Ministry of Education, Culture, Sports, Science and Technology (MEXT), Japan Society for the Promotion of Science (JSPS). I would like to express our gratitude to all those who gave us the opportunity to complete this project.

Contents

Part I
Landscape Design and Urban Biodiversity

Chapter 1
Designing Approaches for Vernacular Landscape and Urban Biodiversity

Keitaro Ito

Abstract Vernacular landscape design and regional planning methods for urban biodiversity were discussed based on examples of interesting landscapes and our design projects and approach methods from landscape ecology. The process and characteristics of designing urban spaces for vernacular landscape and biodiversity were discussed. Public space design usually has limiting conditions. Featured landscapes attract people and are easy to preserve. Therefore, history, culture, topography and ecological matters such as vegetation should be adapted to the design when we think about better landscape design. In this chapter, it is discussed how "ordinary landscape" is situated within the history and nature of an area, and how it should be designed.

Keywords Vernacular · Landscape design · Landscape ecology · Biodiversity · Cultural landscape

1.1 Introduction

We have been designing urban parks, forests, school yards and riverbanks as urban green infrastructure for 20 years. In order to design these spaces, we used landscape ecological approaches, e.g. vegetation, habitat, land use history, biotope map, cultural landscape. When we design these spaces, it is very important to think about the characteristics of the land or spaces. The characteristics of the land or spaces are sometimes called "Fūdo" in Japanese and this term is similar to "Vernacular" or "Milieu" but it is still quite difficult to explain this concept expressed by Tetsuro Watsuji (1979). "Fūdo" was discussed as a phenomenon including nature and culture by Watsuji (1979) and Berque (1992). Watsuji recommended that "We could find the phenomenon of "Fūdo" in human life" (e.g. literature, art, religion,

K. Ito (✉)
Laboratory of Environmental Design, Faculty of Civil Engineering, Kyushu Institute of Technology, Kitakyushu-city, Fukuoka, Japan
e-mail: ito.keitaro230@mail.kyutech.jp

K. Ito (ed.), *Urban Biodiversity and Ecological Design for Sustainable Cities*,
https://doi.org/10.1007/978-4-431-56856-8_1

custom, and so on). He already mentioned the important issue of landscape ecology and design in 1935. Berque (1992) criticized Watsuji's discussion accurately with some examples, seasons, culture, nature protection and destruction in Japan with some contradiction. Relph (2008) pointed out "placelessness" remarkably appeared in Japan less than other countries because this country has the experience of places that was deeply embedded pre-industrialization, and the modern landscape was formed by industrialization and urbanization. Numata (1996) discussed the concept of "landscape", which means the land and structure of biosphere with human activity and culture, we cannot think about "landscape" without excluding the phenomenon of human life. Morimoto and Shirahata (2007) mentioned that landscape potentially includes the information of nature and culture that is the result of interaction between nature and artificial human activity. Kamada (2000) organized the concept of landscape ecology and it should be analysed from the point of view of people's recognition of landscape and the interaction between humans and landscape for understanding total landscape. Washitani and Kito (2007) mentioned that the importance of nature restoration and biodiversity has various functions of ecosystem for stability; therefore, it is very important infrastructure for human beings. Recently, these should be maintained as the "Green infrastructure" (Nakamura 2015). He also mentioned that though forest, river and wetland have various ecological system services, we should preserve their functions for sustainability. Thus, it is a very important issue to preserve biodiversity and landscape as nature natural capital. When we think about future landscape planning and design, it is important to re-think and plan with characteristics of "Fūdo" (Vernacular landscape) including not only nature but also culture. In this chapter, I would like to focus on some examples of vernacular landscape and discuss ordinary landscape. This chapter, in addition to my former discussion about landscape and vernacularity (Ito 2016), I would like to discuss more about the methodology and issues for landscape design through our practical landscape design.

1.2 History of the Place, Vernacularity and Design

1.2.1 A Village in Portugal

Monsanto is a small village surrounded by a castle wall and people live there among huge rocks. This is a special sort of landscape. Some of the houses exist between or under the huge rocks. I felt quite a special feeling when I saw this landscape for the first time. This is a unique landscape that mixes nature with human life in front of me. I felt that it could be a vernacular landscape. This special character attracts people but this is also the ordinary landscape for habitants from a long time ago. I would say we could find some stories in historical context like people have been find niches under huge rocks. Rudofsky (1987) mentioned that there is much to learn from architecture before it became an expert's art. The untutored builders in space and time demonstrate an admirable talent for fitting their buildings into natural

Fig. 1.1 The village, living with huge rocks (Monsanto, Portugal Photo: Keitaro Ito)

surroundings. Instead of trying to "conquer" nature, as we do, they welcome the vagaries of climate and the challenge of topography (Rudofsky 1987).

Relph (2008) suggested a very important point about people's experience in space. He mentioned that geographical space is not objective and different but full of significance for people. Intentionality merely gives direction to experience and the actual experiences are composed of whole complexes of visual, auditory and olfactory sensations, present circumstances and purposes, past experiences and associations, the unfolding sequence of vistas and the various cultural and aesthetic criteria by which we judge buildings and landscapes (Relph 2008).

Monsanto village is one of the places which allow us to have complex experiences and landscape made not only by human intention but also long history and their livelihood (Fig. 1.1).

1.2.2 Hufeisensiedlung (Horseshoes Colony) in Berlin

In Berlin, there is a housing estate called "Hufeisensiedlung" (Horseshoes colony) designed by architect Bruno Taut (Fig. 1.2), built in 1925–33. When I lived in Berlin, I found it was a remarkable landform on the map, so I went to that place. The place has a gradual slope to a shallow pond and trees that were surrounded by low

Fig. 1.2 Hufeisensiedlung by Bruno Taut (Berlin, Germany Photo: Keitaro Ito)

storey houses. And I was wondering why this place is shaped like a horseshoe and read some literature. There used to be many ponds around here and Taut designed the housing with this fact (Suzuki 2002). Now we found this place quite strange because ordinary German style housing is arrayed neatly around there. However, once we know the history and design methodology of this place, the scale and the characteristics of the place as it was, finally we can understand the history of this place. One more interesting thing is the people have been living there and they enjoy taking a walk around the pond and sometimes lie on the grass in the place. It could be said that Taut intended to bring down this pond and landform and combined these as public and nature with people's life. As the result of that the community has formed as time has passed. Suzuki mentioned that throughout the background of Taut's lifetime and his works, his works were affected by garden city movement, "Tomorrow: A peaceful path to real reform" E. Howard (1902). It was one of the most important theories for city planning and influenced later city planning throughout the world.

It was supposed that as the population was increasing rapidly in Berlin, people were obliged to live in graceless apartments apart from nature (Suzuki 2002). Because of these miserable situations in Berlin, the members of the garden city association established "the public architecture union Berlin". Taut was interested in the garden city movement in England when the garden city association asked him to do the planning in Berlin. He thought that an architect should be an artist but also sociologist, economist and strict scientist (Taut 1991). He also mentioned "When we think about Vernacular (Fūdo 風土), we should not rely on only the natural phenomena like rain, snow, temperature, sunlight and seasonal changes". The architect whose design is based on "Fudo" will find the human, customs and the balance of human body (Taut 1974). Thus, I would say that such an attitude as an architect connected his architecture and landscape design by reading topography and nature potential.

1.3 Redesigning an Urban Park for Biodiversity and Ecological Learning

Megurizaka pond is located in Yomiya urban park, Kitakyushu, Japan. The Kitakyushu-city government asked us to redesign this place for urban biodiversity (Ito et al. 2014).Redesigning this park, the history of the region was surveyed. The regional map drawn in 1931 showed us a very interesting description about space and landscape (Fig. 1.3). It was also written that "The twelve-layered lantern parade floats were reflected on the surface of the lotus pond". People could see them from various places in the town. Also they caught the sounds of drums and musical accompaniment that never stop during the Tobata Gion festival period.

These descriptions were very important for us for redesigning the park because it showed us the landform, habitat, lights, food and culture, i.e. total landscape. Hayden (1997) mentioned that "festival and parades also help to define cultural

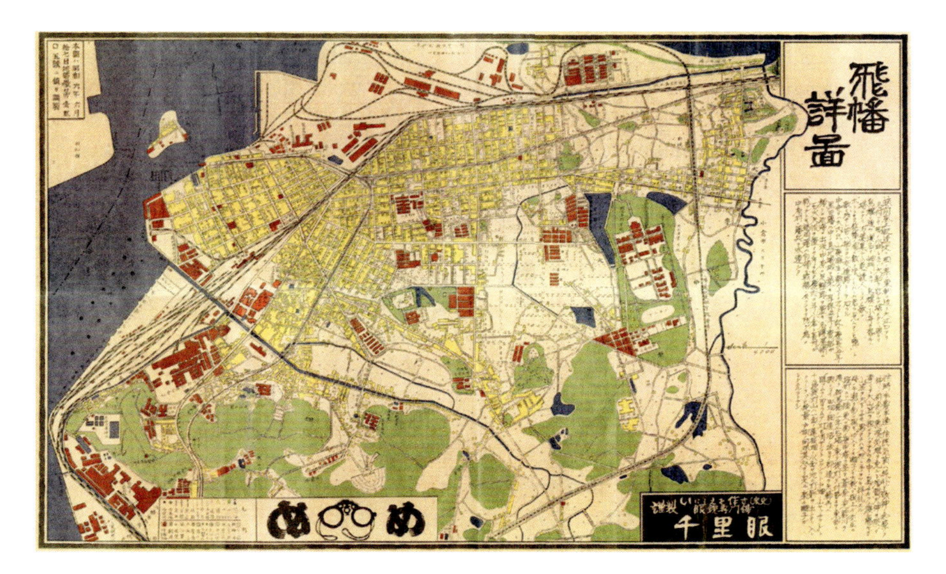

Fig. 1.3 The map around the site in 1931

Fig. 1.4 The design model 1/100 (Keitaro Ito Lab. 2009)

identity in spatial terms by staking out routes in the urban cultural landscape". Therefore, we designed this park taking into account history and biodiversity: native species were selected and planted. And the landform was restored to its former shape based on a topographical map. As the result of this restoration, water flow appeared and never dried up throughout the year. The water was collected from hills behind the pond. A tree (50 cm DBH, *Cedrus deodara*) was cut because of the restoration but it was reused for handmade benches by a primary school and university student's workshop. The trail in the park was barrier-free, i.e. designed for wheel chair and baby car use, and we installed the stones that were used for tram rail 35 years ago as parts of the pavement in the park. Place memory encapsulates the human ability to connect with both the built and natural environments that are entwined in the cultural landscape (Hayden 1997). A strategy to foster urban public history should certainly exploit place memory as well as social memory (Hayden 1997).

Based on these historical planning aspects, the place was designed with regional cultural characteristics and we also tried to design ordinary landscape (Figs. 1.3, 1.4, 1.5 and 1.6).

In this project, a local park has been planned and designed as a green infrastructure. Using domestic plants and trees, the place has been managed by university, primary school and high school students, Ito et al. (2010, 2014), Ito (2016) pointed out the importance of "raising up landscape" through the process of individual landscape design projects (e.g. primary school yard, riverbank and fishway

Fig. 1.5 Process of redesign (2008–2014)

Fig. 1.6 The site after redesign and construction (2016)

restoration). Landscape planners and designers should have an image of the project 50 years later from ecological point of view.

Adopting practices that conserve biological diversity in cities can pay huge dividends for people (Hostetler 2012); in particular, design and management practices that conserve wildlife habitat can provide wildlife watching opportunities for many people in their own neighbourhoods (Hostetler 2012). Furuya (2002) also mentioned that city and architecture is not like stop motion, changing with the times, it would be the dynamic process. I agree with Hostetler and Furuya's opinion that landscape designs should incorporate more ecological principles and I believe that more collaboration between ecologists and designers are needed. Further, when we design urban landscapes, we must think about long-term management and engaging with the local residents in order to minimize negative impacts (Hostetler et al. 2011; Hostetler 2010). For instance, we should think about invasive exotic species (Kameyama et al. 2006) because invasive species could be planted by local residents. These invasives would not only impact the ecological design of the urban landscapes but the invasives would spread into nearby natural areas, decreasing the amount of habitat available for many plant and animal species. From the point of view of creating or preserving landscapes, we should share the information of correct ecological knowledge and exotic species management in the community. In Fukouka, Japan, collaborative design and management of an urban park between

Fig. 1.7 Collaborative management between university students and primary school students

school children and local scientists increased the diversity of native plants and animals and served as a restoration example for future environmental planning in the region (Fig. 1.7).

1.4 Vernacularity and Landscape Design

1.4.1 The Characteristics of the Place and Living Landscape

Touristic places are usually characterized landscape as well as important in historical context. These landscapes are well preserved with time and expense. However, we sometimes find "not living landscape", in other words "not ordinary landscape". For example, I have a slightly strange feeling when I hear a place name such as "Venice in Japan". We find this kind of phenomenon and it is interesting to think about the issue or theme of the identity of the place. Relph (2008) mentioned that the everyday landscape is perhaps more easily understood as all the commonplace objects, spaces, buildings and activities that we accept as comprising the setting for daily routines.

Here, I would like to discuss ordinary, everyday landscape. We, landscape planners and designers usually take part in projects relating to ordinary landscape like park, road and city. Taut mentioned the "Außen Wohnraum" (outer living room). Inside room is usually private space but he thought that outside and garden should be a kind of "Ima" (public space inside house) (Suzuki 2002).

This design philosophy will be effective to create urban landscape and region from the point of view of landscape ecology and environmental design. Thus, Bruno Taut designed in harmony with topography and nature in the 1920s and he taught us the importance of total landscape planning-integrated architecture and landscape design.

Hayden (1997) mentioned that landscape and place matters in his book. "Stalking with stories", a practice of the western Apache, has been recounted by anthropologist Keith Basso. "Learn the names. Learn the names of all these places", insists his Apache guide. "All these places have stories. We shoot each other with them, like arrows". Regional landscape is the result of interaction between nature and humans, sometimes it has interesting stories. Therefore, specific landscape contains history and culture; however, these tend to be lost. It is very important for landscape design field to think about the inheritance of history and culture. Hayden (1997) also mentioned that if there were ways to extend the social portraits of the communities they created into urban public space, it might be possible to make successful projects even more public and more permanent. The places which have long history and keep daily life should have endemic land names and layered history. I suppose it would be important to think about the place with stories.

Sustainability of vernacular landscape has been supporting community people's mind and the region and sharing consciousness as well (Hirose 2016) Thus, we landscape designers should pay considerable attention in each landscape to things such as trees, water, forest and coast, because landscape contains the people's and natural history.

1.4.2 Vernacularity and the Future Landscape

1.4.2.1 Regional Landscape Planning

Characteristic landscape attracts people and such places should be protected. The future issue will be how to preserve ordinary landscape. We should think about relocating these ordinary landscapes in the context of regional history and vernacular landscape and nurturing them for future landscape planning. It is very important to analyse regional landscape which is relevant to their livelihood, for example, forest management, traditional agriculture, and so on (Fukamachi and Oku 2016). Hirose (2016) mentioned that it is important to think about reconstructing vernacular landscape even in modernized and urbanized areas for a sustainable society. Therefore, it is necessary to consider how to incorporate vernacular and endemic landscape into landscape design. Kameyama et al. (2006) discussed how to apply vernacularity by analysing flora for regional planning, especially focusing on exotic species. Natuhara (2006) mentioned the importance of adoptive management, the evaluation of biodiversity and land use. Landscape dynamics were also discussed based on landscape structure analysed by vegetation, land use, time and space

Fig. 1.8 Wild part of Südgelände natur park in Berlin designed by Ingo Kovalik, Andreas Langar and Planning Group ÖkoCon & Planland (Photo: Keitaro Ito 2014)

(Kamada and Nakagoshi 1997). Thus the field of landscape design became important not only in civil engineering, architecture and art but also in crossover ecology.

I suppose that ordinary landscapes are characterized by people's livelihood and endemic nature. These landscapes are formed by people's livelihoods such as agriculture based on nature. Oku et al. (1998) pointed out the importance of landscape planning based on seasonality and regionality. When we plan regional landscape, descriptions of season and climate will be a reflection of regionality and vernacularity. For this reason, designers and planners also need ecological knowledge or collaboration with local ecologists. Thus, when we think about landscape design we should apply incubated knowledge in ecology, landscape ecology and landscape architecture.

From the architectural point of view, the history of the city has been progressing, however ecological thought is undeveloped. On the other hand, ecologists need to have interest in the history of cities and regions (Jinnai 2004). An approach from the fields of history and ecology will be demanded for practical regional planning (Jinnai 2004). Furthermore, landscape from an ecological point of view based on nature and culture will contribute to a sustainable society (Fujita 2016). These day landscape designs and preservation seem to be in a good mutual relationship. For example, Südgelände nature park in Berlin (Fig. 1.8) is a superior design project based on history and vernacularity, because this project has arisen from layered vegetation

Fig. 1.9 Landscape of detention pond in Gainesville, in University of Florida campus, dried two days after the rain (Photo: Keitaro Ito 2020)

diversity, dynamics and history. It was completed by collaboration of specialists of ecology and landscape designers (Kowarik and Langer 2005; Langer 2016).

In cities with lots of impervious surfaces, it is important to think about stormwater treatment and containment because of heavy rainstorms. In Florida, there are planning requirements for stormwater ponds; these ponds are designed to prevent any stormwater running off a site because of the increased amount of impervious surfaces (Hostetler 2012). Stormwater ponds are installed in urban areas to provide the two main functions/ecosystem services of flood control and to improve water quality (i.e. to collect and/or uptake sediments, nutrients, pollutants running off of urban landscapes and prevent them from entering natural water bodies) (Harper and Baker 2007). Nighswander (2019) mentioned that those ecological services can be enhanced with proper pond design and maintenance. Figure 1.9 shows that a detention pond in University of Florida campus, it has flood control function, biodiversity preservation, vernacular landscape as well, because of designing with local plants such as bald cypress (*Taxodium distichum*).

Thus, it will be important to discuss the design methodology about flood control function and urban biodiversity for future cities.

It is a good sign that collaborative projects of ecologists, landscape architects, architects and civil engineers are increasing. Total landscape planning methods have proposed rural and urban planning as a quasi-ecological system in Europe and the

Fig. 1.10 Landscape of detention pond in Gainesville, in University of Florida campus Filled with storm water after the rain (Photo: Keitaro Ito 2020)

USA (Takeuchi 2006). Now we should think of "God is in the detail" again in landscape design process. At the same time, integrated methodology will be possible in landscape ecology because in their fields, they will manage a series of steps including survey, analysis, planning, design and implementation (Fig. 1.10).

1.5 Conclusion

Description by connecting past and present, translation of history, culture and environment for the future will be the role of landscape design. Therefore, the knowledge that has been gained in the field of ecology should be used to create more sustainable landscapes. The knowledge of landscape ecology is easier to apply for landscape design because it usually focuses on applied technology such as land use and planning. Miyagi (2002) mentioned the possibility of landscape ecology and expecting to open not only conservation but also development in terms of preservation in the field of preservation ecology. At smaller scales, Hostetler (2020) proposed landscapes, i.e. individual yards could be eco-attractive in that they not only incorporate ecological functionality into a design, but they also are visually pleasing and they exhibit evidence of human intent and caring.

Kamada (2016) mentioned that this new phase of landscape ecology will have a role of connector between landscape and environment. When we plan and design landscape, the collaboration of ecology, civil engineering, landscape ecology, landscape design, architecture will be required. Then we will be able to discuss green infrastructure based on ecology, history and culture in urban and regional areas. Furthermore, future landscape design taking into account vernacularity, history and culture will be necessary.

Acknowledgement I would like to express especially my thanks to Dr. Mahito Kamada for discussion about ecology and vernacular landscape planning. In the process of designing Megurizaka pond, ecological survey and planning we collaborated with Tenraiji- primary school, Tobata high school, Dr.Tomomi Sudo and students in the lab. of Kyushu Institute of technology, Dr. Tohru Manabe of Kitakyushu museum of natural history and human history and Kitakyushu city government. Concerning local history and cultural studies, thanks to Prof. Ian Ruxton for discussion from a different point of view. And thanks to Dr. Mark Hostetler for discussion from ecological point of view in Florida. This study was supported by Fulbright scholar program in 2019, Fulbright Japan and IIE. Also supported by Kakenhi, Japan Society for the Promotion of Science (JSPS), Fund for the Promotion of Joint International Research (Fostering Joint International Research (B)), 19KK0053, Grant-in-Aid for Scientific research (B)15H02870, Principal Investigator, Keitaro Ito, "Practical study for Environmental Design and space usage based on people's direct experience in the nature" Grant-in-Aid for Scientific research (C) 23601013, Principal Investigator, Keitaro Ito "A practical study on environmental design for children's play and ecological education and nature experiences" Fund for the Promotion of Joint International Research (Fostering Joint International Research (B)), 19KK0053,Interdisciplinary studies for total landscape and ecological education biased on nature experience. I would like to express our gratitude to all those who gave us the opportunity to complete this project. This chapter was written and edited with new information in US based on my article "Vernacular and regional landscape design" (2016).

References

Berque A (1992) Fudo in Japan (trans: Shinoda K). Chikuma Publishing, Japan, 428 pp, in Japanese

Fujita N (2016) Expected specializations for landscape ecologist under the promoting town developments. Jpn J Landsc Ecol 21(1):43–47, in Japanese

Fukamachi K, Oku H (2016) Evaluation of natural resource use and satoyama way of life in the Hira Mountain Range in Otsu City. Jpn J Landsc Ecol 21(1):33–41, in Japanese

Furuya N (2002) Shuffled. TOTO Publishing, Tokyo, 231 pp, in Japanese

Harper HH, Baker DM (2007) Evaluation of current stormwater design criteria within the state of Florida. Florida Department of Environmental Protection, 327 pp

Hayden D (1997) The power of place. MIT Press, Cambridge, 296 pp

Hirose S (2016) Environmental design as an integral part of 'Milieu' formation: through discussion of the notion of 'Milieu' with reference to the research outcomes in the field of humanities. Jpn J Landsc Ecol 21(1):15–21, in Japanese

Hostetler ME (2010) Beyond design: the importance of construction and post-construction phases in green developments. Sustainability 2:1128–1137

Hostetler M (2012) The green leap: a primer for conserving biodiversity in subdivision development. University of California Press, Berkeley, 197 pp

Hostetler M (2020) Cues to care: future directions for ecological landscapes. Urban Ecosyst. https://doi.org/10.1007/s11252-020-00990-8

Hostetler M, Allen W, Meurk C (2011) Conserving urban biodiversity? Creating green infrastructure is only the first step. Landsc Urban Plan. https://doi.org/10.1016/j.landurbplan.2011.01.011
Howard E (1902) Garden cities of tomorrow, urbanplanning.library.cornell.edu. Retrieved 2018-07-02
Ito K (2016) Vernacular and regional landscape design. Jpn J Landsc Ecol 21(1):49–56, in Japanese
Ito K, Fjørtoft I, Manabe T, Masuda K, Kamada M, Fujuwara K (2010) Landscape design and children's participation in a Japanese primary school – planning process of school biotope for 5 years. In: Muller N, Werner P, Kelcey GJ (eds) Urban biodiversity and design. Blackwell Academic Publishing, Oxford, pp 441–453
Ito K, Fjørtoft I, Manabe T, Kamada M (2014) Landscape design for urban biodiversity and ecological education in Japan: approach from process planning and multifunctional landscape planning. In: Nakagoshi N, Mabuhay JA (eds) Designing low carbon societies in landscapes. Springer, Germany, pp 73–86
Jinnai H (2004) Regional design based on ecology and history. Gakugei Publishing, Tokyo, 189 pp, in Japanese
Kamada M (2000) Landscape ecology for inter-relational recognition of landscape and culture. J Landsc Arch 64:142–146, in Japanese
Kamada M (2016) Landscape ecology as a tool for understanding "*Fudo* –the dynamic linkage between environments and human being". Jpn J Landsc Ecol 21(1):57–67, in Japanese
Kamada M, Nakagoshi N (1997) Influence of cultural factors on landscape of mountainous farm villages in western Japan. Landsc Urban Plan 37:85–90
Kameyama A, Kobayashi T, Kuramoto N (2006) Greening with biodiversity hand book- planning and technology for nature preservation. Chijin Publishing, Tokyo, 323 pp, in Japanese
Kowarik I, Langer A (2005) Natur-Park Südgelände: linking conservation and recreation in an abandoned railyard in Berlin. In: Kowrik I, Körner S (eds) Wild urban woodlands. Springer-Verlag, Berlin, pp 287–299
Langer A (2016) From rail to rose. Jpn J Landsc Ecol 21(1):29–32
Miyagi S (2002) A drawn path between landscape and environment. Kagaku, Iwanami Publishing 72(5):545–552, in Japanese
Morimoto Y, Shirahata Y (2007) Landscape design – preservation and creation. Asakura Publishing, Tokyo, 212 pp, in Japanese
Nakamura F (2015) Grey infrastructure to green infrastructure – and adoptive strategy based on nature capital. Forest Environ:89–98, in Japanese
Natuhara Y (2006) Landscape evaluation for ecosystem planning. Landsc Ecol Eng 2:3–11
Nighswander GP, Szoka ME, Hess KM, Bean EZ, Hansen de Chapman G, Iannone BV III (2019) A new database on trait-based selection of stormwater pond plants, FOR347, 1-6, EDIS, IFAS Extension, University of Florida
Numata M (1996) Introduction for landscape ecology. Asakura Publishing, Tokyo, 178 pp, in Japanese
Oku H, Fukamachi K, Shimomura A (1998) Sense of seasonal matters in the rural landscape through a photo contest. J Landsc Arch 61(5):631–636, in Japanese
Relph E (2008) Place and placelessness. Sage Publications, 156 pp
Rudofsky B (1987) Architecture without architect. University of New Mexico Press, 157 pp
Suzuki H (2002) A journey to Bruno Taut. Shinju Publishing, Tokyo, 273 pp, in Japanese
Takeuchi K (2006) Landscape ecology. Asakura Publishing, Tokyo, 245 pp, in Japanese
Taut B (1974) What is architecture? (trans: Shinoda H). Kajima Publishing, Tokyo, in Japanese
Taut B (1991) Nippon (trans: Mori T). Kodansha Publishing, Tokyo, 207 pp, in Japanese
Washitani I, Kito S (2007) Biodiversity monitoring for nature restoration. University of Tokyo Publishing, Tokyo, 231 pp, in Japanese
Watsuji T (1979) Fudo. Iwanami Publishing, Tokyo, 370 pp, in Japanese

Chapter 2
Diversity and Design on a Former Freight Rail-Yard

Andreas Langer

Abstract The Südgelände, originally a freight rail-yard, today is a conservation area and nature-park in which urban-industrial nature is both protected and accessible to the public.

The concept for the nature-park had to address two challenges from its inception: how to open the site to the public without endangering the rich flora and fauna present while concurrently dealing with natural vegetation dynamics which would have led to a complete reforestation of the site in a short period of time.

A concept of limited intervention transformed the already existing tracks into paths. These were complemented by the addition of a metal walkway construction traversing the four hectares of nature conservation area. It provides the general public access to the site without any direct impact on the vegetation.

In order to preserve the immense diversity of flora and fauna a typology of space was defined. The different succession stages characterizing the transformation from rail-yard to wilderness were to be kept and continued by using various maintenance interventions. The remnants of the former train use are still visible.

Keywords Rail-yard · Diversity · Dynamics · Design · Succession · Wilderness · Access

2.1 Introduction

About 50 years of natural succession have converted the Südgelände, a derelict shunting station in the heart of Berlin, into a highly diversified piece of natural urban landscape. Originally a hub of activity, followed by a stretch of more than four decades of almost untouched new wilderness, today the site is designated a nature

A. Langer (✉)
planland – Planungsgruppe Landschaftsentwicklung, Berlin, Germany
e-mail: a.langer@planland.de

K. Ito (ed.), *Urban Biodiversity and Ecological Design for Sustainable Cities*,
https://doi.org/10.1007/978-4-431-56856-8_2

protection site in which urban-industrial nature is both protected and accessible to the public. The site covers an area of about 18 ha stretching over 1.5 km.

2.2 Diversity

The diversity of species at the site has, in principle, developed without human intervention. Originally part of a former much larger freight rail-yard, built between 1880 and 1890, the site was almost completely bare of vegetation. Figure 2.1 gives an impression of what the site looked like in the 30s of the last century, namely a vast area lined with tracks, surrounded by allotment gardens and residential sites. After World War II train service was stopped to most parts of the site leaving it nearly unfettered to natural succession.

The result of more than four decades of natural succession has been a mosaic of habitat types colonizing the plains, the cuttings and embankments as well as viaducts and ramps. The topographically varied character of the site caused by cuttings and embankments is indicated in the cross-section (Fig. 2.2). Besides long-term

Fig. 2.1 View of the rail-yard in the 1930s (Photo: C. Bellingrodt)

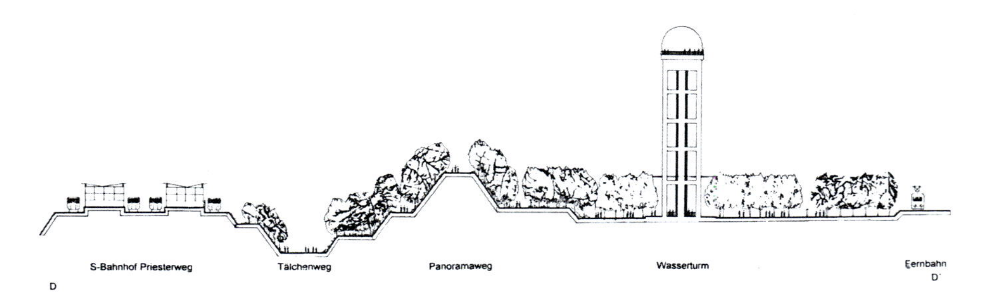

Fig. 2.2 Section across the nature-park (ÖkoCon and planland 1995)

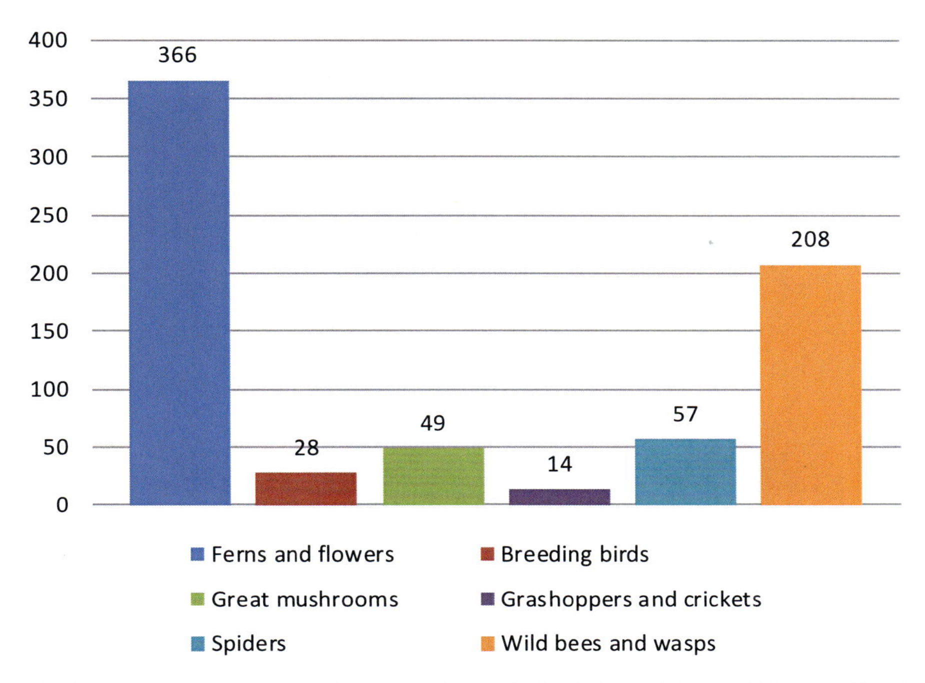

Fig. 2.3 Number of selected species groups of the Südgelände site (Dahlmann 1998; Kowarik and Langer 1994; Saure 1992)

undisturbed development different exposures and therewith ecological conditions are causal for the rich diversity of flora and fauna on the site.

Two comprehensive ecological surveys done last century, one at the beginning of the 80s and the other at the beginning of the 90s revealed a highly diversified piece of natural urban landscape which had been previously perceived as a symbol of decline and negligence, in other words, wasteland. In addition when comparing the two surveys we got a closer insight into the ecological processes due to the ongoing succession.

Figure 2.3 gives you a quantitative impression of the rich diversity of the site.

Behind the figures you find a variety of different habitats with colorful dry lawns, diversified perennials and woody stands. A typical feature of urban habitats is a mixture of indigenous and non-native species creating new and characteristic vegetation. Concerning the woody sites the most dominant native species are Birch (*Betula pendula*) and Poplar (*Populus tremula*), while the non-native ones are dominantly represented by Black locust (*Robinia pseudoacacia*), a species most successful in colonizing abandoned urban habitats.

Especially dry warm open areas are habitats for numerous rare and endangered species like solitarily living bees and wasps. Figures 2.4, 2.5, 2.6, and 2.7 give you an impression of different species you can find on the site.

Fig. 2.4 Centaurea rhenana is one of the characteristical species of the dry lawns (Photo: A. Langer)

Fig. 2.5 Hieracium glomeratum—one of the rare and endangered species of the site (Photo: A. Langer)

Fig. 2.6 Seeds of Bladder Senna (Colutea arborescens) (Photo: A. Langer)

Fig. 2.7 Dasypoda hirtipes a typical bee of the open sites (Photo S. Kühne & C. Saure)

The above-mentioned ecological surveys done in 1981 and 1992 show a remarkable increase in woody vegetation in one decade which doubled in that period (Fig. 2.8). In 1981 only 37% of the Südgelände had been wooded, 10 years later this figure increased to 70%. The results also show that ongoing succession would lead to complete reforestation of the Südgelände in a short period of time. The consequence would be a decline in the characteristic species and plant communities of the open landscapes and a loss of spatial diversity as well.

2.3 Design Concept

Due to the enormous diversity of species and rapidly ongoing succession the design concept for the nature-park had to address two challenges from its inception: the first one being how to open the site to the public without endangering the rich flora and fauna present and secondly whether the natural vegetation dynamic should be influenced or not.

To halt the foreseeable decline of biodiversity the decision was made to create a certain vegetation structure. As a consequence a management system was established to maintain the structure in the long run.

Nevertheless we had to find an appropriate answer to the first question. Especially when taking into consideration the site was officially designated as a conservation area in 1999. The central part is protected as a nature protection site, covering an area of about 4 ha, while the remaining 14 ha are less strictly protected due to its status as a landscape protection site. Both are legal categories of German nature protection.

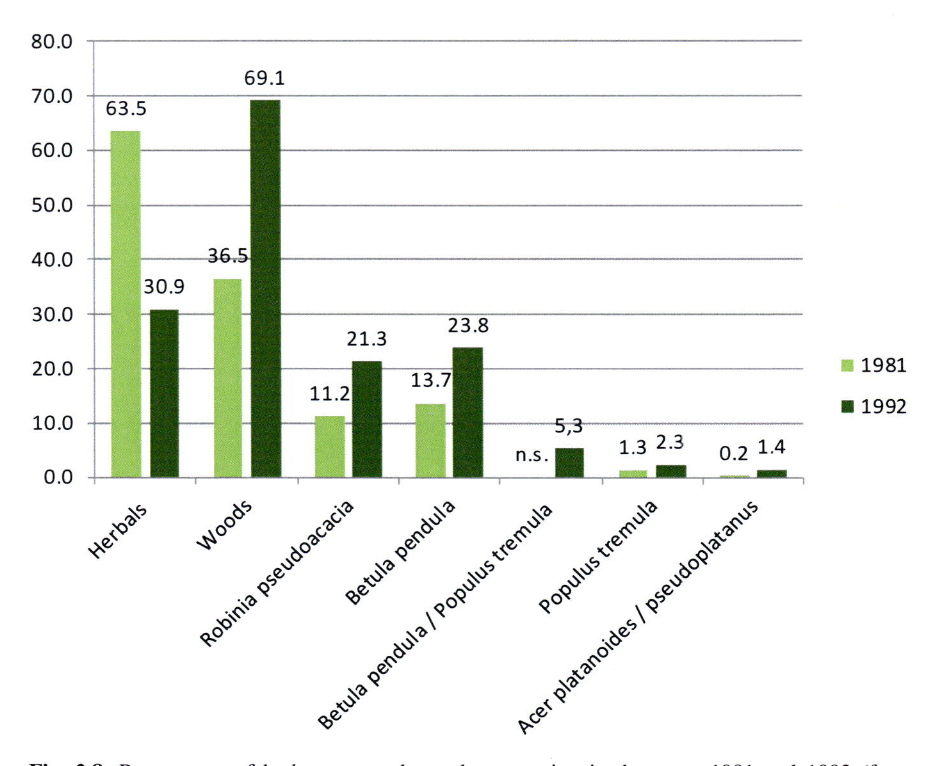

Fig. 2.8 Percentages of herbaceous and woody vegetation in the years 1981 and 1992 (from Kowarik and Langer 1994) (percentages of the named woody species related to the "woods"; n.s.—not stated)

2.4 Design Principles

There are three principles, which are essential to the design concept:

- definition of a space typology,
- access concept,
- preservation of cultural elements

and not to be overlooked is the management concept which has to be seen as an integral part of the overall design concept. The principles together with the management concept do not only secure the diversity of the site, but also keep its special character, namely the amalgamation of nature and cultural elements along with remnants of its former industrial use.

2.4.1 Definition of Space Typology

In order to create different spatial characteristics, lots of trees and bushes were cleared and cut down. Three types of spaces constitutive of the site were defined: clearings, groves, and woody stands. Clearings were opened up and partly enlarged, light and open stands were to be maintained as "groves," while in the "wild woods" natural dynamics were left to proceed fully unfettered (Figs. 2.9, 2.10 and 2.11).

Fig. 2.9 Clearing with shrubs of roses framed by woods making the tracks visible (Photo: A. Langer)

Fig. 2.10 Birch grove (Photo: A. Langer)

Fig. 2.11 Birchwood with water tower—the landmark of the site (Photo: A. Langer)

The result of cutting trees, clearing shrubs and sprouts, and mowing the lawns was a much more transparent spatial structure contrasting shady woody stands, light groves, and sunny open clearings. The interventions in the vegetation structure also emphasize lines of vision, i.e., to the old water tower and highlighted remnants of the past like hidden tracks and the old turntable once again.

The spatial determination of the three types considers both nature conservation as well as landscape aesthetic criteria. The aim has been to demonstrate the transformation from rail-yard to wilderness over time and to make the site more attractive both for rare species of flora and fauna, which are bound to the open sites and for visitors, who can experience a more diverse landscape.

2.4.2 Access Concept

Former preliminary considerations planned to fence in the central, protected parts of the site while giving access to the general public only by means of guided tours (Kowarik et al. 1992). It was clear to us from the very beginning we would have to find a way to open up the site unrestrainedly. A keep-off strategy was not seen as a suitable one for bringing city dwellers in touch with the new wilderness, nor would it make the new wilderness more visible and valuable.

Therefore to make the site accessible to the public a path system was developed that was fundamentally based on the linear structures of the earlier rail-yard. Train tracks were turned into paths by constructing water-bound surfaces between the tracks (Fig. 2.12). Sleepers were used for crossings (Fig. 2.13). The ramps and underpasses that were once used for flyovers are now being used to establish the path

Fig. 2.12 Tracks were turned into path (Photo: A. Langer)

Fig. 2.13 Sleepers were used for crossing tracks (Photo: A. Langer)

system on different levels. In addition a few new connections make circular routes possible.

To access the nature conservation area in the middle of the Südgelände a walkway was proposed. The walkway represents a linking element between the requirements of conservation and the aspiration of the visitors. It fulfills two criteria:

- makes the nature conservation area accessible and at the same time
- avoids direct impact on vegetation.

Fig. 2.14 Sketch of metal walkway (ÖkoCon and planland 1995)

By combining the functions in a single setting—metal walkway with connected observation post and a platform to rest—large areas are protected from intervention, which in turn allows space to be created for undisturbed development.

Our first drafts of the walkway suggested a more or less hidden zigzag-shaped metal and wood construction on different levels integrating lookout posts and places to rest. A railing should prevent people from stepping in to the site (Fig. 2.14). At the end of the discussion process the proposal of the artist group Odious was chosen; a raised metal walkway 50 cm above the vegetation following the old tracks for the most part. The rusty iron construction reflects the former use of the site. The strong linear structure contrasts and highlights both nature and culture. It leads visitors through the nature protection site, keeping them off without any railing (Fig. 2.15).

2.4.3 Preservation of Cultural Elements

On the site we can find a number of selected relics which have been partly secured and even restored including signals, water cranes, and the old turntable (Fig. 2.16). However, the old water tower—the far visible land mark of the site—constructed in the 1920s is still showing its rusty and marked face.

Dilapidated and ruinous buildings were secured by fencing them in or by tearing those parts down seen as posing a danger to visitors.

Technical relicts and nature are different contrasting and communicating layers. By interfering and amalgamating each of them they add to the special atmosphere of the site.

Fig. 2.15 The raised metal walkway makes the nature protection site accessible (Photo: A. Langer)

Fig. 2.16 The old turntable was restored (Photo: A. Langer)

> The ruined building is a remnant of, and a portal into, the past; its decay is a concrete reminder of the passage of time. (. . .) At the same time the ruin casts us forward in time (. . .). The ruin, despite its state of decay, somehow outlives us. And the cultural gaze that we turn on ruins is a way of loosening ourselves from the grip of punctual chronologies, setting ourselves adrift in time (Dillon 2011, p. 11).

An additional cultural enrichment is created by the permanently changing graffiti, works of unknown artists.

The activities on the concrete walls in "Smelly Valley"—the place is cut into the higher leveled surrounding, marked by the smell of the spray-cans during graffiti sessions—add an additional timeline to the site. The sprayers of the day repaint the artwork previously done by their colleagues thus creating a permanent cultural flow contrasting with the slow and inconspicuous pace of ongoing natural succession. A strong cultural heartbeat accompanied by the smell of the spray-cans (Figs. 2.17 and 2.18).

Fig. 2.17 Sprayer at work in the Smelly Valley (Photo: A. Langer)

Fig. 2.18 Graffiti on the walls of Smelly Valley (Photo: A. Langer)

2.4.4 Management

As a consequence of the decision to create and keep a certain structure at the site a management system has been established. Due to nature protection regulation a management plan has to be done. It has to outline the necessary maintenance and development measures for the different biotope types.

The management and development plan is directly linked to the developed spatial structure. The management measures implemented concern the wild woods, the groves, and above all the clearings. From the very outset we decided to leave the wild woods unfettered. The long-term experiment of 50 years of nearly undisturbed succession is to give space and time. Measures are only taken if it is necessary because of safety reasons, i.e., alongside the paths.

The remedies are mostly related to the different open spaces like dry lawns and perennials. Reforestation by expansive woody species like the black locust and aspen is to be prevented. Annual clearing or cutting of the sprouts is especially important with the black locust. Due to the species' capability of binding nitrogen in a very short period of time the species combination changes fostering nutrition rich plants and those which tolerate shady habitats.

The highly diverse dry lawns are to be kept. Every clearing is grazed by a herd of about 50 sheep which are brought to the site for a few days, usually at the end of June. The sheep serve two purposes: they eat up the grass and they also eat the sprouts of woody species keeping the site wild (Fig. 2.19). Additionally, the lawns are to be mown or the sprouts which invade the clearings from the fringes are to be cut.

Fig. 2.19 A herd of sheep maintaining the open grassland (Photo: K. Heinze)

Fig. 2.20 Wafer Ash (Ptelea trifoliata) is to be kept (Photo: A. Langer)

The light and open character of the groves is kept by cutting down trees and shrubs from time to time.

Besides pushing back expansive species others are promoted, especially the different species of roses or some rare species like wafer ash (*Ptelea trifoliata*) (Fig. 2.20). The light-loving species, mostly part of woody stands has to be prevented from being overgrown by cutting the surrounding competitors.

Today management is the key factor in the further development of the park. It has to be seen as part of the design concept while influencing processes in different ways dependent on frequency and intensity of maintenance measures. These are regular and frequent like in the grasslands, infrequent interventions according to need like in the groves and unrestricted succession without any intervention in the woods.

Thus it creates and keeps the different succession stages juxtaposed while maintaining the special character of the site. “The phases of succession that normally would be experienced over the course of time can be experienced here as one moves from one space to another. As one moves through the space the mosaic of succession stages is experienced as a time sequence” (Grosse-Bächle 2005).

Due to the open character of the management approach a monitoring concept was designed. The vegetation development is recorded in reference to the measures. This is the basis for adjusting the intentions and results of the upkeep. Measures always have to be discussed and adapted to changing circumstances.

Management has replaced the previous natural vegetation dynamic. In practice the outlines of the plan are to be reflected and adjusted due to effects of implemented measures.

2.5 Developing Green Axis

Today the nature-park is part of a bigger project changing the vast rail-yard infrastructure into a green infrastructure thus creating a green axis which connects the southern parts of Berlin to the downtown area encompassing Potsdamer square, and one of the largest city parks in Europe, "Tiergarten," a huge park in the heart of the city.

Together with the Gleisdreieck-Park which covers a surface of 32 ha and opened in 2013 and the so-called Bottleneck-Park which was unlocked to the public in 2014 a vision first discussed in the 80s of the last century has come true finalizing the green axis providing different kinds of parks with various qualities along with a wide range of leisure activities.

2.6 Final Remarks

2.6.1 Less Is More

The design concept follows a less is more low intervention concept staging what popped up spontaneously over a period of more than 50 years. It celebrates both the natural and cultural history of the site by keeping existing habitats and spaces and reusing given structures.

2.6.2 Natural Processes

The concept integrates natural processes and thus the dimension of time and therewith of change. The design concept leaves room for spontaneous development on the one hand, while on the other hand spontaneously grown nature is a matter of design.

2.6.3 Urban Nature

The cut out spatial structure showing a change of open spaces with solitary trees and shrubs and woody parts is closely connected to traditional landscape gardens and serves familiar viewing habits. In difference to classical parks the nature-park Südgelände sticks to its urban origin. The traces of the past are still visible, showing that the park is part of the urban fabric thus brightening the perception of urban nature and the conjunction of nature and culture.

2.6.4 Linked

Today the nature-park is linked to additional parks integrated in a green axis running from the south of Berlin to the downtown area thus transforming the former rail-yard infrastructure into a green one.

2.6.5 *Same But Yet Different*

Nevertheless the design concept of the nature-park Südgelände is a unique one integrating what popped up spontaneously on about 18 ha making the new wilderness visible and valuable for city dwellers. Thus opening minds and paving the way for what is called the new wilderness and its integration into public spaces.

References

Dahlmann S (1998) Die Vögel des Südgeländes, unveröff. Mskrpt

Dillon B (2011) Introduction/a short history of decay. In: Dillon B (ed) Ruins, Whitechapel Documents of contemporary art, pp 10–19

Grosse-Bächle L (2005) Strategies between intervening and leaving room. In: Kowarik I, Körner S (eds) Wild urban woodlands, new perspectives for urban forestry, pp 231–246

Kowarik I, Langer A (1994) Vegetation einer Berliner Eisenbahnfläche (Schöneberger Südgelände) im vierten Jahrzehnt der Sukzession, Verh. Bot. Ver. Berlin Brandenburg, 127, pp 5–43

Kowarik I et al (1992) Naturpark Südgelände Bestand - Bewertung – Planung, Unveröff. Gutachten im Auftrag der Bundesgartenschau Berlin 1995 GmbH

ÖkoCon & planland (1995) Natur-Park Südgelände – Entwurf, unveröff. Gutachten im Auftrag der Grün Berlin GmbH

Saure C (1992) Faunistisch-ökologisches Gutachten zur Stechimmenfauna des Südgeländes in Berlin-Schöneberg (Insecta: Hymenoptera Aculeata), Unveröff. Gutachten im Auftrag der Berliner Landesarbeitsgemeinschaft Naturschutz e.V.

Chapter 3
Biodiversity in the Day-to-Day Practice of the Landscape Architect

Frédéric Dellinger, Anne-Cécile Romier, and Ginette Saint-Onge

Abstract Eranthis is a small agency, specialized in landscape architecture, located in Lyon, France. Through four projects, we will try to explain how our work has developed, with tangible ecological designs for urban biodiversity, in many different contexts, at many different scales, and for many different purposes. The projects presented bear witness to this diversity: an historical public forecourt, public spaces in two separate social suburbs, each at a different scale, a privately funded housing complex.

As a minimal requirement within these projects, we wished to successfully deal with two main problematics: the protection and development of urban biodiversity and the adequate management of rain water adapted to each context. Each project held in store a number of problems or successes. In the historical context, we needed to defend a contemporary design among the local authorities, unprepared to such drastic ways of perceiving their urban space, and to convince people in accepting the staging of a new *native* nature. In the social suburbs, the problems laid more in the fears sparked by the city gardener's maintenance concerns and by the possible damages brought to the rain garden plants provoked by abuse from the neighborhood children and the turbulent teenagers that seem to want to damage everything they lay their eyes upon. In the last project, we were pleasantly surprised by the enthusiasm concerning our proposals (fruit and vegetable garden, bat houses, etc.) expressed at first by our client, a social housing corporate executive, and later by the families that have started settling in.

The conclusion remains that these projects are proof that the French (European) society is *prepared* to change, *longs* to change, and deeply *desires* to live in a more sustainable way, gaining more social contact and exploring new limits in community exchange.

Keywords Annecy · Urban nature · Rain water management · Lyon · Rain gardens · Seyssins · Roof gardens

---

F. Dellinger (✉) · A.-C. Romier (✉) · G. Saint-Onge
Eranthis, Landscape Architects, Lyon, France
e-mail: f.dellinger@eranthis.eu; ac.romier@eranthis.eu

K. Ito (ed.), *Urban Biodiversity and Ecological Design for Sustainable Cities*,
https://doi.org/10.1007/978-4-431-56856-8_3

Urban biodiversity and design are at the center of our daily practice. We like to think that all of our projects, with a great design, perhaps even an ecological design, help to develop the urban biodiversity. Today, it is too soon to recognize any success in this matter. In fact, it is still difficult for us to know what will become a success, even a little one!

Today, what we know is that it will take time to see and fully understand the confirmation that we have indeed helped our city to progress on a more sustainable path. But that does not mean we have to wait for results before acting. That is why we accept trying to do our best, adapting our projects to the local circumstances, creating the context where biodiversity, with the help of the local maintenance teams, will develop itself.

Through four projects, we will explain our work on tangible ecological design for urban biodiversity:

The first one concerns the renovation of a public square in an historical environment in Annecy, located in the French Alps.

The second one refers to the public spaces design and parking area in Rillieux-la-Pape, a social suburb of Lyon, France. It will be compared with the third one, in Vaulx-en-Velin, another social suburb of Lyon, but bigger and still under construction. The comparison illustrates how our methodology improves itself with new projects in similar situations.

The last one is a privately funded project in the city of Seyssins, near Grenoble, France, mixing social and private owners, with privatized front gardens, collective fruit and vegetable gardens as well as roof gardens.

These four projects are developed according to their different urban sites and situations each destined to fulfill specific purposes: an historical city center square, street and parking links in social suburbs and, lastly, public spaces within a private living area. In spite of these differences in context, we have tried to apply and use the same methodology in our urban design, counting on our increasing experiences. The result is almost three different designs, each with an ecological scope. We hope to prove by these examples that each context produces its own design.

Before entering into the heart of the matter, let us clarify our approach to some definitions, ideas, and concepts. They have helped us to establish the base of our thoughts and notions, a sort of ambition for a sustainable ideological process. Exposing our definitions will help to understand this approach.

Before and above all, let us also state that we are landscape architects, not ecologists nor engineers. We like to provide our own conceptual interpretation on matters like biodiversity, urbanity, design and ecological design, future cities.

Biodiversity It is the description of a step by step detailed process, involving time, and concerning the understanding of the wellness of a number of species, certain species at a micro life level, and the interaction of plants and animals in general. It takes time to reach the balance in this given situation. We first have to create/work on the life conditions (earth, sun, and climate). Then we can bring in the first species and so on. Our goal is not necessarily to offer protection to species in danger, but just to increase the presence of living organisms. It seems just our chance that plant life is the substance at the heart of our work!

"Urban Character" Is "urban character" as opposed to "natural character". By natural character, we mean the fields, forests, and meadows, not necessarily nature in its wild untouched form. This form of "humanized" nature, found in all types of areas, can actually have a rather low level of biodiversity, but it is not the subject here. In an "urban character" context live people with needs. They are looking for a nice, comfortable, safe area to inhabit. They accept biodiversity, if it does not provide allergies, diseases, or any other form of danger. Let us note that there also exists what we may call a "good" biodiversity (birds, flowers, etc.) as opposed to a "bad" biodiversity (mosquitoes, rats, etc.).

Ecological Design We accept as ecological design a design that uses recycled or recycling materials, with a low carbon footprint and low energy needs for the manufacturing process or for maintenance. The design is closely linked to all forms of eco-city planning.

Ecological City Planning We accept as ecological city planning, the idea that city planning should integrate the following goals:

- the protection and development of urban biodiversity,
- the management and the re-use of rain water,
- the management and the recycling of soils and substrates,
- The search for solutions brought to the needs of its citizens,
- The acceptance that there are limits in the management capacities of today's society,
- The protection and emphasis on culture and local heritage.

Future Cities We define future cities as living areas with high levels of biodiversity, with increased social exchange, with less transportation needs, and with a better balance between living, working, and leisure time.

Public Consultation Meeting or Public Dialogue Public consultation meetings remain still new in France. It is complex to ask people's opinions. The real question concerning them is "what do they really need?" and not "what do they want?" In the matter of public spaces, the goal of a public dialogue is to specify the common interest to all, a common interest that should transcend the personal interest of a few, or one. Design remains something to be discussed between the local authorities, the client, and us. The exercise of public consultations or dialogues demands trust and time. It also asks for a minimum of common cultural basis.

3.1 A Middle Age Castle Front Square

Location: Annecy, the Savoie Region in the French Alps
Client: City of Annecy

Budget:	1,450,000 €
Design Team:	Eranthis landscape architects and team leader/Philippe Buisson landscape designer/Christian Drevet architect/ ECLAR lighting designer/Sitétudes engineers
Beginning of studies:	2008 (after winning the competition)
Start of construction:	2010
Year of completion:	2011 (planting in 2013)
Species planted:	around 4500 plants (5 trees, 170 bushes, and 4300 flowers) coming from nearly 50 different species

3.1.1 The City, the Site Location, the Context Around the Project

Annecy is a small city of about 50,000 inhabitants (200,000 in the whole agglomeration) nestled in the French Alps, at the tip of the lake bearing the same name.

Like other medieval cities in the region, a towering castle seems to still govern the city. But none in the region is as spectacular as Annecy. The castle sits at the frontmost point of a small mountain, on the topmost part of one of its hills and dominates the heart of the medieval city at its feet, as well as the lake.

Before 2008, the castle's forecourt was besieged by the presence of cars: its main function was to be a parking lot. Plant life, in and about the castle, could only be found in the surrounding gardens, whether private or public. As logic would dictate, gardens are much smaller within the heart of the medieval city structure (in the lower part of the city) and larger around the castle where nineteenth century villas sprung up to adjoin the forecourt in what was considered than the old "country side" (Figs. 3.1 and 3.2). All of these gardens are today still rich with city biodiversity (old park trees, birds, insects, small mammals, etc.).

In 2008, we won the design competition to rethink and refurbish the access to the castle for visitors and neighbors, as well as to revisit the forecourt's functionality and design and the castle's general surroundings.

The castle and its outbuildings were built some 500 years ago directly on the limestone structure of the mountain, which is still apparent in some places. In fact, there is hardly any topsoil anywhere on this mountaintop. For the project design, we had to bear this in mind and deal with a topsoil cover included between 0 and 150 cm.

3.1.2 The Project

The idea behind our project was to take advantage of the site's "inconveniences" to form the base for our design. It was decided that it would adopt the mountain's topographical shape and geology (Fig. 3.3) and would organize itself as a central square in the middle of a large surrounding garden, a sort of green connection with

Fig. 3.1 The site as it existed before 2008, a parking

the rest of the city (Figs. 3.4 and 3.5). We proposed a strong contemporary design that would reveal the historical context of the site as well as provide a functional solution to organize the castle's access for tourists and an engineering solution to deal with rainwater management while maintaining the idea to develop, within this newly planned green connection, a new specific biodiversity.

A netting design formed by stone water collectors or gutters offered a structured new outline to the forecourt while guiding the rainwater to the new surrounding gardens (Figs. 3.6 and 3.7).

A mesh, like that of a fisherman's net, draws polygonal forms on the forecourt surface. These forms that overlay the square, a reminder of an early knight's protective armor, which unlike chain-mail was made with small metal polygonal plates (Fig. 3.8) can remind us of the space's medieval heritage. This geometry is reused in the design of the Corten steel furniture accompanying the square: the tree protection shafts, the varied structures for creeping plants, and the three horizontal windows inlaid in the square giving onto the castle's foundation, uncovered by the archeological research team during construction (Figs. 3.9 and 3.10).

At the start, the program was vague. We knew we had to reduce the space allotted for parking and move it over to the sides of the main court in order to preserve a central space for pedestrians. Tourist buses and local residents were to be allowed access to the forecourt.

Fig. 3.2 The site as it existed before 2008, a parking

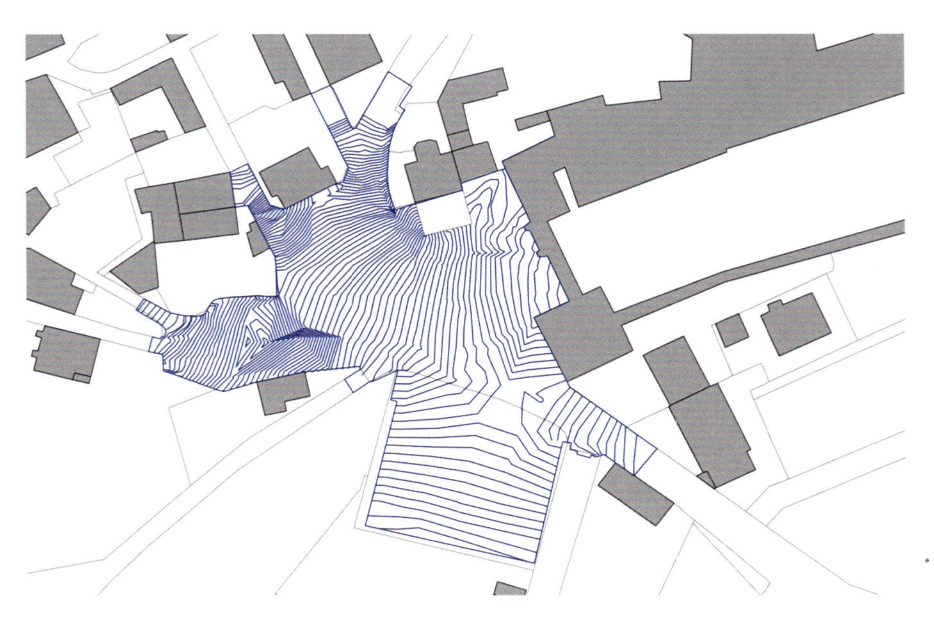

Fig. 3.3 The site topography

Fig. 3.4 Site plan

Fig. 3.5 The newly designed square in 2010—*Albert Videt photo credits*

Fig. 3.6 The stone water collectors or gutter system—*Albert Videt photo credits*

Fig. 3.7 The stone gutters

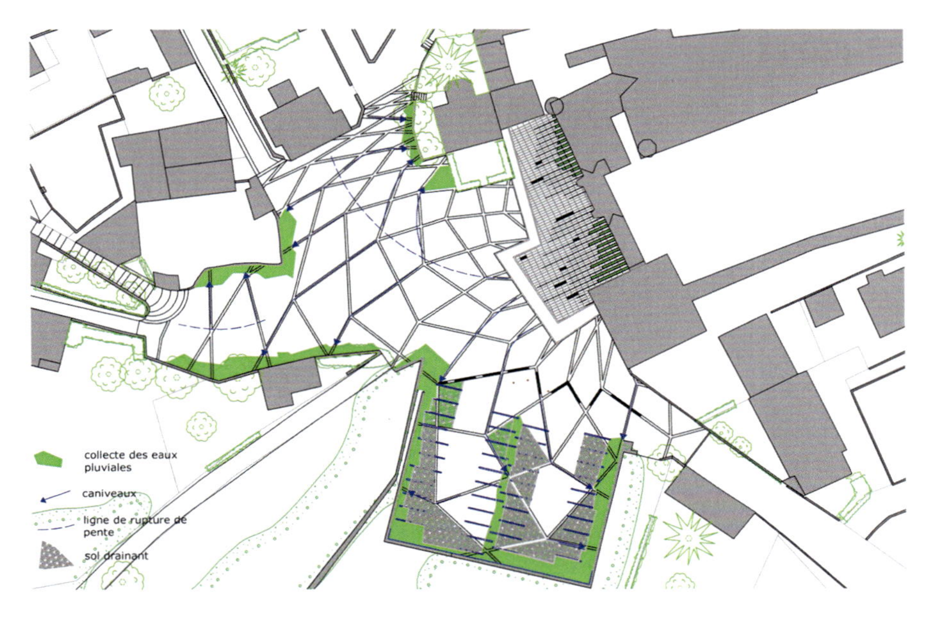

Fig. 3.8 The netting design formed by the stone water collectors or gutters

Fig. 3.9 Furniture accompanying the square—*Albert Videt photo credits*

Fig. 3.10 Furniture accompanying the square—*Albert Videt photo credits*

We have planted nearly 4500 plants in the different surrounding gardens, from nearly 50 different species. These surrounding gardens (nine in all) all put forward diverse agronomical conditions that may go from hot, dry, sunny spots to a wet, undergrowth context.

Plants were provided by two separate local nurseries: one supplying the more "classical" and easily "manufactured" species and a second one supplying the more specific alpine plants and flowers.

Fig. 3.11 The dry and sunny garden, summer 2012—*Albert Videt photo credits*

Annecy's climate is relatively warm despite its location in the middle of the Alps and its altitude (470 m). For example, in 2008, the weather counts were as follows: about 1942 h of sunshine, 900 mm of rain, 26 days of snow, 32 days of storm, and 25 days of fog.

The wet and sunny gardens, as well as the wet and undergrowth garden are rather "classical" in their plant assortment (Figs. 3.11, 3.12, and 3.13): small bushes, climbing plants, and flowers (*Hosta*, *Anemone, Buxus, Berberis, Clematis,* etc.).

By opposition, the dry sunny garden, in front of the castle's main gate, is planted with small plants that normally grow in mountain areas. We chose these plants in concordance with the local nursery. Most of the seeds come from the French Alps Botanical Garden implanted on the Lautaret mountain pass. Ahead of their final resettlement, each seed is planted in the nursery one year in advance. All of these plants were chosen according to their capacity of adaptation to their future agronomic context: hot and sunny in summer, dry and cold in winter. But the choice was still restricted. In fact, in the city of Annecy, there are fewer fog episodes than in the mountains. But in their natural habitat, most of these plants survive with fog humidity not with rain.

Fig. 3.12 The dry and sunny garden, summer 2012—*Albert Videt photo credits*

Fig. 3.13 Summer 2013

3.1.3 Learning Process, Thoughts, and Reflexions

Although this project seemed fairly straightforward, we did end up learning a lot from this plant assortment concept. And on a more social level, we learned about general acceptance to change.

Two main problems revealed themselves during the study period of this project. First, there was the political acceptance of growing plants directly in front of the castle, close to the main gate. Most especially, the city gardeners believed that tourists would destroy the gardens and trample the flowers. Today, years after planting the final elements, no major incidents have been reported. They also feared that the flowers would not survive in such harsh conditions, with great hot/cold variations. Our choices seemed rather judicious, for all species ended up developing themselves quite quickly.

The second problem concerned the local population's acceptance of the new gardens, changing their habits and lifestyle. Some dog owners let their pets do their "business" in the newly refurbished plant beds. They would dig out the newly planted vegetation and leave their waste behind. The more the plant beds would get dirty, the less the city gardeners would clean. And the less respect for the gardens the residents would show.

The question then posed itself: is it possible to do any better? At the first consultation meeting with the population, the questions evoked concerned mainly the parking situation and the noise related to the construction site. No one really saw the need to ever raise a comment about the plant proposition.

We encountered a similar attitude with the city gardeners. We never really succeeded to win the trust of the future maintenance team, and trust is very important for us when designing a public space. Maybe, the problem laid in the need for greater and better communication and exchange?

On the other hand, small French political administrations are rarely adventurous. They prefer to tread existing paths rather than experiment new ideas in projects. It was incumbent upon us, as designers and planners, to prove that our ideas, in theory, worked. In these circumstances, when all was said regarding this major public space, one of the main squares of the city of Annecy, we succeeded in convincing them, theory in hand, to move forward with the project.

Today, construction is done. Plants have settled and grown into their new habitat. Our task now resides in the observation of the evolution of this space. Once a year, during springtime, we revisit and document the square's development and evolution.

3.2 A Public Spaces Renovation in a Social Suburb

Location:	Rillieux-la-Pape, Northeast suburb of the city of Lyon
Client:	The Greater Lyon Area
Budget:	460,000 €
Design Team:	Eranthis landscape architects/Philippe Buisson landscape designer/Arcadis engineer
Beginning of studies:	2008
Start of construction:	2010
Year of completion:	2011
Species planted:	around 4000 plants (50 trees, 600 bushes, and 3350 flowers and bulbs) from nearly 30 different species

3.2.1 The City, the Site Location, the Context Around the Project

Located in the South Eastern part of France, Lyon is today the major agglomeration of the Rhône River Valley. The Greater Lyon Area nears 1.5 million inhabitants and is the second largest city in France, after Paris. It is a dense area centered on an old city heart registered as part of the UNESCO World heritage list. Historical Gallo-roman capital, Lyon remains today a center of culture, finance, and industries (silk, chemical, pharmaceutical).

The city's site has been inhabited for more than two thousand years, taking advantage of the general geography of the area: Lyon sits at the crossing of two important rivers and two hills. Its architecture, today, superimposes large historical areas to new contemporary quarters. In this context, Rillieux-la-Pape is a small village located in the agglomeration's first belt, at the start of a large plateau, in connection with the northern part of the Lyon region.

Like most major French cities, Lyon is surrounded by substantial social living areas, suburbs built between the mid-fifties and the late seventies. Rilleux-la-Pape is one of these suburbs, a "New Town" quarter, like it is considered, built in the 1960s, saw the arrival over the years of mainly residential towers of 5 to 18 stories high. Over 20,000 people live in these buildings today. In all, Rillieux-la-Pape counts a population of 30,000 people. The remaining 10,000 residents can be found in the three old village quarters. Since the year 2000, this waning suburb has launched into refurbishing plans to renovate not only the buildings but the public and common spaces as well. The "rue Bottet" and the synagogue forecourt are part of this general scheme.

In the beginning, our study site consisted primarily of a large asphalt surface area used for parking, some summary leisure areas barely covered with grass and a few old trees (Figs. 3.14 and 3.15), made up of *impromptu* species (*Robinia pseudoacacia, Acer pseudoplatanus*).

Our task was first to organize the site in general, as well as its parking, and to create a new street and small forecourt in reply to the 70's-built synagogue's needs. The project had to insert itself perfectly in the grander scheme, equally being elaborated at the same time for the whole of the district. It went without saying that plants and trees were more than welcome in the area.

3.2.2 The Project

Our idea for the project was to densely organize the needed parking according to each building requirements. This being a space consuming endeavor, we wished to preserve as much as we could of the global space allotted for the introduction of natural areas within the neighborhood and parking facilities. In fact, we succeeded in reserving a small part for planting natural spaces, around 1500 m^2, representing

Fig. 3.14 The site before 2008

approximately 20% of the global project surface area. This action is more than doubled the barely 10% of basic existing grass spots (Fig. 3.16).

To maximize the functionality of the project, we combined the positioning of the newly planted gardens with the implementation of a finely managed topography of all surfaces. This permitted us to collect the rainwater in such a way that the rain would flow directly into the new gardens, transforming them into grand scale rain gardens. These new rain gardens could now manage the rainwater from nearly half the project's global asphalt surface area (Figs. 3.17, 3.18 and 3.19).

In the spirit of maintaining an indigenous natural character for these gardens, we decided upon planting native trees (wild apple, *Alnus incana, Sambucus nigra*, etc.)

Fig. 3.15 The site before 2008

and different varieties of ornamental grass. Nearly all of the tree and bush species present in the gardens have flower and small fruit cycles.

In a more general preoccupation, it was our wish that these new gardens would also be designed to contribute and connect directly with the principal green path, a large pedestrian and bicycle path that draws a line through the entire suburb.

In front of the synagogue, we decided to keep some of the existing ornamental bushes (Fig. 3.20): *Rosa,* etc. The forecourt, now forbidden to cars, was rethought to welcome groups of visitors for religious ceremonies or celebrations. Its surface is made of a light beige concrete embossed with stone incrustations (Fig. 3.21). The stone design recalls certain plant forms, an imprint of a prehistoric fern leaf rib perhaps (Fig. 3.22)?

Amid the new parking spaces, large trees were planted in combination with smaller sized trees upon a ground cover of creeping plants (*Hedera helix, Vinca minor*, etc.), a miniature version of an agricultural hedge.

One of the probing outcomes resulting from the consultation meetings with the local population was the great need for a small children's play area. The project then integrated a small playground replacing a planned small peaceful garden aligned with the forecourt.

The project meant to propose a strong contrast between the strict lines and demands needed for any parking organization and the liberty to offer a new urban natural character to the site and in extension beyond its gardens.

Fig. 3.16 The Rain garden in 2011

Fig. 3.17 The Rain garden in 2012

Fig. 3.18 The Rain garden in 2012

Fig. 3.19 The Rain garden in 2014

Fig. 3.20 Rose beds in front of the synagogue

3.2.3 Learning Process, Thoughts, and Reflexions

In this case again, the consultation meetings with the local residents concerned mainly the parking and the need for a playground. And the local authorities were

Fig. 3.21 The synagogue forecourt

afraid of the maintenance generated by a project so inclined to planting gardens everywhere.

Early on in the project, we had planned to build the parking in evergreen tiles. At the last minute, afraid of the long-term maintenance, the local authorities wished to transform them in asphalt. But, we did manage to save the rain garden.

The city gardener, for his part, feared that children would eat the wild berries and get ill or that the *Sambucus nigra* berries would "splash" some of the parked cars.

This social suburb is rather known for its turbulent teenagers that seem to want to damage everything they lay their eyes on. Having been we ourselves "contaminated" by these fears, influenced certainly by the local social climate, we reduced the presence of a number of species.

Today, we must admit that the project works. Its functions and its new constructions are respected, by young and old! Even the bulbs (*Eranthis, Narcissus*, etc.), which spring up in summer, are full of life, well protected from rabbits and field mice, numerous in the many green areas nearby.

3.3 Creation of New Public Spaces in a Social Suburb

Location:	Vaulx-en-Velin, Eastern suburb of the city of Lyon
Client:	The Greater Lyon Area
Budget:	14,000,000 €

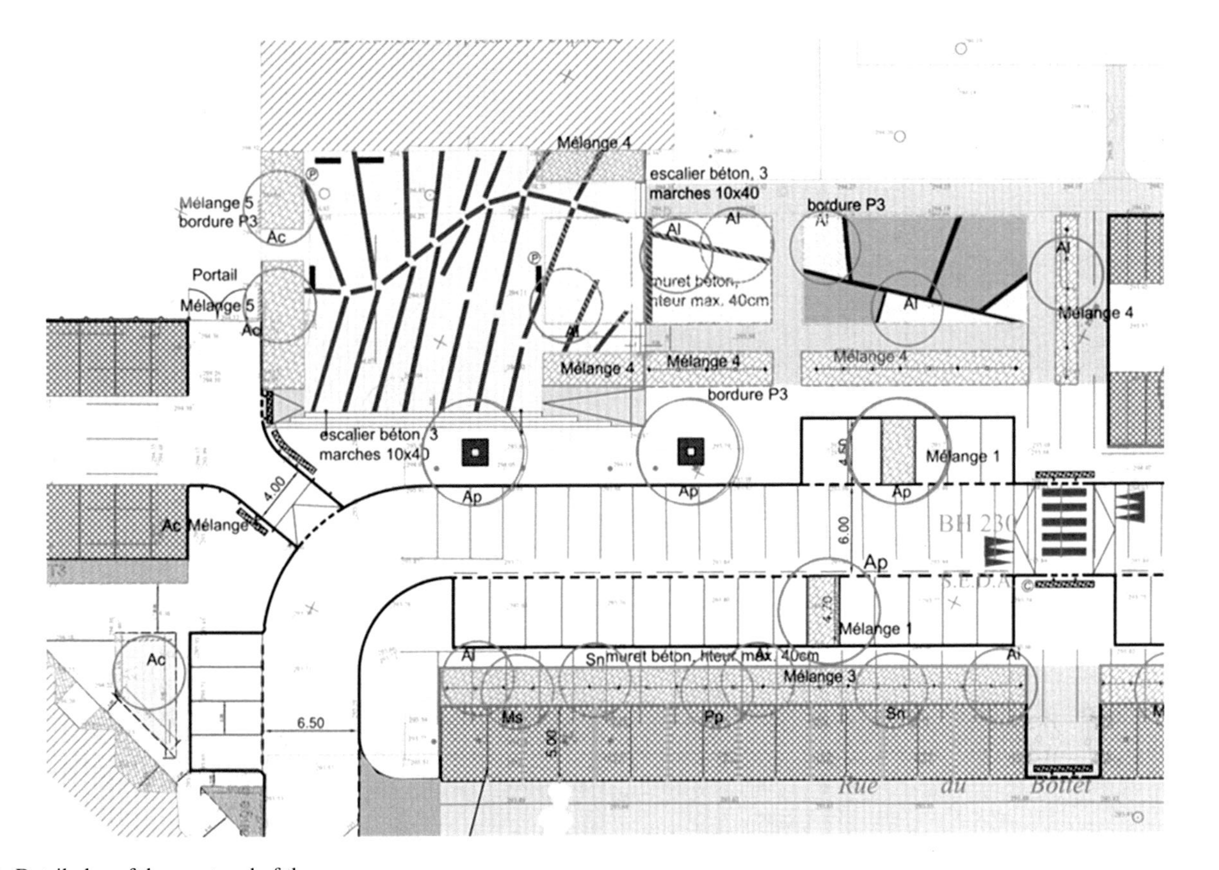

Fig. 3.22 Detail plan of the courtyard of the synagogue

Team: Eranthis landscape architects/Marc Pelosse Architect (Lyon)/François Gschwind lighting designer/Sitétudes engineers
Beginning of studies: 2008
Start of construction: 2010
Year of completion: 2018
Species planted: around 15,000 plants (160 trees, 3500 bushes, and 12,000 flowers) from nearly 76 different species

3.3.1 The City, the Site Location, the Context Around the Project

Vaulx-en-Velin, in the eastern part of the agglomeration, is another social suburb of Lyon.

Located in the alluvial plain of the Rhône River, this area was well known from the beginning of the nineteenth century for its market gardening and its industry of artificial silk.

In 1970, almost at the same time as Rilleux-la-pape, began the construction of a new massive housing district which saw the arrival of over 35,000 new inhabitants (Figs. 3.23, 3.24, and 3.25).

Fig. 3.23 The site before 2010

Fig. 3.24 The site before 2010

In 1981, and then again in 1990, after many signs of social tensions, Vaux-en-Velin suffered many violent riots. To help reverse this tendency, the political powers in place decided, from 1993 on, to change the face of the city, hoping this would change the city's image and by extension the city's way of life. They demolished the old shopping mall, symbol of the last riot (In 1990, the mall was badly vandalized and was set on fire).

By the year 2000, the city had built itself a new city center following a grand project covering most of the city surface (Bernard Paris city planner and Alain Marguerit landscape architect). They kept the old city hall and certain other public buildings but created new streets, a public park, a new square following a "traditional" French urban design (a large formal clean perspective organized by a tree "allee"—all of the same species—flanked by new constructions reflecting a homogeneous and controlled architectural style).

In 2010, our team was chosen to work on the center's reconstruction. We proposed a more sustainable ambitious project, focused on rainwater management, biodiversity, and the quality of life in this district (local climate, noise, pollution, etc.). For the design, we kept in step with the formalism of the original designers. Maintaining coherence in the choice of plants, of materials and colors, even going beyond our design preferences seemed important and essential. We accepted the fact that the superimposition of too many styles of design in such a confined and rather small city center would be counterproductive.

Fig. 3.25 The site before 2010

3.3.2 The Project

The project consists mainly of green infrastructures and rain gardens (Fig. 3.26), using permeable materials (evergreen pavement for parking spaces).

It is a long-term project that involves the simultaneous construction of both the public spaces and the residential buildings. The city has planned constructions for the equivalent of 14,000,000 euros, from the start of operations in 2010 to project completion in 2020.

Fig. 3.26 Site plan

Fig. 3.27 Proximity garden—*Jérémie Cormier graphic design*

Our project involves the design of three small public squares, a large parking for the city hall, a proximity garden, and over 1.5 km of street strips (from Figs. 3.27, 3.28, 3.29, 3.30, 3.31, and 3.32).

With this project, we saw a great opportunity to implement in the city center a new type of experimental treatment of rain water by the use of plants chosen for their cleaning and filtering capabilities. By chance, the Greater Lyon's tree department and water works department backed our experimental project and pushed for its construction, the local authorities lacking any foresight in the matter. A treatment of the underground water is made through the tree substrate. Plants present on riverbanks have an inherent capacity to support droughts as well as floods. These plants were selected and included in our project for their capacity to survive in polluted environments through root filtration of the rain water and the physical properties of the substrate.

Twelve different tree species are used (*Alnus, Salix, Gymnocladus, Robinia, Quercus, Morus*, etc.) around 20 bush species, 40 flower species (flowers, climbing plants, bulbs, and grass), and 4 types of bamboos.

Piezometers (for water pressure measures) and probes are installed at the same time as planting occurs. The aim of this monitoring is to study the cleaning results as well as the growth rate of the plants. To provide us with dependable information, the urban waste management department will test new alternatives to salt and glycol for winter ice road treatment.

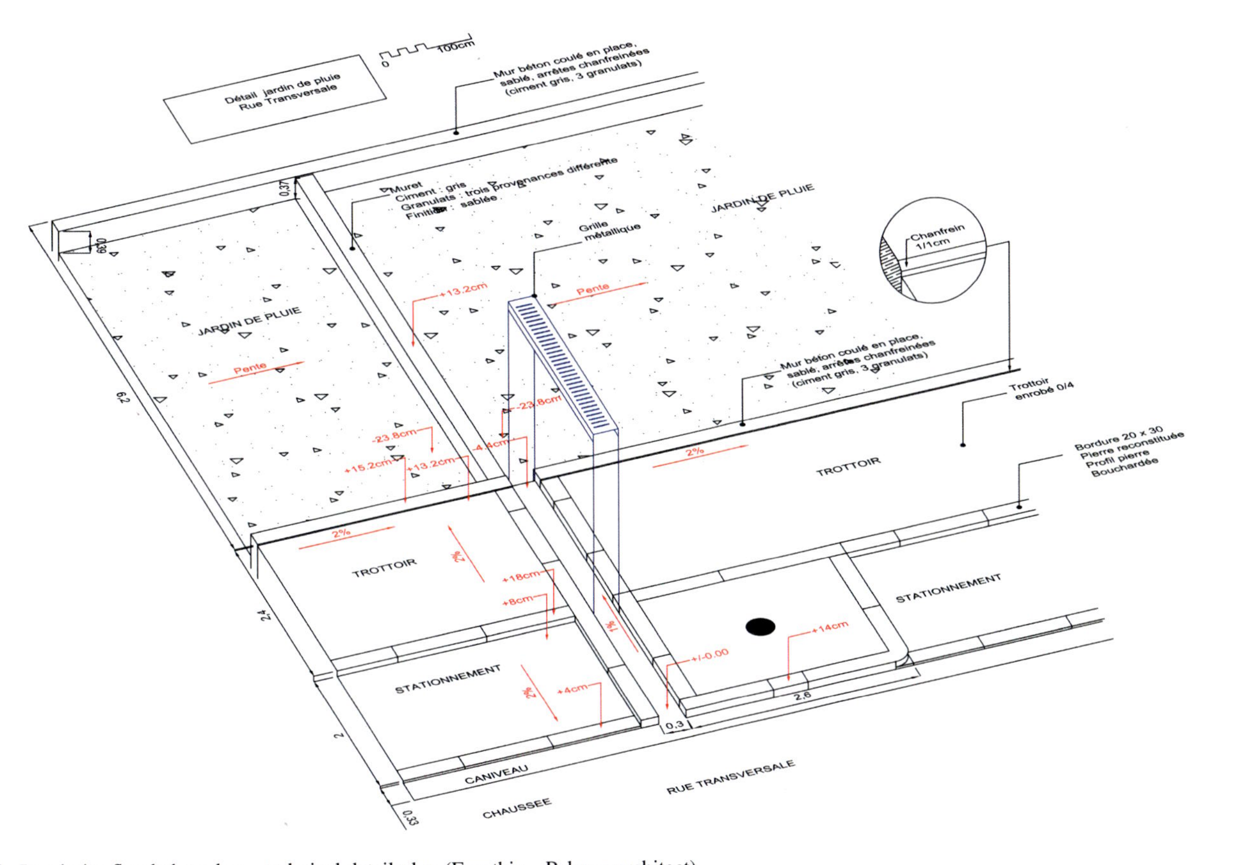

Fig. 3.28 Proximity flooded garden—technical detail plan (Eranthis—Pelosse architect)

Fig. 3.29 Street rain garden

A plant nursery, as well the University of Lyon, has accepted to become our partners in this joint venture. The plant nursery studies the tree growth. For each species, certain individuals will be planted in this "rain water cleaning system," in a classical planting pit (a mix of stone and soil), and others in the nursery. The University will analyze the quality of the underground water and the soils.

For us, this project has a second major goal: it must be exemplary in the matter of local soil management and re-use. Vaulx-en-Velin being located in the large Rhône River Valley, its soil is made up of a system of imbricated layers of clay, gravel, and sand. In this region, the urban ground cover structure that constitutes the main city surfaces has an approximate thickness of about 80 cm, composed of many materials. But when in construction, our rain gardens, including the underground drainage system (gravel wrapped in a geotextile), as well as our tree planting pit, need almost *2m* of depth. With this depth, we come across diverse materials. In agreement with the different contractors, we have organized a sort of "soil bank" on the last lot to be built. So, during the excavation, the different types of materials are directly sorted by nature, (artificial melting soil, clay, sand, gravel or rich soil) and set aside for re-use during the project's construction. We need street structure, sandy rich soil, rich soil mixed with stones and then compacted (70% of stones with a diameter of 80–100 mm), gravel for the drainage system, etc. The contractor has an obligation to give priority to the re-use of the material selected and stocked previously on site and to complete with new material only when the supply cannot be met. This way we expect to save on costs and to minimize truck transportation, source of general and noise pollution for the local residents.

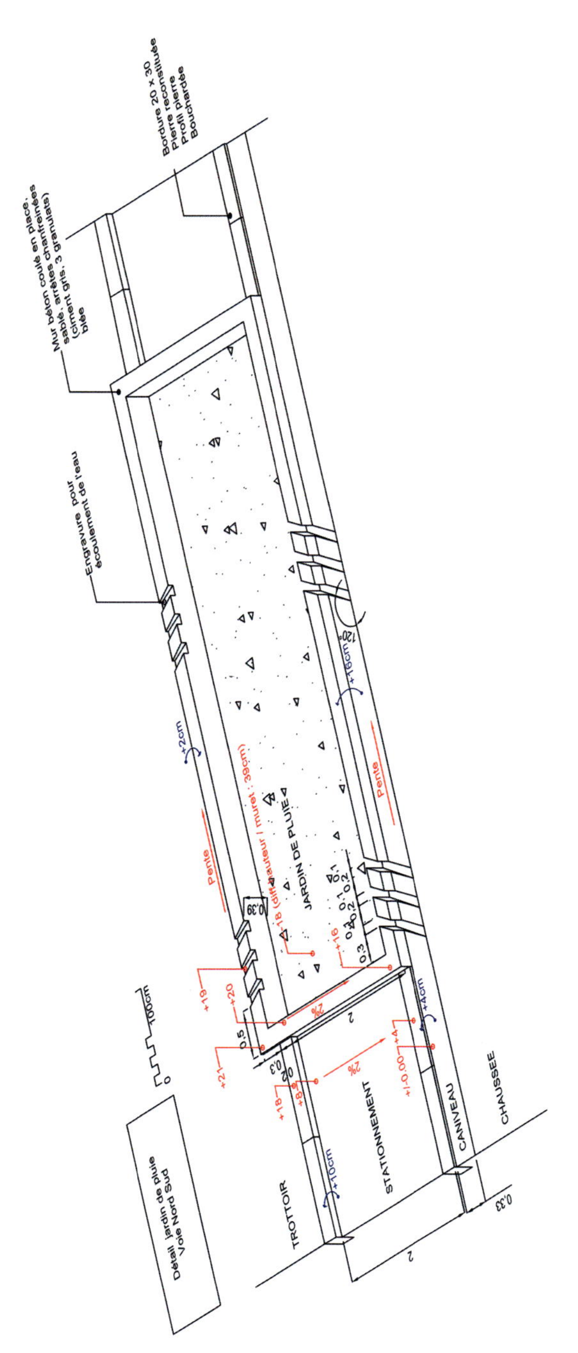

Fig. 3.30 Street rain garden—technical detail plan (Eranthis—Pelosse architect)

Fig. 3.31 City hall parking—*Jérémie Cormier graphic design*

Fig. 3.32 Public square—*Jérémie Cormier graphic design*

At this day and time, the second stage of our project is under construction, whereas the building projects have run into certain delays. These constructions are only partially under way at the present time.

3.3.3 *Learning Process, Thoughts, and Reflexions*

The project succeeded in bridging the operational gap between the traditional departments managing all public streets (rain water dept., green spaces and tree infrastructure dept., road building dept., urban waste management dept., City of Vaulx-en-Velin lighting dept., City of Vaulx-en-Velin gardening dept.) and a new storm water/rain water management system: underground water treatment through tree substrate. The ecological design and planning deployed for this project did not only demand to cross-reference the team's expertise, but also forced our society's mind set, and especially our management department's mind set, on changing their long-term views and expectations of public spaces.

Learning from our previous experience in Rillieux-la-Pape, we proposed a wider plant diversity. This confirms that, with an increased experience (and testing), we gain in self-confidence, in power of persuasion and a sort of courage to experiment beyond.

For any gained technical knowledge, it is too soon to know what will come of this project. We have, however, managed to confront our technical theories with the contractors building the project and as is usual with an experimental project, plans and details are not carved in stone. They have gradually been fuelled and nourished with the technical knowhow drawn from the contractors themselves.

3.4 A Small Social Living Area: 55 Flats in Seyssins

Location:	Seyssins, suburb of the city the Grenoble, French Alps
Client:	Grenoble Habitat (Social investors)
Budget:	660,000 €
Team:	GTB architects/Eranthis landscape architect/ING Clim, Kaena, Soraetec, and Editec engineers
Beginning of studies:	2010 (winning competition)
Start of construction:	2011
Year of completion:	2013
Species planted:	around 9000 plants (55 trees, 2100 bushes, and 6845 flowers and bulbs) from nearly 90 species

3.4.1 *The City, the Site Location, the Context Around the Project*

Seyssins is a rich suburb of the French city of Grenoble. Grenoble is located at the confluence of two rivers, the Drac and the Isere. The city is surrounded by three big mountains, the Chartreuse, the Belledonne, and the Vercors. Most people enjoy skiing on these mountains in winter. Some consider Grenoble, the biggest city in the

Fig. 3.33 The site before the construction

Alps, as its capital. The city is well known for its university, high-tech industry, and its eco-friendly layout.

They are 680,000 inhabitants in the Greater City of Grenoble, some of which live within the municipality of Seyssins (7000 inhabitants). Seyssins is more precisely located between the Drac River and Vercors mountains. Well exposed on the sunny side of the mountain (important in the alps), it makes the most of its beautiful view on the Grenoble city center and the other mountains.

Even more precisely, the project site is niched between the last tramway station and a large green space reserve, formerly a fruit and vegetable garden. This garden was moved closer to the project site to accommodate the new constructions (Fig. 3.33).

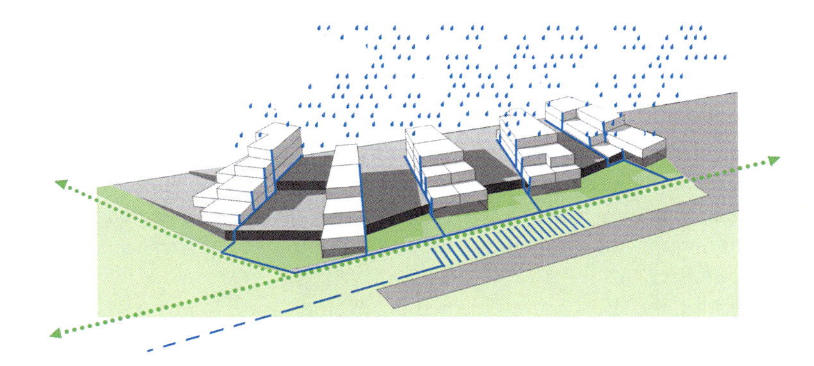

Fig. 3.34 Site plan

Fig. 3.35 Rain water management concept

3.4.2 The Project

The project concerns the building of 55 flats in an old family garden lot. As the landscape architects of the winning competition team, we designed the building's integration in the natural mountainous landscape which included rain water recycling systems for the gardens, extensive rooftop gardens, and fruit and vegetable gardens (Fig. 3.34). The project site is divided between private owners (one-third of the flats) and a social housing organization (the other two-thirds of the flats).

The housing is made up of five slender buildings sharing a long underground parking (Fig. 3.35). Each building offers a bike shed, a small winter garden with a mailbox, and an orange tree. Between the buildings, a small private garden (front garden, south of each flat) and a common garden are used to cover the parking spaces. The access to the flats requires going through this common garden, planted with fruit trees.

Each flat has its own private access, like a small house, many of which are two stories high. On the eastern side of the site, between the building and the public road, maintained for bike and pedestrian purposes only, we organized a common open field with a small playground, a small forecourt, and vegetable gardens.

In this case, obtaining more biodiversity is a goal strongly connected with the site management and the social organization of the future residents, especially in terms of "living together."

From the start, the site was well located to develop a local biodiversity: the project sits directly on the border of an extensive green area, a nearby fruit and vegetable garden and a wetland area (also used as an infiltration area for rainwater).

The privacy for the terraces is provided by pergolas and plants creeping up the sides of big wood containers. In these containers grow climbing plants (*Clematis, Rosa,* etc.), but also flowers (*Gaura, Lavandula*, etc.) and herbs (*Thymus, Rosemary,* etc.). This project has a distinctive feature rather rare in France: all plants and gardens, whether private or public, have all directly been planted *before* the future residents move in. It is clearly explained to all, private owners as well as the social tenants, prior to their moving, that living in this housing complex involves that everyone has to take part in the caring of these plants and gardens (Figs. 3.36, 3.37, 3.38, and 3.39). For the extensive and semi-extensive green roof gardens, we have

Fig. 3.36 View on terraces

Fig. 3.37 View from south east

selected seeds and plants from low-maintenance dry and rocky meadows from natural areas in the Vercors, a mountainous site set against the village.

To the seeds, we added small plants. Fate would have it that the nursery used by the garden contractor, a nursery connected to the Lautaret botanical garden, was the same as the one used in Annecy (Figs. 3.40 and 3.41).

Every flat on the ground floor has an access to a small front garden. The rest of the available space is dedicated to the bike and pedestrian path (Fig. 3.42) and fruit and vegetable gardens. These fruit and vegetable gardens occupy small terraces and are managed directly by willing and motivated residents. This design choice to involve the residents in the future garden management has for consequence the necessity of an organized public dialogue. Everyone must participate to decide how things should be done.

In the central building, a large locale is symbolically rented to most gardeners. Rain water, collected in four tanks set in the site, is also symbolically sold to them. The rain water surplus is steered into a succession of small ponds, the last one located in the public space garden nearby (Figs. 3.43 and 3.44). The garden management specification book forbids all chemical treatment.

The proximity between the gardens, bushes, and wetland areas offers a better chance to develop biodiversity. The gardens were also invested with some bird and bat houses (Figs. 3.45 and 3.46).

The ambition to live differently *with, in, and around* nature while still living *in* the city served as the foundation to all the commercial information and marketing when the flats were to be sold. At the same time, gardening interest and respect for nature were part of the selection process for all the social housing flats. Beginning September 2013, tenants and residents alike started to move into their new homes

Fig. 3.38 Roof garden for the collective terrace

and we still meet with them to explain and share experiences about the management of their gardens.

To give life to these gardens, we planted 9000 plants from 90 different species: 55 trees from 19 different species, including 15 varieties of fruit trees, 2200 bushes from 30 species, and 6900 flowers from 41 species, including 380 bulbs of 9 different species and climbing plants (12 species).

For the roof gardens, we used seeds as well as 3500 small plants, from 40 species.

In this project, we tried to create a melting pot of native and horticultural plants, ornamental and producing plants, grass species, and herb species (fruits trees and bushes). Our intention was to try and respect a sort of "natural plant association" an "eco systemic association." All these plants were selected for their food potential for insects, birds, and small mammals.

3.4.3 Learning Process, Thoughts, and Reflexions

We entered this project with many hopes and desires, many things to test and surprisingly our client, a social housing corporate executive and our partner, the architectural team leader seemed to agree with most of them! A second surprise came when the families who bought these quite expensive flats, just as the families chosen for social reasons, seemed thrilled with the project and simply waiting for a more ecological living environment. We met with all of them as they were moving in, and most of them were "just smiling," happy to be moving into these new flats. When looking at them, the social flats quite resemble the private flats, to the point that from

Fig. 3.39 Roof garden for the collective terrace

Fig. 3.40 Planted semi-extensive roof garden

the outside, it is impossible to distinguish the social housing building from the private housing building.

Ironically, the desire to live there was strangely mixed with some old fears: the fear of mosquitoes, of bats, of the water quality, of their own capacity to take care of the vegetable gardens, of having to deal with potential crisis when sharing the fruits, etc. (Figs. 3.47 and 3.48).

Fig. 3.41 Planted semi-extensive roof garden and then 2 years later (Photo Mr Billic)

Fig. 3.42 Passage between two gardens

Today, discussions are still needed, ever evolving. The residents have now been living in their new homes since moving in 2013/2014, prepared to take care of the gardens. They meet regularly and almost everyone knows each other! We are ourselves impatient to see how people will keep getting involved, how the project itself will evolve in time as much for the social integration, as for the biodiversity development.

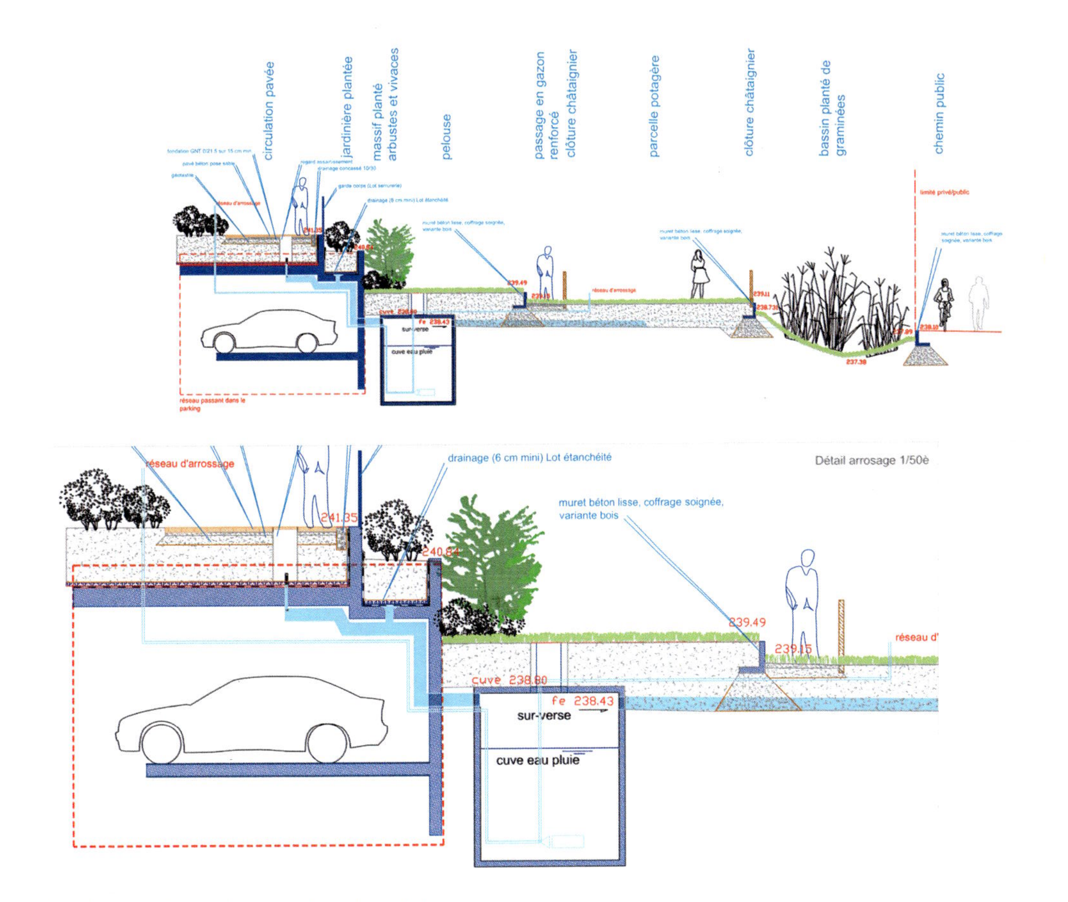

Fig. 3.43 Garden and parking section with rain tank–technical drawnings

Fig. 3.44 Rain garden at the front entrance

Fig. 3.45 Bird houses located in the common terraces

Fig. 3.46 Bat houses

Fig. 3.47 Vegetable gardens

Fig. 3.48 Vegetable gardens

Meanwhile, all of these projects represent a great opportunity to learn to be patient and learn that the ambition linked to biodiversity in project design can help create a peaceful social environment.

The Seyssins real estate project illustrates, just as all of the three other projects, that most of the sustainable and ecological character in our projects constitutes the hidden part of the iceberg. Our designs just follow the appetite our generation seems to have that desire for nature and simplicity. The pragmatic integration of sustainable goals in the design process, from the first sketches on, explains perhaps that we forgot that all these gardens are rain water "machines," a source of new biodiversity in the heart our cities, a local "air conditioner," and a powerful catalyst of social link.

Let us also remember that the refurbishing or creation of public or common spaces needs to involve the living being, whether it be human, animal, or plant. Today, urban living does not mean breaking away from any form of nature (water, flora, and fauna) or our intimate relation with these elements. The return to more permeable grounds, the sharing of garden allotment spaces, the welcome of a new biodiversity at every building's feet, at every building's top: this return of nature in our towns and our neighborhoods creates a new landscape for our cities.

Chapter 4
Edible Landscapes: Relocalising Food and Bringing Nature into North London

Joanna Homan

Abstract As a response to restore scarcity and climate change, urban design needs to shift decisively towards more localised food production and increased biodiversity. The area of North London around Manor House is an interesting example of an urban setting which has many burgeoning grassroots and third sector projects which demonstrate this principle. Furthermore, there is enormous potential for social enterprises to be established that will ensure a greater economic and social resilience for the area. Forest gardening is an approach that tackles both food scarcity and biodiversity issues. Foraging is a simple way to unlock the existing food potential of the area.

Keywords Food scarcity · Urban food security · Forest gardens · Peal oil · Climate change · Urban biodiversity · Foraging

4.1 Food Scarcity: The Background

4.1.1 Food Scarcity Due to Peak Resources

London is dependent on outside sources of food. In fact 80% of London's food is imported (Boycott 2008) and 'has only three or four days' stocks of food should there be any disruption to supply (Jones 2010)'. This makes it hugely vulnerable to *peak resources*, meaning resources such as fossil fuels, water and trace elements:

- As fuel price rises, food prices will also rise and food will become less available to Londoners.
- As additives to artificial fertilisers such as phosphorus become harder to obtain, artificial fertilisers will become more or less available and more expensive.

J. Homan (✉)
Finsbury Park, London, UK
e-mail: tt@jo.homan.me.uk

K. Ito (ed.), *Urban Biodiversity and Ecological Design for Sustainable Cities*,
https://doi.org/10.1007/978-4-431-56856-8_4

- With a growing population, polluted water supplies and decreasing summer rains, water will also become more of a scare resource.

Some of the impacts of *peak resources* on food production could include:

- A much greater proportion of people's income would be spent on food.
- Hunger/decreased health/starvation/civil unrest—perhaps government attempts to feed large numbers of people.
- A movement of people from cities into the countryside where food production becomes more people-powered rather than fossil-fuel powered.
- Less food waste—it would no longer be possible to gain free waste food in community projects or continue the practice of 'skipping'—where individuals go to shops and markets at the end of the day to collect and eat food which is destined for landfill sites.
- A necessity and desire for many more people to grow food locally—leading to increased health (Healy 2006).
- A necessity and desire for food growers to harness natural fertilisers and grow crop which are less demanding of nutrients.
- A necessity and desire for food growers to grow drought resistant crops.
- More integrated gardening designs—e.g. one which intelligently integrate human waste and rainwater harvesting.

Of course it is impossible to know what would happen in a capital city like London. It is highly likely that central government would do what it could to ensure there was a food supply to keep its citizens fed. However, there is a limit to what even the most determined government can do if the resources simply are not there.

Even without this happening, there is already a situation where, due to a lack of real choice, some areas are regarded as fresh food desert (Bowyer et al. 2006). This report demonstrates that the poorest people in urban city centres end up paying the highest price for a limited range of less healthy foods. Locally produced on fresh fruit and vegetables simply are not available—just the more processed, over-packed and ultimately costly convenience foods.

4.1.2 Food Scarcity Due to Climate Change

Climate change is predicted to pan out as follows (Crawford 2010c):

- Warmer temperatures on average, over the year—London would be as hot as Southern France (1960 temperature). The average London temperature is already about 1.8 degrees higher than in the 1960s.
- In wintertime, the average night temperature will increase more than the average day temperature. But in summertime it will be the other way around—days feeling proportionally warmer.
- Fewer frosts and temperatures may not drop enough to 'chill' certain plants—leading to lower yields.

- Increased instances of erratic, heavy rainfall along with rising sea levels will increase the chances of flooding. This could lead to increased soil erosion; soil destabilisation; loss of top soil; less land available for growing; salination and/or contamination of soil.
- Rainfall will increase in the winter and decrease in the summer and summers will be less humid. There will be less cloud cover in the summer but slightly more in winter.
- Increased solar radiation will extend the growing season in London and increased carbon dioxide will help healthy plants grow bigger.
- Desalination of the oceans caused by melting ice caps will affect the density of water. This will disrupt the circulation of oceanic waters, manifested as the Gulf Stream. This will in turn affect the weather patterns that currently keep the South Western UK warmer.
- Soil will be generally less moist because of the combination of changes to rainfall, temperature, evaporation, wind speed and solar radiation.
- Changes to temperature and humidity will have an impact on insect lifecycles, allowing potentially harmful insects and diseases to thrive.

This could have these effects on food growing:

- Food growers growing annual plants in monocultures could lose their entire crop in one hit. This could lead to certain food staples being unavailable. For example, the 2012 London apple harvest was virtually non-existent because of prolonged heavy rainfall coinciding with the pollination window.
- Plants grown indoors will be more prone to pests and diseases—some of these pests could survive outdoors. Again, plant monocultures would be more vulnerable.
- A necessity and desire to grow drought resistant plants, e.g. deep-rooted perennial plants and to prioritise rainwater harvesting in any project designs.
- A necessity and desire to grow plants that do not require winter chilling.
- A necessity and desire to grow food polycultures which would spread the risk of plants responding poorly to extreme and unpredictable weather or sudden insect-population explosions.
- A necessity and desire to plant more trees to protect soils and shelter smaller plants.
- A necessity and desire to protect, preserve and build soils, e.g. by covering or planting with shallow-rooted plants.

4.1.3 Food Scarcity Due to Lack of Biodiversity

Pollinating Insects. It is widely accepted that insects are essential as pollinators for many of our food crops. Insect numbers are depleted due to a range of different factors, mostly man made, to the point that there is now a significant pollination deficit, especially where food is not grown in biodiverse setting (Tirado et al. 2013).

A healthy food growing space will have an abundance of airborne and other pollinating insects.

Biodiversity—other species that live above the ground. In addition, having a range of arthropods (e.g. minibeasts such as insects, spiders, centipedes, woodlice), bird and amphibians existing within a garden system leads to a more balanced ecosystem which has enough predators to ensure no specific species becomes over-dominant. In plant monocultures it is very easy for one species to dominate and destroy a food crop because:

- The plants in a monoculture are competing with each other for the same nutrients, water and light and are therefore stressed. In this weakened state, the plant is less able to deter predatory species, such as burrowing beetles or fungi.
- Plants in a monoculture are easy for predators to spot because they are physically near each other and conveniently laid out in rows.
- Food growers usually isolate monocultured plants in earth deserts which provide no habitat for other species which would otherwise prey upon the 'pest' species.

Biodiversity—other species that live below the ground when considering the biota living in soil, it is also clear that having a range of species enhances the growing potential of any site. A rich soil biota is only obtained when soil remains largely undisturbed from one season to the next, is protected from the weather and from compression (e.g. being walked on) and is fed using plenty of organic matter. The practice of frequent digging associated with monocultural food growing kills many beneficial soil species, dries out soil, releases carbon to the atmosphere and also reduces soil fertility. This further weakens the plants and increases their vulnerability.

Lack of biodiversity within urban settings leads to plenty of other problems for humans that are related to mental health, e.g. increased stress levels or nature deficit disorder. There are established links between being in nature and good mental health. The UK mental health charity, Mind, found that participating in outdoor activities, 'provides substantial benefits for health and wellbeing' Mind (2007). Furthermore, urban biodiversity is now identified as extremely important for supporting certain species populations whose survival is under threat in the supposedly natural setting of the countryside (Davies et al. 2012; Fuller et al. 2012). Back gardens across the UK make up 3% of the total land available (Cannon et al. 2005) and when public spaces are included, this makes for a reasonable amount of potential food growing land.

Although these issues are beyond the scope of this article, it is worth remembering that biodiversity and food availability are issues which sit within a wider context. Likewise it is important to realise that creating or developing projects that increase biodiversity and food availability will have enormous additional benefits to local communities and society at large. These benefits may even be more important than the professed aims of the projects—food scarcity and biodiversity. To put this in a different way, developing a community's resilience to the external shocks of climate change and peak resources by developing community cohesion may be of greater significance than obtaining a yield of food.

To summarise this section: We need a food growing system that is an intelligent response to the challenges of climate changes, resource scarcity and poor biodiversity.

4.2 Forest Gardening >>> Food Production and Biodiversity

Forest gardening is growing food in a *polyculture* that includes trees. 'Polyculture' simply means that several plants are being grown together, as opposed to a 'monoculture' where just one crop is being grown—i.e. the current norm. The name, forest gardening, is slightly misleading because it implies a growing approach which is dominated by trees. In actual, forest gardens emulate the growing conditions found at the *edge* of a forest or woodland. They do not have to cover acres of land or include trees if space is not available. They are beautiful, productive and intrinsically resilient because they mimic what is found in nature.

There are seven main features of a forest garden which are outlined here.

4.2.1 Plants Are Grown in a Polyculture

Natural woodlands are robust systems which have evolved and stood the test of time. They are polycultures, containing a mixture of plants. Forest gardens mimic this because there are several advantages:

- Plants are not competing with each other so much for light, water, etc.
- Plants will be multifunctional. For example, a Siberian Pea Tree will provide shelter for smaller plants, provide edible peas and fix nitrogen from the air, which can then be used by other plants. As well as having an edible function, forest garden plants can yield firewood, poles, tying materials, medicine, soaps, dyes, honey, mushrooms, sap and spices.
- Plant biodiversity is essential when there are unpredictable weather patterns because the risk of crop failure is significantly reduced—there are so many plants some are bound to survive of thrive. Martin Grawford's two acre forest garden has over 500 species in it (Crawford 2010b).
- Plants in a polyculture are a living, collaborative system which includes the soil biota—see Sect. 4.2.7. Plants and soil biota support each other by passing around nutrients, camouflaging each other against predators and attracting pollinating insects.
- The huge range of habitats within a polyculture ensures there are enough different minibeasts, birds and amphibians to ensure no single species becomes over-dominant. Martin Grawford's forest garden was found to contain a higher range of species than a native woodland of a similar age (Crawford 2010a). This means forest gardens encounter far fewer problems with pests and diseases.

4.2.2 *Plants Are Grown in Layers*

Where a woodland meets an open space, the sudden availability of light enables plants to grow at every height. At the highest level are the trees which offer shelter from strong winds to younger plants as well as protection from the fiercest solar radiation. Air beneath the branches is nice and humid and so the trees, or the *canopy layer*, create a safe space for other plants. Beneath the canopy layer the plants gradually diminish in size: going from smaller trees, through shrubs and herbaceous perennials to groundcover plants and annual plants. Cutting across the layers in a forest garden are climbing plants such as kiwi, hops and grape vine. Below ground level, the roots of different plants reach different depths which means that a wider range of nutrients is being captured by the plants and they are not competing with each other in the way that would be found in a monoculture (see Fig. 4.1).

Growing plants in layers makes the most efficient use of the physical space available—both its height and width. It allows a mixture of plants to grow to their optimal height in way that contrasts with planting rows of annuals which will need to compete for the same light, water and soil space.

Growing plants in layers has another benefit to do with the timing of the uptake of water and energy from soil. When plants come into leaf in the spring, they have a sudden demand for water and energy as they put on this new growth. In forest

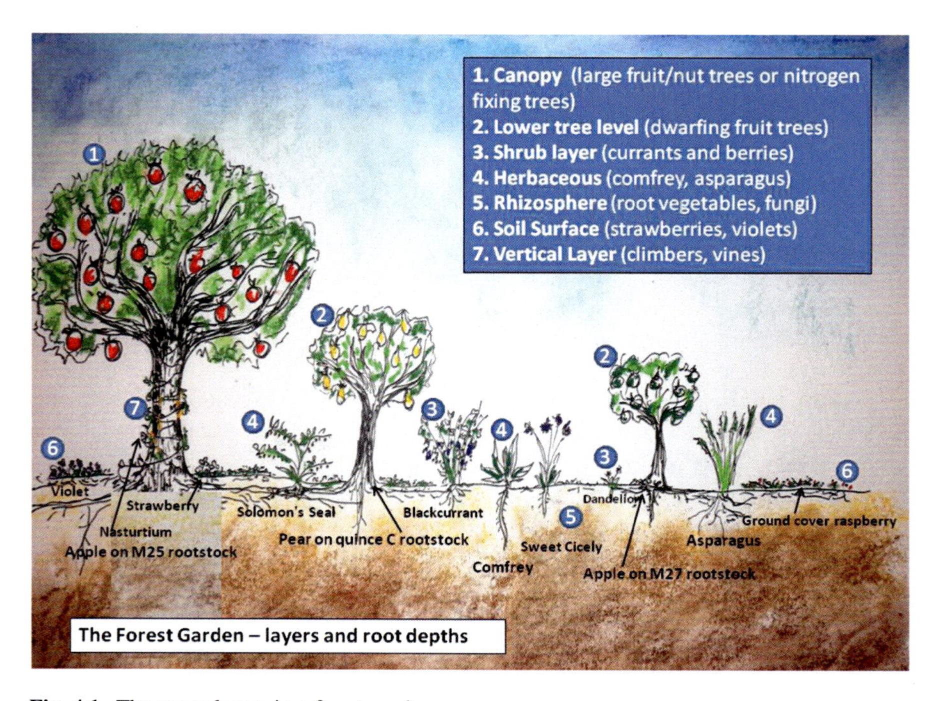

Fig. 4.1 The seven layers in a forest garden

gardens the shrub layer (and some of the groundcover layer) comes into leaf before the canopy layer, thus spreading out that demand. For example, blackcurrant and berry bushes will come into leaf before apple or oak trees. This makes it more likely that each of these plants' needs will be met—contrasting again with what happens in a monoculture (Whitefield 1996). In addition, the yields from different plants are also spread out over the year—so the workload and plant yields can be more comfortably managed.

4.2.3 Perennial Plants Are Often Used

Herbaceous perennials are unwoody plants such as rhubarb or artichoke, that die back to ground level in the winter time, but which have root systems that survive over winter. Every spring, they magically appear to come back to life because all their energy was stored in their root system. Shrubs, bushes and trees do exactly the same thing except that they retain a woody structure over winter, instead of rotting back to ground level. The fact that these plants grow back each year is what makes them '*perennial*'. In fact, all of the layers in a forest garden are perennial, except for the *annuals*, which are usually profligate self-seeders that need very little maintenance.

Annual plants have a life cycle that lasts for one year. Every year they must start from scratch: seeds must germinate, grow a root system, a plant structure and then leaves, flower and fruit. Annual plants need a lot of energy and light to be able to achieve this rapid growth. Perennial plants have a life cycle of more than two years. Trees are perennials and some of them live for hundreds of years. Other perennials may only be productive for three or four years.

The distinction between annual and perennial plants is important to understand because it is one of the main differences between forest gardening and conventional gardening. Where conventional food growing is based on growing annuals, forest gardens mostly use low-maintenance perennial plants. There are several advantages to this, especially when considered in the light of climate change and resource depletion.

- Perennial plants start off every year with a root system and, in the case of trees and shrubs, a physical framework. This means that they get a head start over annual plants. Perennial plants with edible leaves (for example, Small Leaved Lime trees, or Wild Garlic) are producing leaves during the so-called hungry gap where conventional gardeners are still waiting for their annual crops to deliver a yield.
- Because perennial plants have less need to grow quickly they have lower energy demands than annual plants.
- Perennial plants are usually deeper rooted than annuals. This makes them more likely to tolerate dry conditions and withstand extreme weather conditions, e.g. strong winds.

- It is less work growing perennial plants because they do not have to be replanted every year. This lack of digging also has huge benefits for soil biodiversity—see below.

4.2.4 Forest Garden Plants Have an Increased Nutritional Content

Growing plants in an organic polyculture produces plants that are much higher in proteins and minerals. Shiva and Singh (2011), in their study of Indian farms growing plants in a monocultures and polycultures, found, amongst many other things, the following results: (Table 4.1).

4.2.5 Non-native and Unusual Plants Can Be Used

As the climate changes there is an opportunity for forest gardeners to experiment with growing unusual and non-native plants. Some plants are simply seen as unusual because they have fallen out of fashion—for example, Medlars and Service Tree fruits. Other unusual forest garden plants are well-known plants such as Fuchsia and Ice Plant that people simply do not realise are edible (see Fig. 4.2). There is an enormous number of non-native plants that are now beginning to thrive in the UK. They have many advantages such as disease resistance, larger fruit, better flavours and greater yields. Some plants are different versions of traditional UK plants, such as non-native Hawthorns and American Elder. Non-native Hawthorns are easy to grow and produce fatter, tastier fruit than the UK varieties. The American Elders have a much longer flowering season that the UK elder and, if planted in isolated, will not be pollinated and therefore keep on producing flowers. As such these plants offer plenty of exciting potential.

Table 4.1 Comparing the nutritional content of monocultures and polycultures (after Shiva and Singh 2011)

	Protein (kg)	Carbohydrate (kg)	Fat (kg)	Total energy (kcal)
Average production of nutrients from organic mixed farming	240	833	66	4,914,270
Average production of nutrients from conventional mono cropping	116	785	23	3,711,475

Fig. 4.2 Less well-known edible plants—Medlar (*Mespilus germanica*), Service Tree (*Sorbus domestica*), Fuchsia (*Fuchsia*), Ice Plant (*Hylotelephium spectabile*)

4.2.6 There Is Rarely Any Bare Earth

In forest gardens the soil is always covered with something. This could be groundcover plants such as Clover or Strawberries. It could also be cardboard, woodchips, leaves, plants that have just been weeded, newspaper, straw, or some other semi-permeable but opaque membrane such as carpet or plastic sheeting.

When dead organic matter is used as groundcover—also called a 'mulch'—there are huge benefits to the growing system:

- water remains in the soil because it does not evaporate away so quickly—unlike with bare earth
- new seed growth is suppressed—they cannot grow without light. This means there is less competition to new plants. For young trees this is really important as they grow twice as fast when there is no competition (Crawford 2010d)
- soil is kept slightly warmer, enhancing plant growth
- the much eventually rots down, feeding the soil

The use of groundcover is one of the reasons why forest gardens are easier to maintain because there is less time spent weeding, watering and feeding plants.

4.2.7 The Soil Is Biodiverse

The main reasons why soil is more healthy in forest garden are as follows:

- less digging, so soil biota left to flourish, e.g. arthropods, bacteria, mycorrhizal fungi and saprophytic fungi. Not only does the soil biota act as a carbon sink, they are also like mini bags of compost, trapping nutrients that would otherwise be washed away by rain (see Fig. 4.3).
- soil is less compressed by footfall or machinery, so soil structure is kept healthy. Plant roots need air pockets in the soil to be able to metabolise the

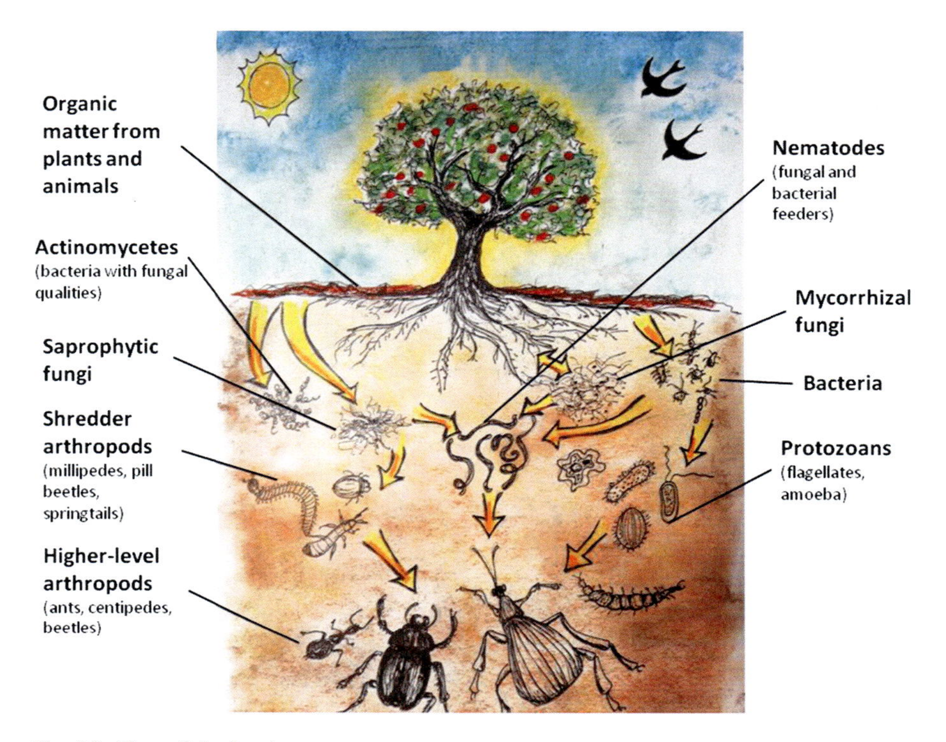

Fig. 4.3 The soil food web

elements in the gases, as well as enough space to be able to physically push through soil.

- soil is fed naturally by mulching which also enhances the soil structure. There needs to be lots of soil biota to process the mulch as it is added.
- mycorrhizal fungi living in the soil support all the other plants. Underground networks of fungi transport nutrients from areas of high concentration to areas of low concentration. This amazing process happens because fungi have a symbiotic, mutually beneficial, relationship with plants. The fungi grow into the fine root hairs of plants and literally exchange nutrients for sugars, which the fungi are unable to generate for themselves. When left alone, soil will be full of these fungi which will fertilise the plants completely gratis.

To summarise, forest gardening is a more efficient and resilient approach to food growing that encourages biodiversity.

4.3 Foraging: Coming from Biodiversity

Foraging is the practice of harvesting plants growing naturally in a biodiverse setting. There are hundreds of common plants that have edible leaves or fruits which can be selectively picked to create an additional food source. Well-known examples of foraged plants are blackberries, apples, mulberries, acorns and sweet chestnuts. People know less about the edible leaves that are available in spring and the edible flowers that come a little later.

All of the plants grown in conventional agriculture are cultivated varieties derived from plants that still grow in the wild and can be foraged. The plants that can be foraged change throughout the year and there is a growing body of individuals in North London who are capable of teaching people how to recognise, eat and cook foraged foods. As long as plants are not overly picked, there is an extant abundant source of food. Some of the plants, such as nettle, ground elder, three cornered leek, fat hen, mallow and chickweed grow so enthusiastically that it is hard to imagine them every being over-harvested!

It is rare to find any land that is completely un-managed by humans in a city, i.e. complete wilderness. For example, brambles must be cut back, paths must be kept clear, wind breaks and hedgerows are often planted and trees are often thinned out or planted. It is therefore entirely possible that nature reserves and public amenities can be managed intelligently so as to develop the foraging potential of any given site.

To summarise, foraging is an outcome of biodiversity which can produce food for humans with relatively little effort.

4.4 Manor House, North London: The Background

Manor House is an interesting and useful area to examine when considering food growing and biodiversity. It is an area of extremes.

It includes four electoral wards in the London boroughs of Hackney and Haringey. They are among the most deprived wards in the UK, with pockets of severe deprivation. The English Indices of Deprivation measure relative levels of deprivation in small area of England called Lower Layer Super Output Areas. It reports that 29% of people in Haringey are in the most deprived Super Output Areas, meaning that about one-third of households in the area are in the top 10% most deprived areas in the country. Furthermore, the Mental Health Needs Index shows that wards in the Manor House area have higher than average levels of admission for mental illness—the worst 10% in the country.

Characteristics of the area include (Manor House Development Trust, Manor House PACT Final Bid Document 2012):

- A high percentage of unemployed people and people claiming benefits

- Low income families and a high level of mobility in and out of the area
- Isolated and/or vulnerable individuals
- Distinct ethnic minority communities
- Public open spaces, with barriers to connectivity and potential for collective action
- Higher than average (for London) levels of students and older people
- High density housing and Victorian houses that have been converted into flats
- Two major 'Red Route' roads cutting though the area and high levels of localised pollution

Most local residents do not have access to a garden or to land they can freely cultivate. Waiting lists for nearby allotments are closed 'due to excessive demand', showing there is a un-met need for people to have access to food growing opportunities. Haringey Council's sustainable food strategy notes 'Would be home-growers often lack the technical skills needed to grow food'.(Haringey 2010) This shows the need for easily accessible horticultural learning opportunities.

There is a £1 billion regeneration project going on in Woodberry Down estate with the estate's population planned to double from 5000 to 10,000 over the next ten years. This means there is lots of building work going on and limited potential to develop some more community spaces. Funding was recently gained to improve the quality of life along one of the main roads, Green Lanes.

Furthermore additional funding was also obtained (£1 million over 2013–2015) to help protect vulnerable residents from the effects of climate change (Manor House Development Trust, Manor House PACT Final Bid Document 2012). Part of this money will be spent on horticultural training and one of the project's aims is to secure two community spaces which are under local control or ownership.

The Manor House area has some really strong natural assets and amenities, such as West and East Reservoirs, Finsbury Park, Clissold Park, the New River walk, Islington Ecology Centre and Railway Fields. Dotted around the area there are several local environmental groups and community organisations offering a range of food growing and biodiversity projects. When looked at in combination it is possible to see that it would not take much effort to pass the tipping point that would enable significantly better outcomes, in terms of biodiversity and local food production.

To summarise, the Manor House area of North London has a complex mixture of needs and possibilities.

4.5 Manor House, North London: Examples of Resilience

See Fig. 4.4.

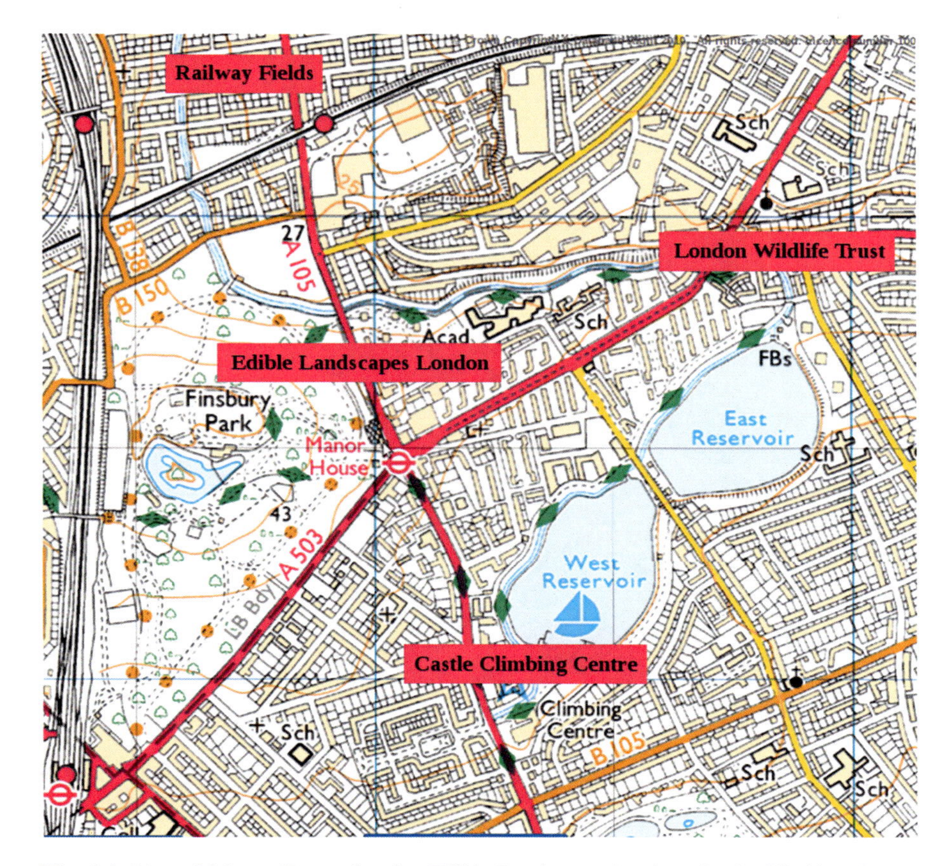

Fig. 4.4 Map of Manor House showing Edible Landscapes London, Castle Climbing Centre, Railway Fields and London Wildlife Trust

4.5.1 Edible Landscape London

This volunteer-led project is a highly accessible training centre and plant propagation site that operates on forest garden principles. Edible Landscape London, known as 'ELL', aims to help Londoners grow more of their own food by propagating low-maintenance edible plants which are a range of methods such as planting seeds, grafting or taking cuttings. ELL has been running since September 2010 and so far over 1200 plants have been supplied to over 90 community run food growing projects (ELL Propagation and Plant 2013). Example of projects are estates, schools, community gardens, hospitals, churches—everywhere that people are growing food in public settings. Plants are propagated in raised beds and other site features are a rainwater harvester, compost heaps, separating compost toilet, pond and a large forest garden with over 200 species in it (see Fig. 4.5). As well as supplying plants, ELL's mission is to educate and train people about forest gardening. People are

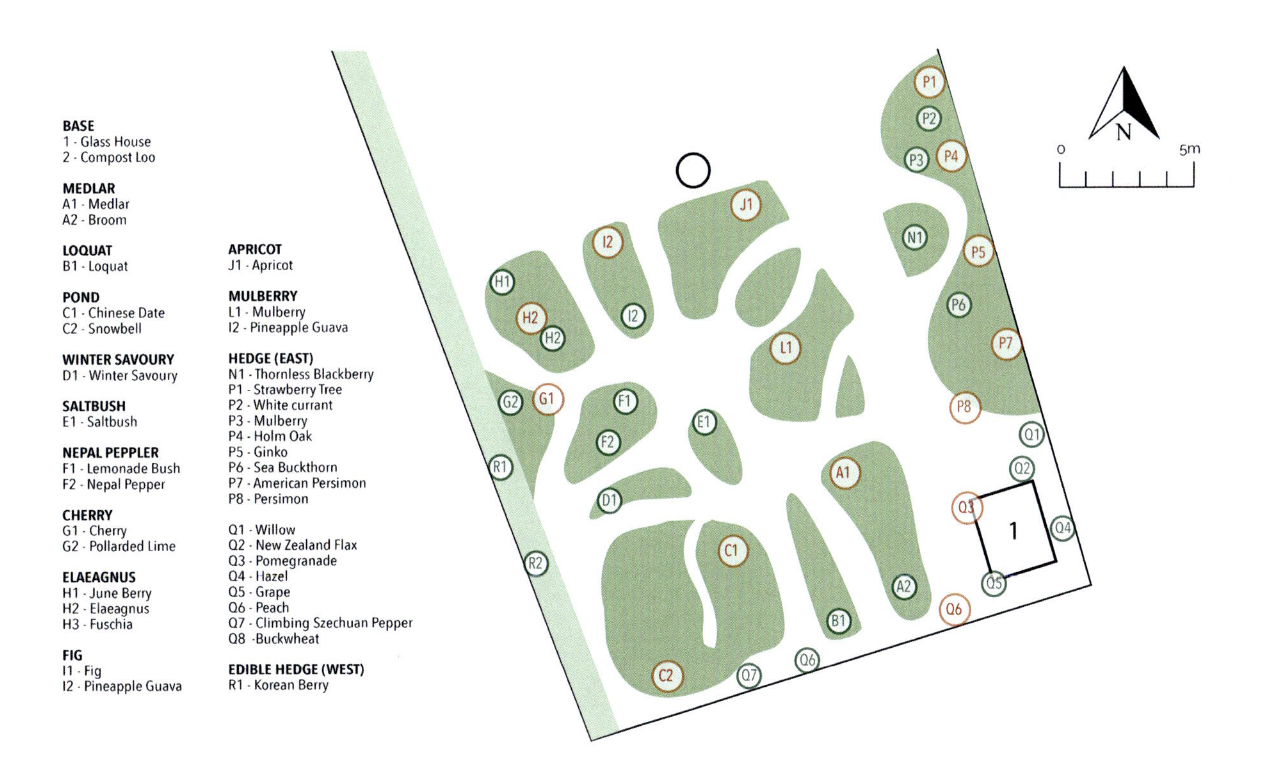

Fig. 4.5 Map of ELL site

taught how to recognise the plants, which parts are edible, how to propagate them, how they are grown in a forest garden and even how to cook with them. There are many informative signs which will allow the casual park user to learn all about forest garden plants.

ELL runs many horticultural training sessions. 120 people were taught in the second year by project volunteers. Course attendees could pay in cash or by volunteering. Following a successful funding bid to the lottery, which was in partnership with many local organisations, ELL scaled up its training efforts in 2013 to include accredited training. People could continue to pay for training by volunteering, which has been a useful way of gaining the time and experience of very good quality volunteers. One of the training courses was brand new; Creating a Forest Garden. The accreditation for this ground-breaking, 20-week course was developed in partnership with the Permaculture Association and enabled over 50 students to gain a nationally recognised qualification. Furthermore, ELL has joined the Permaculture Association's LAND network (the Permaculture Association), which is a useful mechanism for spreading the learning opportunities more widely. LAND centres throughout the UK will be able to run the training pioneered at ELL. There are plans to add a classroom which will allow ELL to scale up its training programme.

Because ELL runs as a forest garden, there are always plenty of flowering plants for the bees which are based on site and cared for by another site stakeholder. There is also a high diversity of wildlife on site but no site survey has yet been carried out.

ELL operates on a site owned by Haringey Council. ELL uses about 300 square metres of the site which is found near the Manor House entrance to Finsbury Park. ELL runs regular work days which are open to anyone who has an interest in growing and eating food, and these workdays include a shared meal. Becoming a volunteer has minimal paperwork associated with it and the project's location within a busy park means that the project is well-stocked with volunteers.

ELL has strong links with many local organisations and a willingness to collaborate with other like-minded groups. Children from the local secondary school work on site and will support the planting of a new showcase bed inside Finsbury Park itself. The recent Food Forest Project was an outreach project that saw the creation of forest gardens at three local community venues: an old people's daycare centre, a drop in centre for mentally unwell people and a residential site for male refugees. Additional sites have been located and it is very likely that outreach will continue to be an important strand.

4.5.2 *Castle Climbing Centre Gardens*

The Castle Climbing Centre is an extremely successful indoor climbing centre business which has a robust and clearly thought through environmental policy (environmental sustainability and the castle 2013) (The Castle Climbing Centre). The Castle is found just south of the busy Manor House junction and is very near

Fig. 4.6 Castle Climbing Centre Gardens

West Reservoir. The centre is working its way through an ambitious project plan which includes some outstanding work in the outdoor areas in the Castle Grounds (The Castle Climbing Centre). One of their aims is to 'Develop a permaculture based garden on our land that promotes biodiversity and produces food for the Cafe'. (The Castle Climbing Centre 2013) They are well on the way to achieving this (see Fig. 4.6).

The Castle is very accessible with any visitors able to simply walk into the reception desk, sign in and visit the grounds. There are several celebratory and activity days during the year as well as regular volunteer days. Gardening volunteers can earn themselves climbing credits as well as free refreshments from the Castle Cafe. Work in the 1 acre garden began in 2009 and now the grounds are multifunctional, with several concurrent, complementary projects up and running:

- allotments for local residents
- growing space for local vegetable delivery box scheme, Growing Communities (Joe S)—in fact, the Castle is a collection point for Growing Communities customers
- growing space to supply herbs and salads for the Castle Café
- bee hives
- composting systems
- forest garden, including a pond
- herb collection and drying areas
- cob oven under a green roof shelter made from cob and roundwood
- mushroom growing

It describes itself as a "vibrant green space teeming with wildlife" and it is very popular with local residents and the climbers. Indeed it won the Sustainable City award for Sustainable Food in 2013.

Other local projects know about what the Castle is doing and it is well linked in with the London—wide organisation, Capital Growth. There are at least three paid members of staff working part time on the project and the project benefits from the full support of the financially successful and ethical Castle management.

4.5.3 Railway Fields

This two acre site is based north of ELL, just off Green Lanes. It is a long thin site of mostly woodland managed by the conservation charity, The Conservation Volunteers (TCV) since 2002. This site is bounded by the New River, a rail line, Green Lanes and residential back gardens.

Railway Fields is devoted to promoting an awareness and appreciation of wildlife. It has around 600 children visiting each year as part of their education programme. Events taking place through the year, and conservation volunteering every month. It is open on weekdays, mainly (see Fig. 4.7). As well as an active Friends group, Railway Fields has paid staff who work in partnership with Haringey Council to deliver some of their conservation management work across the borough. Because the land used to be a goods depot, it has some contamination and it is not possible to grow food on site or forage the wild plants. However, fast-growing fruits such as blackberries are considered safe to eat.

Although small, the site has many habitats: woodland; scrub; meadow; pond and marshland. There are over 200 species of wild plants on site, including the unique Haringey Knotweed discovered in 1987, which is a remarkable cross between the Japanese Knotweed and the Russian Vine. Twenty-one kinds of butterfly have been recorded and more than sixty species of birds have been observed since Railway Fields first opened. As such, Railway Fields is an important local amenity acting like a biodiversity bank for the local area. The biodiversity is monitored jointly by TCV and Haringey Council (Haringey 2013a, b).

4.5.4 London Wildlife Trust

London Wildlife Trust (LWT) has a base next to the New River and East Reservoir just to the east of Manor House Junction. LWT is a London-wide organisation that wants London to be city where all people treasure wildlife and natural spaces. They try to achieve this by campaigning to protect threatened land, managing land appropriately and communicating their aims with as many people as possible (London Wildlife Trust 2013).

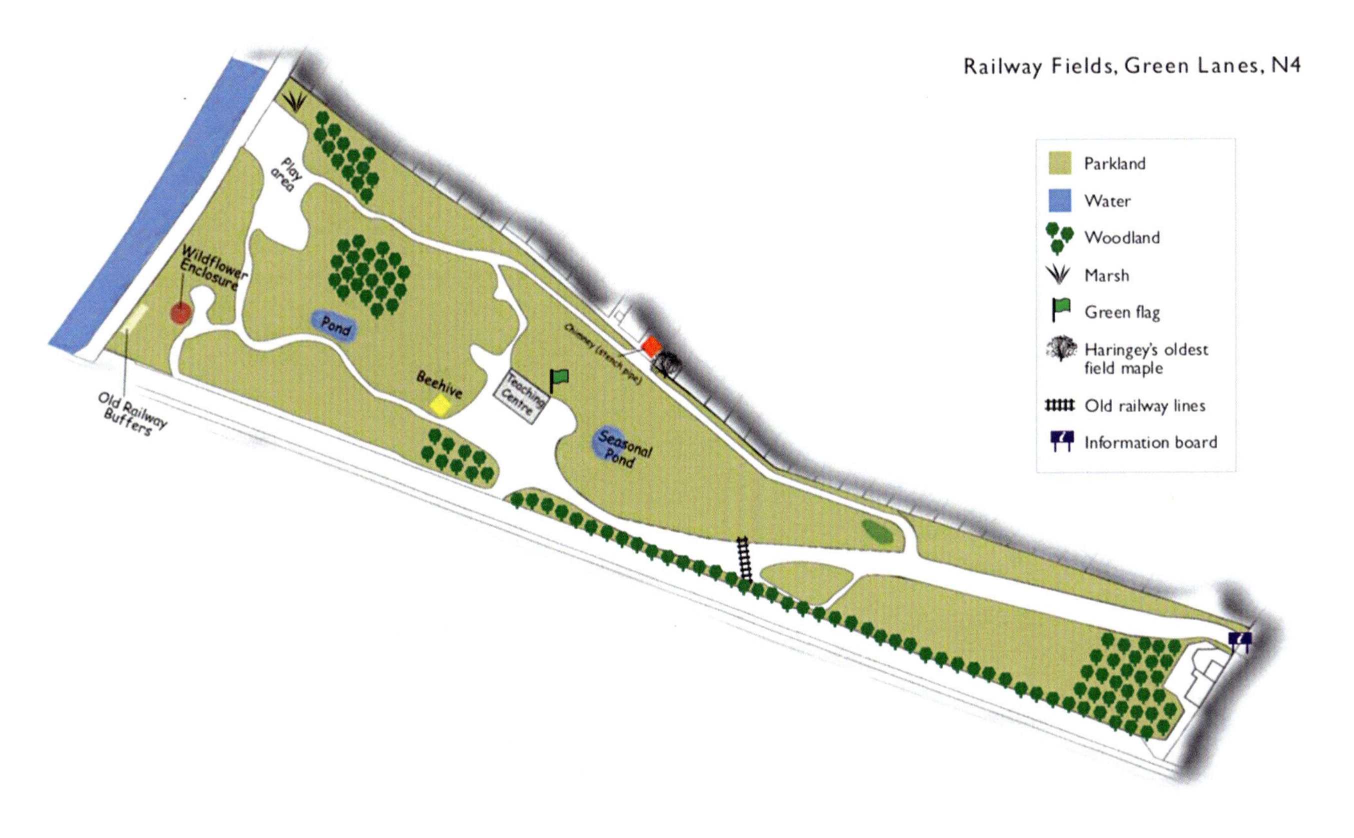

Fig. 4.7 Map of Railway Fields

Fig. 4.8 View of East Reservoir from London Wildlife Trust

LWT East Reservoir has a building with a large teaching space and about 17 ha of land which demonstrates a range of habitats such as canalised river, wildlife pond, open water, reedbeds, summer meadow, woodland copse, stag beetle sanctuary, grassland, scrub and community garden. LWT East Reservoir has a number of strategies for communicating with people such as working with local school children, making space available to local residents who wish to grow food, having regular volunteer days, running wildlife walks and giving people land management skills that could enable them to get work. Some of the activities people can get involved in are 'bird watching over the reservoir, arts and craft workshops, landscape design and construction, wildwalks, pond and river dipping and mini-beast hunts along the nature trail (The wildlife Trusts 2013)'.

LWT worked in partnership with Thames Water, London Borough of Hackney and Berkeley Homes to make the area around East Reservoir more accessible and educational, leading to the opening of the flagship project, Woodberry Wetlands, in 2016—by Sir David Attenborough. New floating islands, bat and bird habitats, an orchard, new reedbeds and hedgerows all increased the biodiversity of the area (London Wildlife Trust 2013). At present, wildlife such as Great Crested Grebe, Common Tern, Reed Bunting, Emperor Dragonfly, Speckled Wood Butterfly are commonly found (see Fig. 4.8).

4.5.5 The New River

The New River is a natural amenity and wildlife corridor running through the Manor House Area and is managed by Thames Water. It is a man-made river built in 1613 to bring fresh water into London and it is part of London's 'Capital Ring' of walks. The relatively undisturbed parts of the New River are host to large amounts of wildlife, notably Lady's Smock, River Water-Dropwort, Stream Water Crowfoot, Dragonflies and Kingfishers (Haringey 2004a, b, c).

Because there are a good number of wild plants growing along the New River, there have been regular wildlife and foraging walks, which set out from the Castle Climbing Centre. During these walks people learn about wildlife and also how to safely identify edible wild plants. This empowers people by giving them an awareness that there is fresh, edible food freely available. It also connects people with the area more and with their neighbours (Manor House PACT 2014).

4.5.6 *The Reservoirs*

Manor House is very lucky to have two beautiful reservoirs of 21.25 ha in size, described in London Borough of Hackney's Sites of Importance for Nature Conservation as follows:

> Two small reservoirs surrounded by built-up areas, being fed by the New River from chalk springs. Of interest mainly as a haven for waterfowl. Wintering tufted duck numbers reach national significance at times, particularly in cold weather when the high through flow of water ensures ice-free conditions. Substantial numbers of moulting tufted duck also spend the summer here. The reservoirs have formerly held important numbers of wintering pochard and smew, but these have since declined. They still attract significant numbers of gadwall, and small numbers of mallard and other waterfowl throughout the year, while regular passage species include common waders. The east basin, an operational water supply reservoir, is the most important of the two for waterfowl, being less disturbed than the west reservoir, which is used for water sports. The reservoirs are also valuable for amphibians, supporting substantial populations of smooth newt and common toad (Hackney 2012).

The New River feeds into the East Reservoir at the northern end, where LWT has its centre. The New River Walk runs along the northern sides of the reservoirs to the Castle Climbing Centre. The walk is gated at some points to prevent cycling but is otherwise fully accessible.

4.5.7 *Finsbury Park*

Finsbury Park is a fully accessible, 46 ha (115 acre) park (Haringey 2013a, b) managed by Haringey Council. At present it is dominated by mown grass but has a few biodiverse areas such as woodland, hedgerow, neutral grassland, mature trees, a lake and amenity grassland. Species found in the park include Lady's-Smock, Lesser Chickweed, Parsley-Piert, Common Whitlowgrass. There is a closed-off road looping running around the park, many tree-lined footpaths, sports facilities and play facilities. Recently the Woodland Trust supplied a large number of native species which were planted along one of the park's boundaries. There is a community garden at Manor House lodge near the already mentioned Edible Landscapes London. There are plans to increase biodiversity still further by creating wildflower meadows in the northern part of the park and allow grass to grow longer in other parts of the park (see Fig. 4.9).

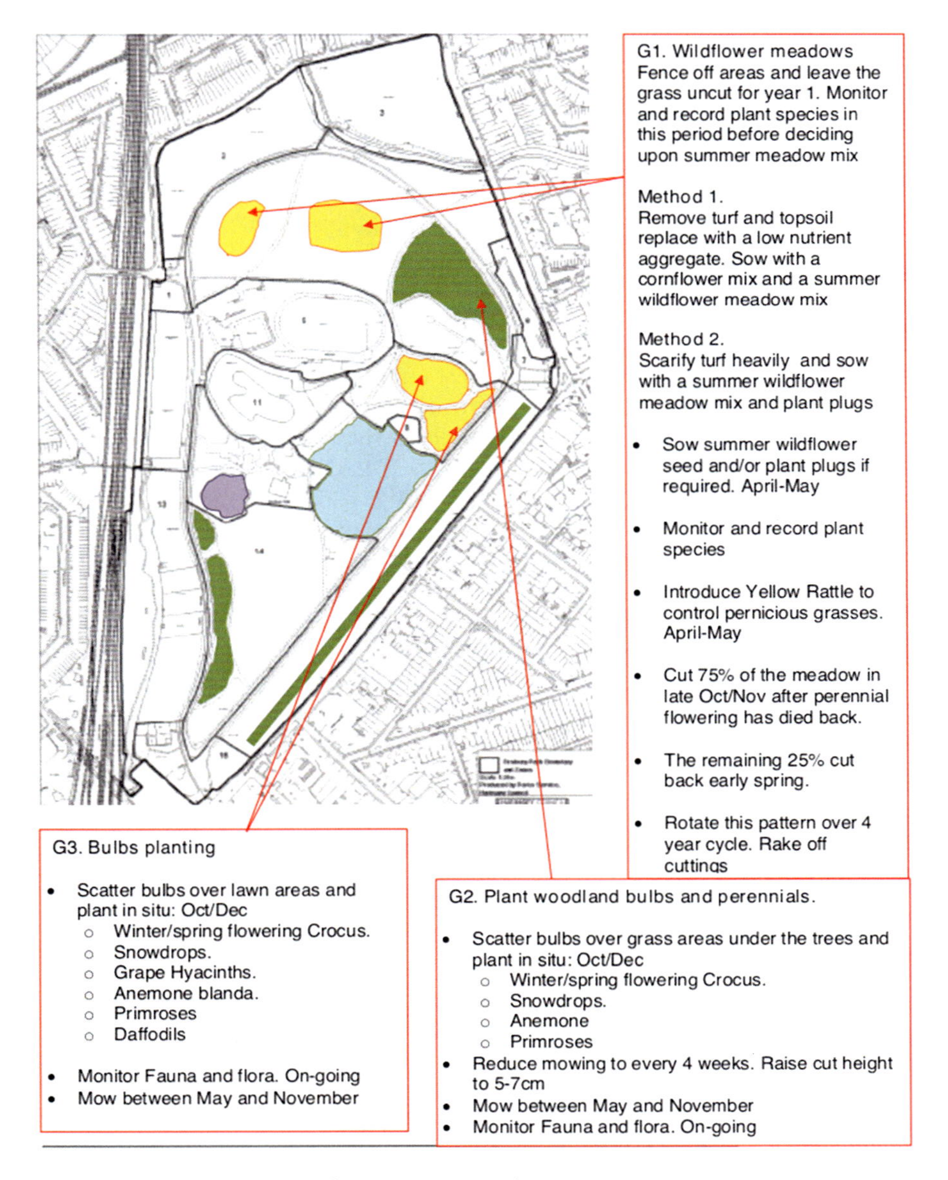

Fig. 4.9 Map of Finsbury Park from Management Plan

4.5.8 Projects and Amenities in the Wider Area

Just outside the Manor House area there are several foods growing and biodiversity projects or sites which further strengthen the food growing culture in this part of North London. These sites are ordered going clockwise from south east of the Manor House junction.

Growing Communities, based in Stoke Newington, runs a patchwork farm of food growing sites that provides Hackney residents with a food box scheme and runs a weekly farmer's market in Stoke Newington. Their food comes from 25 small scale organic farmers, some of whom are in London. They describe themselves as 'building community-led alternatives to the current damaging food system. We're working out how to feed urban populations in the face of climate change and peak oil (Growing Communities)'. Growing Communities supports individuals wishing to set up similar schemes in their area and they also run an apprenticeship scheme that helps people earn money by being food growers. People can also volunteer them—approximately 120 people volunteer with them each year.

Abney Park Cemetery is a 13 ha woodland memorial park and Local Nature Reserve, managed by the Abney Park Trust (Abney Park 2004). It is accessed by Stoke Newington Church Street and runs many wildlife, woodland management and woodland crafts activities. Some of the trees on site are ancient, around 60 being classified as 'veteran' (Abney Park 2004) (WOODLANDTRUST).

Clissold Park is a Hackney-run (Hackney 2012) park which has many amenities and natural features to recommend it. There are deer and several domesticated animals on site, as well as plenty of wildlife in the ponds. The wooded edges to the park are home to a good range of species. Growing Communities has a small but well-sign posted patch in the park and there are regular compost giveaway organised by Hackney Council.

Islington Ecology Centre (ISLINGTON) is a council-run, 2.8 ha nature reserve with a purpose-built resource and training centre. It is home to 244 species of plants, 94 species of birds and 24 types of butterflies with pond, woodland and meadow habitats. It is located south of Finsbury Park and there are plans to link the park to the nature reserve in the park, in the latest planning document relating to the area around Finsbury Park train station (Islington 2013). There is a weekly work day and the centre has good links with local community groups, playing host to the popular Gillespie Park festival (Islington 2012) each year.

Parkland Walk This is another wildlife corridor which runs westerly, away from Finsbury Park. This ex-railway line is managed by TCV and the Friends of Parkland Walk (Friends of The Parkland Walk) and is extremely popular with local walkers and cyclists because it provides a beautiful and functional off-road route linking Finsbury Park to Crouch Hill and Highgate. It is also popular with local foragers who enjoy, amongst other things, the blackberries, hogweed, cherries, plums, elderberries and ground elder. In terms of wildlife, Bats, Muntjac (deer), Teasel, Rosebay Woodpecker, Tawny Owl, Slow Worm and White-Letter Hairstreak (butterfly) have been found (Haringey 2004a, b, c).

To summarise, when looked at in combination, there is a good range of projects in North London which are already offering the following:

- **forest gardening**
- **plant propagation**
- **social enterprises food growing/selling**
- **wild food foraging**
- **pockets of biodiversity—wildlife 'corridors' and 'banks'**

4.6 Recommendations and Opportunities

4.6.1 Local Authorities

Each of the local authorities in North London (London Boroughs of Haringey, Hackney and Islington) could adopt the practical recommendations made in Gemma Harris's well written document on how parks could relatively simply be made into productive and educational spaces (Michelle 2013). In this document she explains how areas of 'My Local Park'—fences; ornamental beds; 'dog-free' zones; and display beds—could relatively easily be changed into laid hedges, productive trees, community orchards and demonstration beds. She suggests this could be achieved in three years, but any progress in that direction could gradually transform Clissold or Finsbury Park. The design and planning process that has been applied so successfully at the Castle could be applied in the park.

In addition local authorities could support any of the ideas listed elsewhere in this section.

4.6.2 Social Entrepreneurs

There are the following gaps in the market that could be investigated for their suitability as successful social enterprises.

Forest gardening and foraging It would be relatively straight forward to modify and adapt existing biodiverse settings so that they produce more crops that are edible and palatable for humans. The fresh produce could be harvested and sold. The Castle and ELL could be developed as outlets for this locally produced food.

Eco tourism if the level of interest from European and Asian tourists is anything to go on, there is a sizeable untapped market of eco-tourists who are fascinated to visit urban food growing, permaculture and forest garden projects. At present these determined people will compile their own itinerary of activities or exploit the existing relationships they happen to have with local people. It may be that someone local could establish a livelihood by supporting eco-tourists by creating local tours and setting up visits to local projects. The relative proximity of these projects would add to the chances of success of this venture.

Training There is a clear interest from local people in training in horticulture, forest gardening and in craft activities, e.g. bee-keeping, cob building or carpentry. In addition, cooking and eating are hugely popular—especially when it involves unfamiliar and unusual foods, such as foraged leaves or flowers. A training social enterprise with contacts with local skilled tradespeople could probably create a livelihood in managing this process. It is the experience of ELL that it is far easier to gain income from training than selling relatively low-value items such as potted plants—unless the plant sales person was able to sell other things.

Food enterprises While food is still relatively cheap there is an abundance of food that is currently going to waste. There is huge potential for one or more people to process waste and foraged food into a wide range of long shelf-life products such as jams, chutneys, pickled vegetables, fermented vegetables, fruit leathers, infused oils and vinegars, dried herbs, dried wild leaves, teas and spices. Perhaps these could be sold at an outlet such as the Castle Climbing Centre or ELL. The castle is already producing its own herbal tea, amongst other things. In addition, dried ingredients and oils could be bought in bulk and sold.

Other foods such as soups and fruit smoothies could also be made, ideally with a strand of production going straight to vulnerable local residents.

Delivery As fuel prices rise any of the burgeoning food hubs—ELL or the Castle Climbing Centre could become the sites for a bike delivery company. Fees paid by middle income people could subsidise deliveries of food to vulnerable locals—disabled, housebound and elderly people. There is already a community owned cargo bike based at the Redmond Resource Centre for a trial period.

4.6.3 Researchers

Biodiverse settings such as Railway Fields or London Wildlife Trust or parts of Clissold and Finsbury Park could be the site for research plots. There is much research that needs to be done about forest gardens. For example:

- how much more of its own food could Manor House produce?
- what are the most productive polycultures? Measure yields from different combinations of plants in different habitats and locations.
- what plants deliver the highest yield of calories/trace elements for the lowest effort?
- what forest garden foods taste the best and how can they be cooked/processed to maximise flavour, minimise embodied energy (cooking, storing and packaging) and minimise the food miles of the ingredients?
- how much more productive and biodiverse is a polyculture compared to a monoculture?

- what are the most efficient watering systems, in terms of water used and time spent?
- are reeds and nettles (and other fibrous plants that are growing abundantly in Manor House) useful or viable plant fibres for the textile industry?
- what opportunities are there for creating biochar (Wikipedia 2014) in this part of London?
- what coppicing opportunities are there and could they generate enough income to create a livelihood?

4.6.4 Risk

There are a number of reasons why food security and biodiversity might not be satisfactorily addressed in Manor House:

- project managers are so busy they do not have time to network with each other
- because this cohesive approach is being led by volunteers in an un-hierarchical way, it could be haphazard and incoherently delivered
- there are a lot of potential sites and it may be difficult to pull them all in the same direction.
- it might not be possible to find enough volunteers to run the ambitious research projects suggested above
- there might not be a 'market' for any of the social enterprise ideas suggested above
- the food security, social or environmental situation may deteriorate so quickly there is not enough time to adapt

What sometimes happens, however, is that haphazardly put together projects that are dependent on personal contacts and relationships between individuals—rather than 'professionally' obliged individuals working 'efficiently'—can lead to excellent outcomes so long as the project intention is clear and correct. It is worth starting with some small collaboration and then building on those.

4.6.5 Conclusion

Although these are challenging times, they are also extremely exciting if we are prepared to adapt a little and make more of what we already have. We may need to spend more time growing food and alter our eating habits but there will be big benefits for this, such as improved physical and mental health.

Key points for the Manor House area, as an example:

- there are many local opportunities to increase biodiversity

- some 'natural' settings could be managed towards becoming productive forest gardens or at the least, foraging sites
- any biodiverse setting will have foraging opportunities
- there are many ways in which the projects in Manor House could collaborate to provide or develop training, community networking opportunities and social enterprises that are related to food security
- there is a strong extant body of knowledge and experience in the Manor House area as well as a proven enthusiasm for this approach
- lessons learned and project outcomes can be shared and extended using the existing network of organisations such as lottery project partners and the Permaculture Association.
- there are many ways food production can be re-localised in Manor House

References

Abney Park (2004). http://www.abney-park.org.uk/. Accessed Aug 2013

Bowyer S, Caraher M, Duane T, Carr-Hill R (2006) Shopping for food: accessing healthy affordable food in the three areas of Hackney. http://openaccess.city.ac.uk/489/. Accessed Sep 2013

Boycott R (2008) London's Achilles Heel. http://www.theguardian.com/commentisfree/2008/oct/01/food. Accessed Aug 2013

Cannon AR, Chamberlain DE, Toms MP, Hatchwell BJ, Gaston K (2005) Trends in the use of private gardens by wild birds in Great Britain 1995–2002. Vol 42(4), pp 659–671. http://onlinelibrary.wiley.com/doi/10.1111/j.1365-2664.2005.01050.x/full. Accessed Aug 2013

Crawford M (2010a) Creating a forest garden, working with nature to grow edible crops. Green Books, Devon, p 22

Crawford M (2010b) Creating a forest garden, working with nature to grow edible crops. Green Books, Devon, p 27

Crawford M (2010c) Creating a forest garden, working with nature to grow edible crops. Green Books, Devon, pp 33–39

Crawford M (2010d) Creating a forest garden, working with nature to grow edible crops. Green Books, Devon, p 71

Davies ZG, Fuller RA, Dallimer M, Loram A, Gaston KJ (2012) Household factors influencing participation in bird feeding activity: a national scale analysis. PLoS One 7:e39692. http://kevingaston.com. Accessed Aug 2013

ELL Propagation and Plant, World Domination Yet? tab https://docs.google.com/spreadsheet/ccc?key=0AkKXpP9KC5PkdDdoRW1nS0p5cjk0R3Z6Y3BGMWR0Y1E&authkey=CLeWh88B&authkey=CLeWh88B#gid=5. Accessed Aug 2013

Fuller RA, Irvine KN, Davies ZG, Armsworth PR, Gaston KJ (2012) Interaction between people and birds in urban landscapes. Stud Avian Biol 45:249–266

Growing Communities. http://www.growingcommunities.org/

Hackney C (2012) Hackney_SINCs. http://www.hackney.gov.uk/sites-of-importance-for-nature-conservation.htm#.UgdbiRLJAiQ, p 6. Accessed Aug 2013

Haringey C (2004a) Haringey's Biodiversity Action Plan. http://www.haringey.gov.uk/biodiversity_action_plan.doc, p 44. Accessed Aug 2013

Haringey C (2004b) Haringey's Biodiversity Action Plan. http://www.haringey.gov.uk/biodiversity_action_plan.doc, p 46. Accessed Aug 2013

Haringey C (2004c) Haringey's Biodiversity Action Plan. http://www.haringey.gov.uk/biodiversity_action_plan_2009-2.pdf. Accessed Aug 2013
Haringey C (2010) Sustainable Food Strategy 2010–2015: sustainable food at the heart of every day life. http://www.haringey.gov.uk/index/environment_and_transport/going-green/sustainable-food/foodstrategy.htm. Accessed Aug 2013
Haringey C (2013a) Finsbury Park Management Plan 2013–2016. http://www.haringey.gov.uk/finsbury_management_plan_2013.pdf, p 9. Accessed Aug 2013
Haringey C (2013b) Railway Fields Local Nature Reserve: Management Plan 2013–2016. http://www.haringey.gov.uk/railway_fields_management_plan_2013.pdf. Accessed Aug 2013
Healy B (2006) How Cuba survived its oil shock from green left weekly. http://www.powerofcommunity.org/cm/content/view/28/49/. Accessed Sep 2013
Islington G (2012) Community Gears up for Gillespie Park Festival. http://www.islingtongazette.co.uk/news/community_gears_up_for_gillespie_park_festival_1_1509090. Accessed Aug 2013
Islington C (2013) Finsbury Park Town Centre Supplementary Planning Document – Draft for Consultation. http://www.islington.gov.uk/publicrecords/library/Planning-and-building-control/Publicity/Public-consultation/2013-2014/(2013-06-28)-Finsbury-Park-Town-Centre-draft-SPD.pdf, p 28. Accessed Aug 2013
ISLINGTON. http://www.islington.gov.uk/services/parks-environment/parks/islington_nature_reserves/gillespie/Pages/default.aspx. Accessed Aug 2013
Jones J (2010) Cultivating the capital: food growing and the planning system in London. http://legacy.london.gov.uk/assembly/reports/plansd/growing-food.pdf. Accessed Aug 2013
London Wildlife Trust. http://www.wildlondon.org.uk/about-london-wildlife-trust. Accessed Aug 2013
London Wildlife Trust. http://www.wildlondon.org.uk/woodberry-wetlands. Accessed Aug 2013
Manor House Development Trust, Manor House PACT Final Bid Document (2012). http://www.transitionfinsburypark.org.uk/ManorHousePACT. Accessed Aug 2013, p 25
Manor House PACT (2014). http://www.manorhousepact.org.uk/page/pact-wildlife-and-foraging-walks. Accessed Aug 2013
Metropolitan Improving life together. http://www.metropolitan.org.uk/
Michelle (2013) Haringey Parks. https://dl.dropboxusercontent.com/u/62113333/Haringey%20Parks%20for%20Michelle.docx. Accessed June 2014
Mind (2007) Ecotherapy, the green agenda for mental health. http://www.mind.org.uk/assets/0000/2139/ecotherapy_executivesummary.pdf, p 2. Accessed Aug 2013
Shiva V, Singh V (2011) Health per acre: organic solutions to hunger and malnutrition. http://www.navdanya.org/attachments/Health%20Per%20Acre.pdf. Accessed Aug 2013
The Castle Climbing Centre. http://www.castle-climbing.co.uk/sustainability. Accessed Aug 2013. http://www.castle-climbing.co.uk/garden. Accessed Aug 2013
The wildlife Trusts. http://www.wildlifetrusts.org/reserves/east-reservoir-community-garden. Accessed Aug 2013
Tirado R, Simon G, Johnston P (2013) A review of factors that put pollinators and agriculture in Europe at risk. http://www.greenpeace.org/eu-unit/Global/eu-unit/reports-briefings/2013/130409_GPI-Report_BeesInDecline.pdf, p 40. Accessed Sep 2013
Whitefield P (1996) How to make a forest garden. Permanent Publications, Hampshire, p 5
Wikipedia (2014) Biochar. https://en.wikipedia.org/wiki/Biochar. Accessed June 2014
WOODLANDTRUST. http://www.woodlandtrust.org.uk/en/why-woods-matter/what-are-they/types/veteran/Pages/veteran.aspx#.UgeABRLJDqc

Chapter 5
Landscape Design and Ecological Management Process of Fishway and Surroundings

Keitaro Ito, Tomomi Sudo, Kazuhito Ishimatsu, and Hayato Hasegawa

Abstract Ecosystem Services (ES) are the ecological characteristics, functions, or processes that directly or indirectly contribute to human well-being. Green Infrastructure (GI) is characterized by its multiple benefits. In Japan, GI is defined as infrastructure and land use planning which enhances regional and national sustainability. The ecological functions and use by people in the study site were limited. Consequently, the Ministry of Land, Infrastructure, Transport, and Tourism (MLIT) asked the Laboratory of Environmental Design at Kyushu Institute of Technology (KIT), to design a new river bank. This project started in 2008, and the restoration plan was developed in cooperation together with local residents. We also considered how to implement GI in local community and could provide ecosystem services for local residents. In the 1960s, the river was a familiar environment for people. Children who lived close to the river had the opportunity to learn about nature from the river; however, the present river bank covered in concrete does not provide natural experiences for children. Preserving areas such as wildlife habitats and spaces where children can play is a crucial issue nowadays. The planning site had the potential to be a place for children to learn local ecology. Therefore, we designed the riverbank fishway not only for nature restoration but also as a place for children's ecological learning. This chapter demonstrates the process of GI construction on the river Onga estuary in Japan to contribute to regional biodiversity conservation and provide ecological learning for children. It also should be noted that the purpose of urban landscape design or planning is to connect nature and people's daily lives. In other words, an integrated approach in terms of "nature rehabilitation" and "lifescape" will be essential to create vernacular (i.e., domestic and functional) places in the future.

K. Ito (✉) · T. Sudo · H. Hasegawa
Laboratory of Environmental Design, Faculty of Civil Engineering, Kyushu Institute of Technology, Kitakyushu-city, Fukuoka, Japan
e-mail: ito.keitaro230@mail.kyutech.jp

K. Ishimatsu
Department of Civil Engineering, National Institute of Technology, Akashi College, Akashi-shi, Hyogo, Japan

K. Ito (ed.), *Urban Biodiversity and Ecological Design for Sustainable Cities*,
https://doi.org/10.1007/978-4-431-56856-8_5

Keywords Nature restoration · Green Infrastructure · Fishway · Wildlife habitat · Children · Riverbank · Lifescape · Vernacular

5.1 Introduction

Conservation of ecosystem services (ES) is recognized as a critical issue for sustainable development. ES are the ecological characteristics, functions, or processes that directly or indirectly contribute to human well-being, i.e. the benefits that people derive from functioning ecosystems (Costanza et al. 1997; Millennium Ecosystem Assessment 2005). In order to promote conservation and provision of ES, a Green Infrastructure (GI) approach has been developing in recent years. GI is characterized by its multiple benefits for human beings. In Japan, GI is defined as infrastructure and land use planning which enhances regional and national sustainability by providing multiple functions such as improving wildlife habitat, regulation of the climate, and a variety of benefits for human well-being (MLIT 2015; Green Infrastructure Association 2017). GI can include various natural and semi-natural green spaces (Town and Country Planning Association 2012), and a wide range of objectives has been put forward for green infrastructures such as improving human health and well-being, urban esthetics, biodiversity conservation, water management, sustainable land management, climate change mitigation and adaptation, job creation, and urban regeneration (Chenoweth et al. 2018). This chapter demonstrates the process of GI construction on the river Onga estuary in Japan to contribute to regional biodiversity conservation and to provide ecological learning for children.

5.2 The Study Site

The study site is located in the River Onga estuary in Fukuoka Prefecture, Kyushu, Japan (Fig. 5.1). The River Onga has a total length of 61 km and a catchment area of 1026 square km. The population size in the area surrounding the river is around 670,000 people, and the population density is around 650 per square km. The area surrounding the river is composed of mountainous area (80%), agricultural area (14%), and residential area (6%).

Lowland riparian habitats in Japan have been lost mainly because of the remarkable shift of increasing population size and the influx of social assets into the areas that have occurred since the medieval era, and in modern times only tiny portions of relict remain along rivers (Washitani 2001). The Onga river has contributed to local society, economics, and culture over the centuries. After the Second World War, urbanized areas have dramatically expanded in the area surrounding the river. These significant changes have resulted in an increased risk of flooding. Consequently, rising demands for flood control as well as for water supplies for industries and cities have led to an intensification of river regulation, dam construction, and water

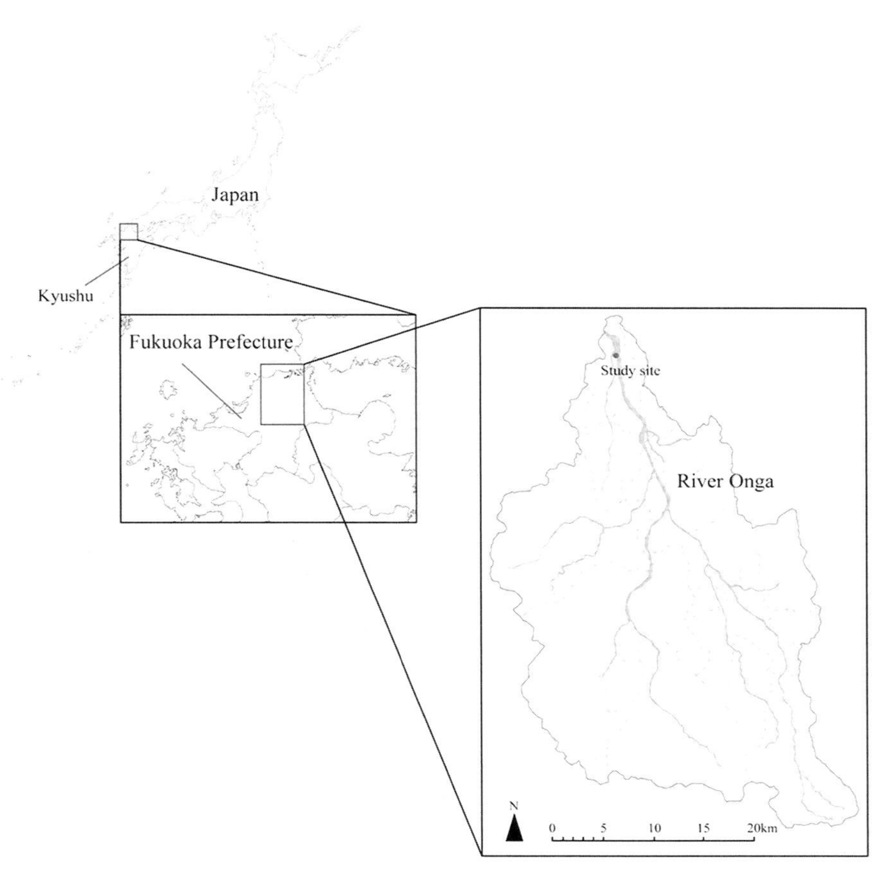

Fig. 5.1 Location of River Onga

abstraction throughout this region (Yoshimura et al. 2005), as well as in the Onga river area. A weir was constructed on the River Onga estuary in 1983 (Fig. 5.2). Beside the weir, a fish ladder was also built in order to mitigate the adverse impacts of the weir construction on fish migration. Also, the area surrounding the fish ladder was completely covered in concrete (Fig. 5.3). The ecological function and use by people were limited in the study site. Consequently, the Ministry of Land, Infrastructure, Transport, and Tourism (MLIT) asked the Laboratory of Environmental Design at Kyushu Institute of Technology (KIT), to design a new river bank. This project started in 2008, and the restoration plan was developed in cooperation with local residents. The construction started in 2010 and was completed in 2013.

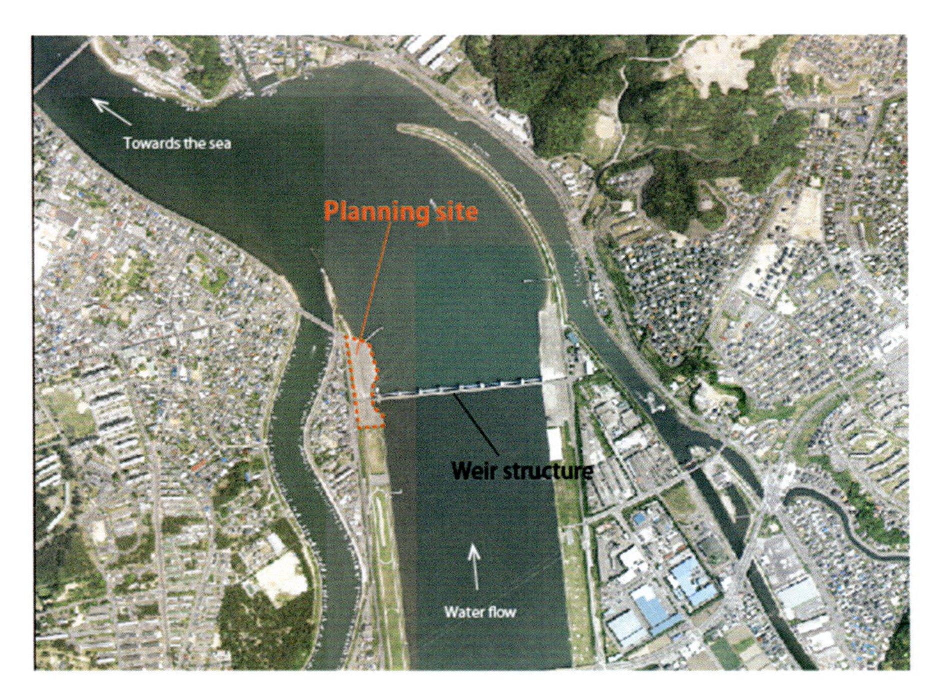

Fig. 5.2 Aerial photograph of the study site

5.3 Design Concept

The blocking effect of dams and weirs needs to be mitigated, especially in midstream and downstream sections, because these artificial structures are responsible for the decrease of diadromous fish species and they have negative ecological effects on upstream river systems (Yoshimura et al. 2005). There was the concept of a fish ladder; however, scientists have recently revealed that it does not work well for the river ecosystem. For example, fish passage facilities have been built predominantly on the main stems of large rivers; however, fish generally use tributaries rather than main stems of large rivers to spawn (Ovidio and Philippart 2002). In addition, several past studies suggest that once fish ascend to upstream reaches, individuals fail to return through the fishways to downstream reaches (Agostinho et al. 2011). If so, fishways would work only as one-way routes, considering that many fish species pass through these facilities on their upstream movements. In the study site, the fish ladder was designed for Japanese eel (*Anguilla japonica*) and Ayu (*Plecoglossus altivelis*). The fish species which could go up from downstream to upstream through it were quite limited. For these reasons, creating a new riverbank fishway which reconnects the ecological network downstream to upstream was established as an essential goal in the project (Ito et al. 2020).

Fig. 5.3 River bank covered by concrete before nature restoration (2008)

In addition, we considered how the place could provide cultural ecosystem services for local residents. In the 1960s, the river was a familiar environment for people. Children who lived close to the river had the opportunity to learn about nature from the river; however, the present river bank covered in concrete does not provide natural experiences for children. Preserving areas such as wildlife habitats and spaces where children can play is a crucial issue nowadays. Play in nature during childhood is an essential experience in learning about the structure of nature while environmental education has been afforded much greater importance in primary and secondary school education in Japan since 2002 (Ito et al. 2010, 2016). It is also important to consider physical activity as it relates to the multiple demands of childhood and adolescence associated with physical growth, biological maturation, and behavioral development (Strong et al. 2005). In fact, Fjørtoft and Sageie (2000) found that the stimulation of inventiveness and creativity, and the possibility of discovery are directly related to the number and the kind of features in the natural landscapes. Hence, natural landscapes represent potential grounds for playing and learning, and this has to be taken into serious consideration for future policy and planning of outdoor grounds for children (Fjørtoft and Sageie 2000; Haug et al. 2010). The planning site had the potential to be a place for children to learn local ecology. Therefore, we designed the riverbank fishway not only for nature restoration but also as a place for children's ecological learning (Fig. 5.4).

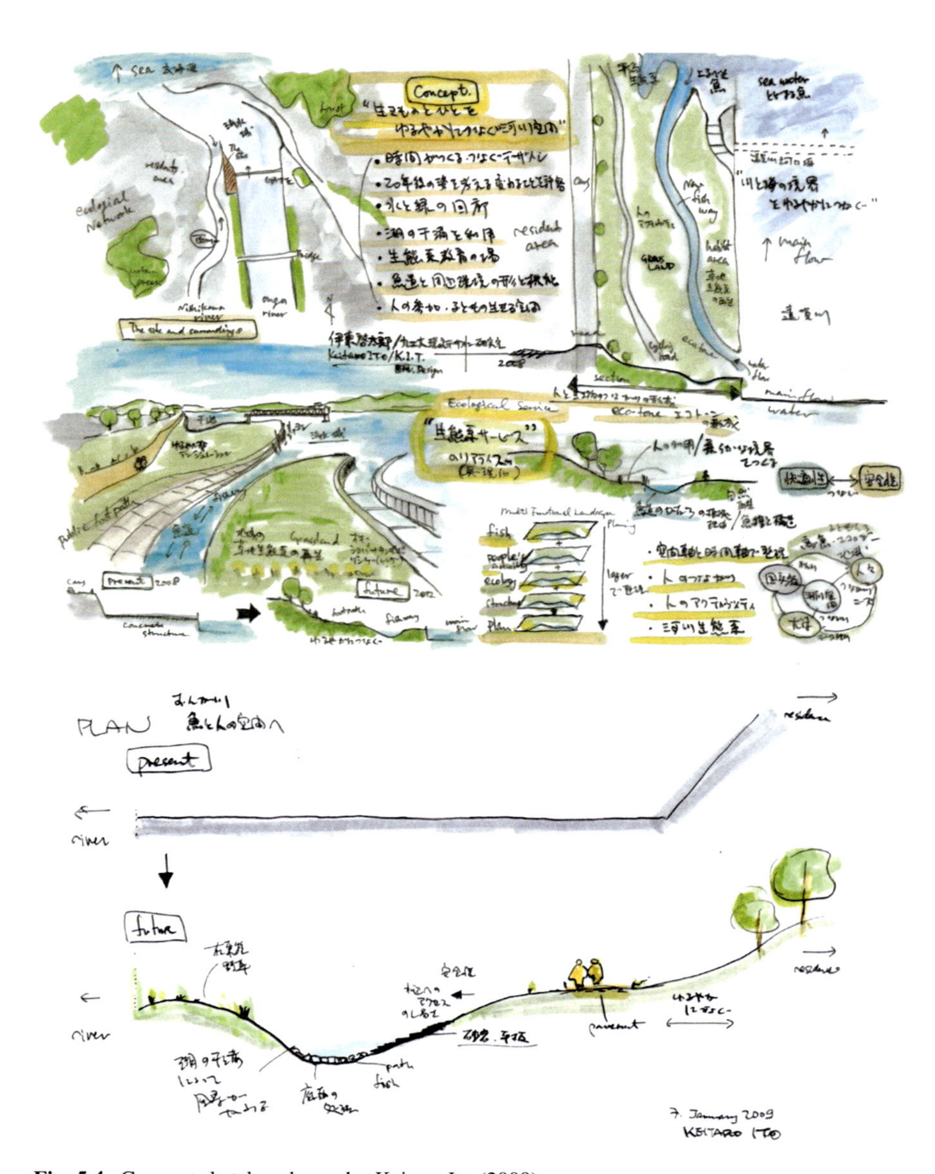

Fig. 5.4 Concept sketches drawn by Keitaro Ito (2009)

5.4 Design Process

5.4.1 Basic Design

At the first stage of this project, the presented image of the fishway was trapezoid shaped. I thought it might have a function for fishway, however it would not

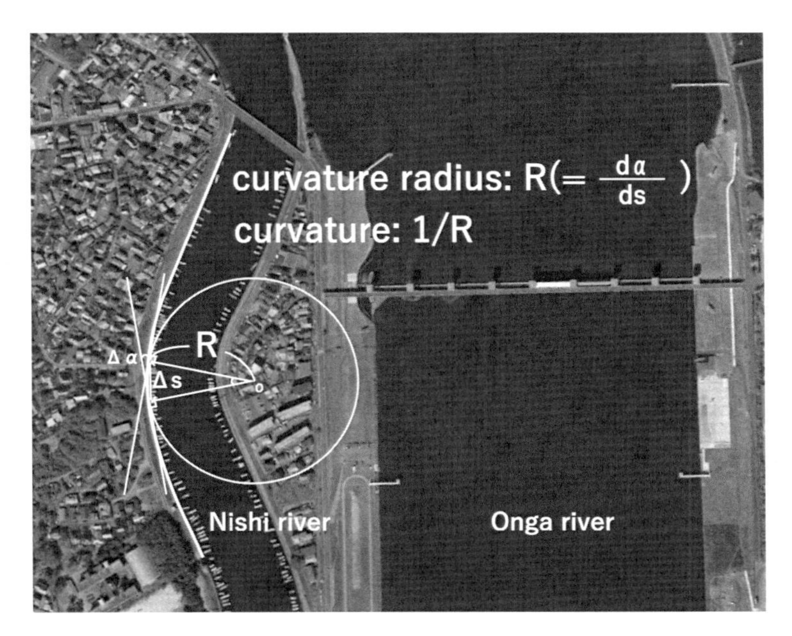

Fig. 5.5 The curvature of Nishi-river next to the planning site that was used for designing the shape of the fishway

be beautiful for regional landscape. Therefore, when I determined the shape of the fishway for local landscape and ecological functions, I used the curvature of Nishi-river which flows next to Onga river. It was difficult for designing the shape of the fishway, so I quoted the curvature of Nishi-river because that natural river shape would reflect the landscape of this site (Fig. 5.5). And after discussing current issues using a 1/200 scale model, we designed the shape of the new riverbank fishway by using a 1/200 scale study model. The tide and brackish water are distinguishing phenomena at the river mouth zone, and very important for many water creatures to build their habitats. The new riverbank fishway was, therefore, designed so that it can take advantage of the phenomena in terms of water biodiversity conservation and landscape design as well.

Concerning accessibility to water area for the users, the levee crown was connected to the water area by a gentle slope. The irregular slope angles were designed to consider children's play space based on Gibson's affordance theory (1979). Besides, many variously-sized stones were randomly placed within the riverbank fishway in order to create diverse water space not only for creating water creatures' habitats but also for encouraging user's activities. Furthermore, the water depth is strongly affected by the tide. Thus, this small change would be beneficial for enhancing water biodiversity, and local children can observe and learn the abundant water ecosystem with easy access to the water surface. Finally, the area at the end of the lower riverbank fishway was designed for a wetland (a tidal flat) which can attract not only water creatures but also birds (e.g., wading birds).

5.4.2 *Participation of Local Residents*

It was expected that local people would become familiar with this riverbank fishway before completing this renovation work by the fact that four workshops took place from 2009 to 2013 in order to share the design concept and process with them (Fig. 5.6). The client, MLIT, is going to cede the riverbank fishway to a local government, Ashiya-town, after finishing the renovation work. Therefore, the local government and people have to manage this space in the future independently. Thus it should be noted that they knew the reason the renovation work was conducted for future management works. The attendees were from Ashiya-town government; local companies, Ashiya-Higashi Primary School, local clubs, etc.

5.5 Results of the Restoration Project

5.5.1 *Landscape Changes After Construction*

The concrete was removed, and a riverbank fishway and the surrounding environment were constructed (Fig. 5.7). Within the riverbank fishway, variously-sized stones are allocated to be hiding places for living creatures, and the water depth of the riverbank fishway can be regulated due to tide (Fig. 5.8a). A creek tidal flat was designed as a habitat for benthic living creatures and it had diversified environments that are born by tidal variations (Fig. 5.8b). According to the fish survey conducted in 2016, 813 individuals consisting of 36 species were captured in the riverbank fishway. Additionally, this renovated site is not only designed for the enhancement of biodiversity but also for the creation of public space for local people. Therefore, we had to design the most basic artificial structures for the users avoiding any disturbance of the natural landscape (Fig. 5.8c, d).

5.5.2 *Ecological Learning for Children*

After the restoration of the new riverbank fishway and the surrounding environment, it became possible for Ashiya-Higashi Primary School to practice their ecological education. The school is located near the study site. The school and its pupils have participated in the design process from the beginning. The ecological learning is for fourth-grade pupils in their curriculum named “Learning the local ecology; people, water and lives in Onga River” in the Period for Integrated Studies. The ecological education program is implemented by cooperation between Ashiya-Higashi Primary School, MLIT, Ashiya-town, and the Laboratory of Environmental Design, KIT. The pupils learned about the relations between the Onga River and their daily life for a year by studying the local ecology. They captured and observed fish in the

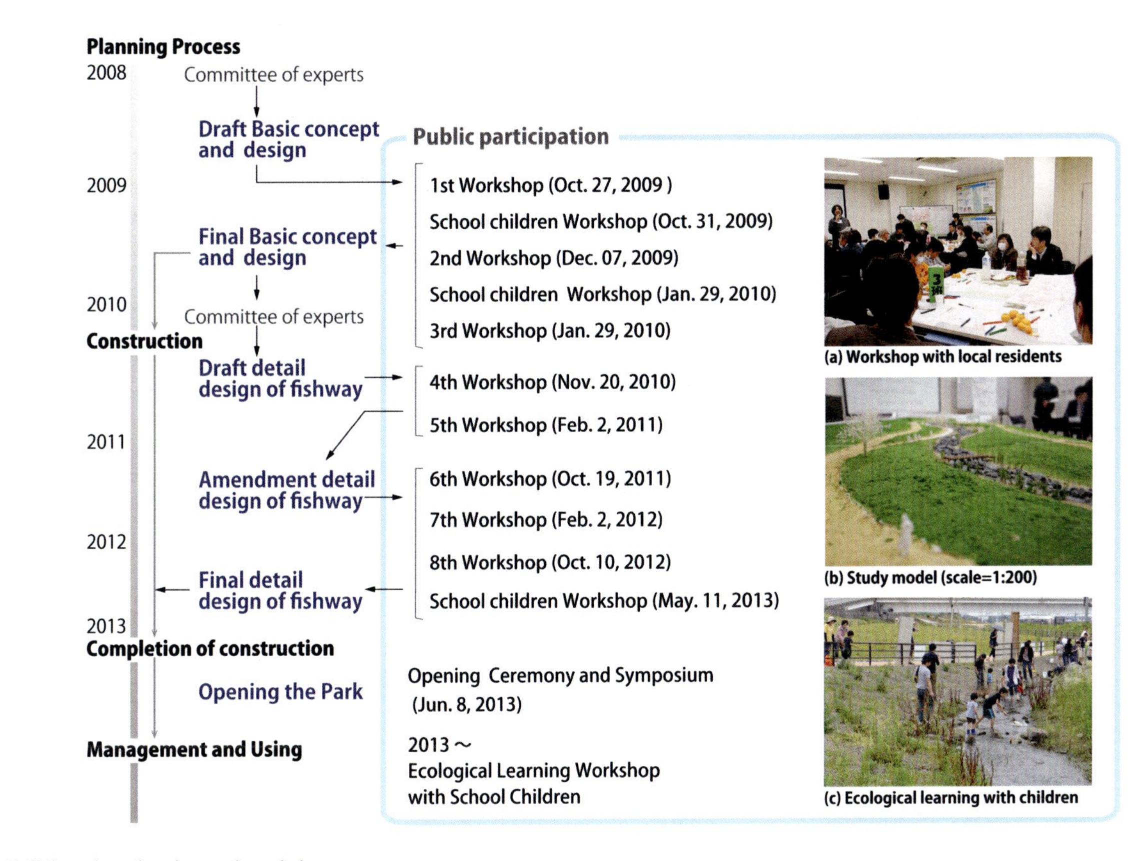

Fig. 5.6 Collaborative planning and workshop process

Fig. 5.7 Completed river bank and fishway (2013)

riverbank fishway and identified the species (Fig. 5.9). After identification, they learned the ecological characteristics: habitat, lifecycles, food, etc., of each species both in the books and from the specialists. Through these experiences, they learned the river ecology and the function of the riverbank fishway for the conservation of local ecosystems (Fig. 5.10). Their learning outcomes were presented by poster in a symposium to their parents, local residents, the staff of local governments, and other grade pupils.

5.6 Discussion

5.6.1 Ecological Design for GI

Until recently, the land in Japan had developed in various ways for economic growth and disaster prevention since the end of the Second World War. As a result, although Japan could become a significant economic country in the world, it is argued that the level of natural disasters slightly tends to be worse. Why? It can be safely said that the reason is the substantial destruction of nature through land development. That is to say, the current ecosystem services in Japan are incredibly fragile. Consequently,

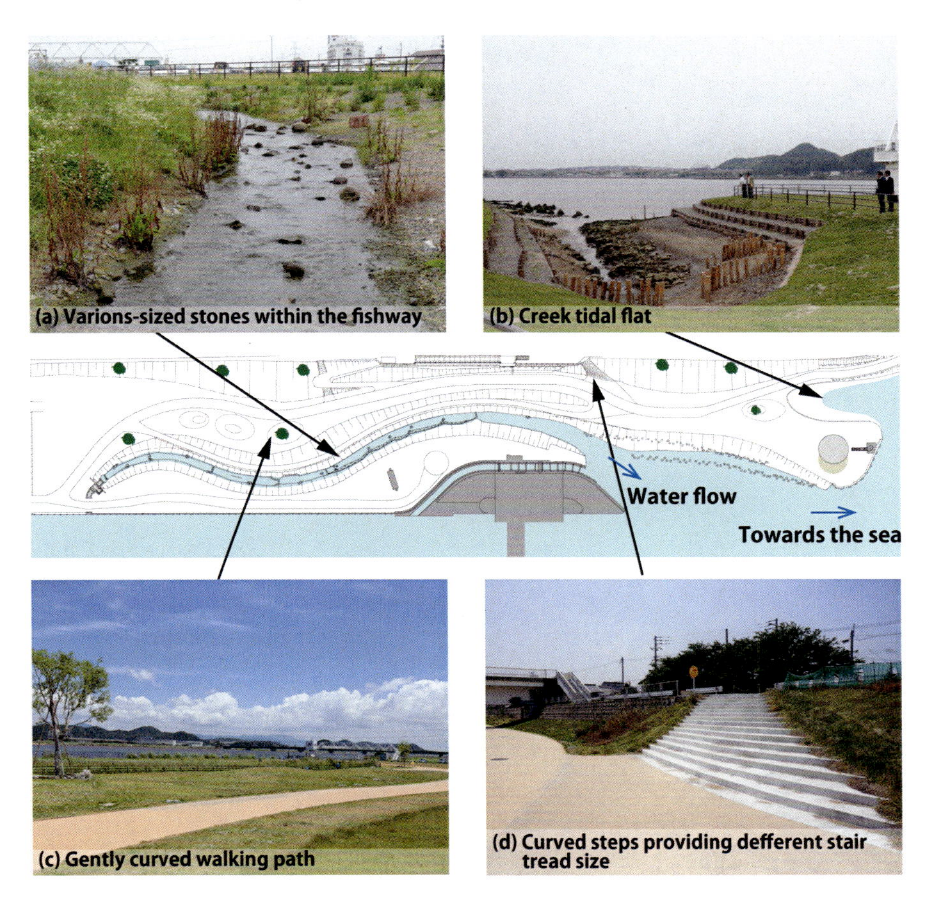

Fig. 5.8 Designed environment of river bank fishway and surroundings. (**a**) Various-sized stones within the fishway. (**b**) Creek tidal flat. (**c**) Gently curved walking path. (**d**) Curved steps providing different stair tread size

a massive number of projects of nature restorations have been taking place all over Japan. Nakamura (2004) distinguished nature rehabilitation from restoration. Nature restoration is to rebuild structures of past strong ecosystems; nature rehabilitation is to install only the functions of past ecosystems instead of their structures. It is generally challenging to conduct nature restoration in urban areas because of the lack of open space; nature rehabilitation is thus more realistic. Notably, for nature rehabilitation, the enhancement of biodiversity is one of the most critical aspects of ecological design. Nassauer (2012) mentioned that "rather than attempting to return to a more natural order, mimicking nature, or compromising between human desires and the limits of nature, ecological design invites the invention and realization of new, resilient landscapes that visibly embody societal values, thoughtfully

Fig. 5.9 The ecological survey at the fishway with Ashiya-Higashi Primary School pupils and KIT students

Fig. 5.10 The group discussion about their findings through the ecological education class in local primary school

incorporate our best knowledge of environmental processes, and are adaptable to surprising change."

The creation of the riverbank fishway project can be a rehabilitation which installs multiple functions such as wildlife habitat, reconnecting ecological network, as well as providing recreation and education. As long as people need a stable water supply, the study site cannot return to the environment before the weir construction is built.

The project shows how the replacement of single-use infrastructure by GI promotes biodiversity conservation. The location of the study site is a critical environment for the watershed aquatic ecosystem, where the sea and the river are connected. The essential goal of the project was to reconnect the ecological network, which was separated by the weir structure. A more detailed analysis will be conducted; however, the result of the fish survey in 2016 shows that the riverbank fishway is functioning. In addition, the fishway is very important not only for fish but also benthic macroinvertebrates to maintain their life cycles (Rawer-Jost et al. 1999), so some of them can also become key index species. According to McKinney et al. (2011), in spite of their decreased size, the wetland areas continue to provide habitat for many birds and may in fact even have greater importance because of the resources they provide to remaining species. Wetlands used to exist everywhere at river mouth areas in Japan; however, they have currently become rare or small due to concrete embankment construction.

As a result of the project, the study site became an important place for wildlife. At the same time, it means that people can encounter wildlife in that place. The riverbank fishway, wetland, and lawn that were created in the study site can be an important place for an opportunity to observe ecosystem as an ecological education site for local children as well as for conservation of the local ecosystem.

5.6.2 *Participation: Building a System to Promote the Use of GI*

To build a system for GI, the programs vary according to the regions, states, and landscapes in which they occur but some conventional planning approaches will include involvement of stakeholders, establishing a mechanism for making decisions, development of a clear mission, and engagement of the public throughout the process (Benedict and MacMahon 2006). In this project, the cooperation of national government (MLIT), an academic (the first author), and student members of the Laboratory of Environmental Design of KIT, local residents, Ashiya Primary School, local clubs, etc., made possible the holding of workshops in the process of the design concept and future management planning.

The participants provided their valuable opinions for landscape design in the process of the project, and decided to create a place both for nature restoration and for children's ecological learning. The importance of public participation in this project can be seen for contact with nature, local problem solution, and recognition and realization of their interests and demands. For nature restoration, concerns about the habitats and biodiversity conservation have been considered. Moreover, the users' activities after the restoration have been included for the nature contact and awareness. Suebvises (2018) proposed that experiencing nature may decrease in urban communities, and it is suggested that encouraging people to be more connected to nature may be essential to increase human well-being in cities. Miller suggested that more effort should be expended in broadening the value and relevance of nature in the public mind by raising awareness of ecological realities that provide the context for people's lives. The concerns about the habitats, biodiversity conservation, and the users' activities have been considered in this project. Nature awareness activities are needed in order to allow people to interact with nature directly and perhaps for children to develop deeper connections to nature in childhood (Bixler et al. 2002). The involvement and practical implementation with public participation, including children, in the planning and design process can drive the motivation and interaction with nature afterward in this project. Rosenzweig (2003) has proposed that there may be a need to modify natural places that are already dedicated to human activities to become more natural. The participation of citizens in the landscape design process strengthens the effectiveness of their new ideas adjusted to the particular local problem situation (Lin et al. 2014) and their needs. In this project, the planning and management of the study site have been conducted with the local public participation consideration for the use of GI.

In addition, how to utilize and manage after restoration is also as important as the physical changes in nature rehabilitation. The project team/stakeholders include the staff of MILT (as the professionals of Onga river estuary and weir), professor and student members of the Laboratory of Environmental Design of KIT (as landscape ecological designers), and teachers from Ashiya Primary School (as educators). The role and effort of this partnership supports and develops the ecological education program in Ashiya Primary School through many workshops engaging a wide range of perspectives from individual fields.

It is vital to promote the importance of well-being and learning local ecology. The ecological education program in Ashiya Primary School is established by the installation of a curriculum which builds strong relationships in the connection between humans and nature. The riverbank fishway plays a highly significant role for direct experiences by catching fish, observing them, perceiving their habitat, and how they live in the water. This curriculum also provides opportunities to explore and learn, in a practical sense, how the environment of organisms has changed before and after restoration at the study site. From the ecological psychology perspective, all animate beings learn about the environment and about their own competencies in effecting change through their actions simultaneously (Heft and Chawla 2006). The children can conceive the development of perceiving, thinking, and acting within their individual environment relationship through direct experience in the study site. It also needs an evaluation of the relationship between the riverbank fishway and the users' activities. Fjørtoft and Sageie (2000) found that a diverse natural landscape had the qualities to meet the children's needs for a varied and stimulating play environment where the composition and structures of the landscape were conducive to different play functions.

In summary, the participation of people in the design process before and during the nature rehabilitation can connect the activities of future utilization; promote nature experiences, and consequently, biodiversity conservation management of the study site. As a physical setting, the issue of designing an accessible, open, and natural space and environment is important, and for a social setting, the educational philosophy and practices encouraged activities of children engaged in exploring their environment (Sudo et al. 2019). After nature rehabilitation, environmental learning could support enhancement of the children's awareness of sustainable development by restoring to the original natural environment.

5.7 Conclusion

It is important to renovate the urban open space as a Green Infrastructure that provides ecosystem services in the urban areas. Although Green Infrastructure has a wide range of definitions, it has a lower impact for energy and material than Gray Infrastructure does and produces additional benefits such as providing wildlife habitats or amenity space (Chenoweth et al. 2018). In the future city planning, urban landscape designers and planners must consider green spaces as places for people, especially in urban areas. It is required to build a system approach to balance a wide array of interests and needs in the planning (Yoshii et al. 2015). Mitchell et al. (2015) stated that consolidated green infrastructure of urban areas should be designed so as to maintain the supply of abundant urban ecosystem services. At this study site, the renovation work was not only for the enhancement of biodiversity but also for people using it, so it functions as Green Infrastructure. The implementation of GI construction in this project has been integrated into promoting multifunctional ecosystem services and the enhancement of local sustainability of the study site. Heft and Chawla (2006) claimed that sustainable development is also the most pressing challenge of the new century. Moreover, its possibility depends on nurturing the children who recognize the action of human environmental connection and who can imagine themselves as being participants in achieving this goal. However, lack of outdoor space to play in, fear of violence in public spaces, the longer working hours of parents, and the artificial nature of most playgrounds have helped create the present-day situation in which young children have gradually lost contact with nature (Herrington and Studtman 1998). Practical implementation is necessary for developing urban natural spaces—not just green but with substantial quality and promoting the multiple functions needed to contribute to the quality of children's natural experience with an understanding of children–nature interaction and its value (Sudo et al. 2019). Landscape planning, as a "learnscape" should embrace the five senses, not only sight but also touch, taste, hearing, and smell (Ito et al. 2010). It is thus vital that present-day planners and landscape designers consider "Landscape" as an "Omniscape" (Arakawa and Fujii 1999; Numata 1996) in which it is much more important to think of landscape planning as a learnscape, embracing not only the joy of seeing but exciting the five senses as a whole. Therefore, it should also be noted that the purpose of urban landscape design or planning is to connect nature and people's daily lives. In other words, the approach in terms of "nature rehabilitation" and "lifescape" will be essential to create vernacular places in the future.

Acknowledgments We wish to express our gratitude to all those who made the writing of this paper possible. We would like to appreciate the cooperation and support of all staff members of MLIT, Ashiya-town, and the construction consultancy companies who have been involved in this project. We are greatly indebted to Prof. Ian. Ruxton about discussion on historical point of view and english expressions. We would like to show our appreciation to Dr. Yuichi Ono, Dr. Mahito Kamada and Dr. Tohru Manabe about discussions from earlier stage of this project. We also thank to Takuya Ito, Yuji Shin, Kosuke Ide, Yuta Tanaka, Masayoshi Yamamoto, Naruki Baba, Natsuko Ayukawa and all students in Keitaro Ito's laboratory in Kyushu Institute of Technology, and the children, the teachers, and parents of the Ashiya-Higashi Primary School. This study was supported by Kakenhi, Japan Society for Promotion of Science (JSPS), Grant-in-Aid for Scientific Research (C) (No. 23601013) in 2011–2013, and Grant-in-Aid for Scientific Research (B) (No. 15H02870) in 2015–2019.

References

Agostinho CS, Pelicice FM, Marques EE, Soares AB, de Almeida DA (2011) All that goes up must come down? Absence of downstream passage through a fish ladder in a large Amazonian river. Hydrobiologia 675:1–12

Arakawa S, Fujii H (1999) Seimei-no-kenchiku (Life architecture). Suiseisha, Tokyo (in Japanese)

Benedict AM, MacMahon TE (2006) Green infrastructure; linking landscape and communities. Island Press, Washington, DC

Bixler RD, Floyd MF, Hammitt WE (2002) Environmental socialization quantitative tests of the childhood play hypothesis. Environ Behav 34:795–818

Chenoweth J, Anderson RA, Kumar P, Hunt WF, Chimbwandira JS, Moore LCT (2018) The interrelationship of green infrastructure and natural capital. Land Use Policy 75:137–144

Costanza R, d'Arge R, Groot R, Faber S, Grasso M, Hannon B, Limburg K, Naeem S, O'Neill VR, Paruelo J, Raskin GR, Sutton P, Belt M (1997) The value of the world's ecosystem services and natural capital. Nature 387:253–260

Fjørtoft I, Sageie J (2000) The natural environment as a playground for children: landscape description and analyses of a natural playscape. Landsc Urban Plan 48:83–97

Gibson J (1979) The ecological approach to visual perception. Houghton Mifflin Company, Boston

Green Infrastructure Association (2017) Green infrastructure. Nikkei, Tokyo

Haug E, Torsheim T, Sallis JF, Samdal O (2010) The characteristics of the outdoor school environment associated with physical activity. Health Educ Res 25:248–256

Heft H, Chawla L (2006) Children as agents in sustainable development: the ecology of competence. In: Spencer C, Blades M (eds) Children and their environments: learning, using, and designing spaces. Cambridge University Press, Cambridge, UK, pp 199–216

Herrington S, Studtman K (1998) Landscape interventions: new directions for the design of children's outdoor play environments. Landsc Urban Plan 42:191–205

Ito K, Fjørtoft I, Manabe T, Masuda K, Kamada M, Fujiwara K (2010) Landscape design and children's participation in a Japanese primary school: planning process of school biotope for 5 years. Willey-Blackwell, Oxford, pp 441–453

Ito K, Sudo I, Fjørtoft I (2016) Ecological design: collaborative landscape design with school children. In: Murnaghan A, Shillington L (eds) Children, nature, cities. Routledge, New York, pp 195–209

Ito K, Sudo T, Hasegawa H, Ishimatsu K, Shiote K, Mitsuhashi S, Ono Y, Fukaura T, Shimada T, Izumi D and Toyokuni N (2020) The process of constructing the fishway of Onga river as Green Infrastructure, Landscape ecology and Management, Vol. 25, 2-5 (in Japanese)

Lin BB, Fuller RA, Bush R, Gaston KJ, Shanahan DF (2014) Opportunity or orientation? Who uses urban parks and why. PLoS ONE 9(1):e87422

McKinney RA, Raposa KB, Martin RM (Cournoyer) (2011) Wetlands as habitat in urbanizing landscapes: patterns of bird abundance and occupancy. Landsc Urban Plan 100(1):144–152

Ministry of Land, Infrastructure, Transport and Tourism (2015) National Spatial Strategies. http://www.milt.go.jp/common/001100233.pdf
Mitchell MGE, Suarez-Castro AF, Martinez-Harms M, Maron M, McAlpine C, Gaston KJ, Johansen K, Rhodes JR (2015) Reframing landscape fragmentation's effects on ecosystem services. Trends Ecol Evol 30(4):190–198
Nakamura F (2004) Nature restoration: analyses at regional, catchment and local site scale and how to realize the concept of nature rehabilitation. J Jpn Soc Reveget Technol 30(2):391–393 (in Japanese)
Nassauer JI (2012) Landscape as medium and method for synthesis in urban ecological design. Landsc Urban Plan 106:221–229
Numata M (1996) Keisoseitaigaku: introduction of landscape ecology. Asakura Shoten, Tokyo (in Japanese)
Ovidio M, Philippart JC (2002) The impact of small physical obstacles on upstream movements of six species of fish. Hydrobiologia 483:55–69
Rawer-Jost C, Kappus B, Böhmer J, Jansen W, Rahmann H (1999) Upstream movements of benthic macroinvertebrates in two different types of fishways in southwestern Germany. Hydrobiologia 391:47–61
Rosenzweig ML (2003) Win-win ecology; how the earth's species can survive in the midst of human enterprise. Oxford University Press
Strong WB, Malina RM, Blimkie CJR, Daniels SR, Dishman RK, Gutin B, Hergenroeder AC, Must A, Nixon PA, Pivarnik JM, Rowland T, Trost S, Trudeau F (2005) Evidence based physical activity for school-age youth. J Pediatr 146:732–737
Sudo T, Shwe YL, Hasegawa H, Ito K, Yamashita T, Yamashita I (2021) Nature environment and management for children's play and learning in kindergarten in urban forest, Kyoto, Japan. In: Ito K (ed) Urban biodiversity and ecological design for sustainable cities. Springer, Berlin
Suebvises P (2018) Social capital, citizen participation in public administration, and public sector performance in Thailand. World Dev 109:236–248
The Millennium Ecosystem Assessment (2005) Ecosystems and human well-being: synthesis, a report of the millennium ecosystem assessment. Island Press, Washington, DC
Town and Country Planning Association (2012) Planning for a Healthy Environment Good Practice Guidance for Green Infrastructure and Biodiversity
Washitani I (2001) Plant conservation ecology for management and restoration of riparian habitats of lowland Japan. Popul Ecol 43:189–195
Yoshii C, Yamaura Y, Soga M, Shibuya M, Nakamura F (2015) Comparable benefits of land sparing and sharing indicated by bird responses to stand-level plantation intensity in Hokkaido, northern Japan. J For Res 20:167–174
Yoshimura C, Omura T, Furumai H, Tockner K (2005) Present state of rivers and streams in Japan. River Res Appl 21:93–112

Part II
Landscape Management for Biodiversity in Urban Area

Chapter 6
Nature in the Cities: Places for Play and Learning

Ingunn Fjørtoft, Tomomi Sudo, and Keitaro Ito

Abstract The natural environment has been a site to play for many children. However, the children's physical play environments and facilities for play are changing, and opportunities for free play in stimulating environment are declining. Previous studies have shown the value of natural environment for children's development in different ways: developing motor skills, more harmonious and imaginative play, improving quality of playing and perception of landscapes. This chapter introduces how natural outdoor environments affect children's play activities and development. Children perceive environments as functions: functions to move, functions to build, functions to hide, to play, etc. Hence, physical environments afford such challenges to children in different ways, and the more diverse, the more challenging. Children interpret landscapes as functions to play and they operationalize the affordances as an awareness of the environments and their functional meaning into action. Natural features are important qualities of play environments, and they allow a wide range of learning opportunities not available from other playground options. Therefore, natural environments afford many opportunities for free play and physical activity for children. Thus creating green playgrounds in urban areas is important for children's physical, explorative, and imaginative play and should be of high priority for future cities. Green playgrounds must be considered as an essential quality of cities and urban planning should provide public open spaces capable to perform ecosystem services as functions for children.

Keywords Children · Green playgrounds · Affordance · Physical activity · Creative play

---

I. Fjørtoft (✉)
Faculty of Humanities, Sports and Educational Science, Department of Sports, Physical Education and Outdoor Studies, Campus Notodden, University of South-Eastern Norway, Notodden, Norway
e-mail: Ingunn.Fjortoft@usn.no

T. Sudo
Faculty of Civil Engineering, Kyushu Institute of Technology, Kitakyushu, Fukuoka, Japan

K. Ito
Laboratory of Environmental Design, Faculty of Civil Engineering, Kyushu Institute of Technology, Kitakyushu-city, Fukuoka, Japan

K. Ito (ed.), *Urban Biodiversity and Ecological Design for Sustainable Cities*,
https://doi.org/10.1007/978-4-431-56856-8_6

6.1 Introduction

The natural environment has traditionally been a site for play and physical activity for many children, and children a generation ago had access to wildlands and used them for exploring, challenging, and exercising skills to master a defiant landscape and unforeseen situations (Karsten 2005). However, the children's environments and facilities for play are changing, and the opportunities for free play in stimulating environments are declining (Clements 2004). There is a growing concern that children are becoming more sedentary in their adolescent lives, and scenarios predict enervated health perspectives later in life from an inactive adolescence. Lack of physical activity during childhood is a serious risk factor related to healthy lifestyle (Andersen et al. 2006; Biddle et al. 2004; Jones et al. 2013; WHO 2020).

When children have the opportunity to play in natural and green environments, how will this affect their play patterns and their physical activity and motor development? This has gradually become a highly valued approach and the topic has been increasingly subject to environmental research. Previous studies by Hart (1979), Moore (1986), Moore and Wong (1997), Rivkin (1990, 1995), Titman (1994) and others described the value of complex environments and wildlands for children, and how children perceive and experience wildlands and places of their own domain. Focus has been directed on learning effects from a natural environment and its impact on children's development. Studies by Grahn et al. (1997) and Fjørtoft (2000, 2001, 2004) have described how natural environments in parks and forest areas influenced the children's use of these environments in different ways: by moving on rugged ground and by climbing trees and rocks the children's motor fitness was affected. The studies showed that the children became more harmonious, and imaginative, and playing itself improved in quality. Playing in the parks also influenced their perception of space. A growing body of evidence indicates that activity-friendly environments may influence physical activity in children (Ferreira et al. 2007; Fjørtoft 2001, 2004; Sallis and Glanz 2006; Olivieira et al. 2014).

Landscape ecology analyses show a relation between landscape characters and children's play and the contextual relation to motor development (Fjørtoft and Sageie 2000; Fjørtoft 2004, Fjørtoft et al. 2010; Ito et al. 2014). We are gradually beginning to understand how landscapes and environmental features can contribute to increase play and physical activity in children. Behavior-environmental approaches to physical activity place emphasis upon mutual and independent relations among the individual, behavior, and the environment (Fig. 6.1). Physical environments are settings where activities occur and thus comprise physical and social contexts of behavior (Fjørtoft and Gundersen 2007; Sallis and Glanz 2006).

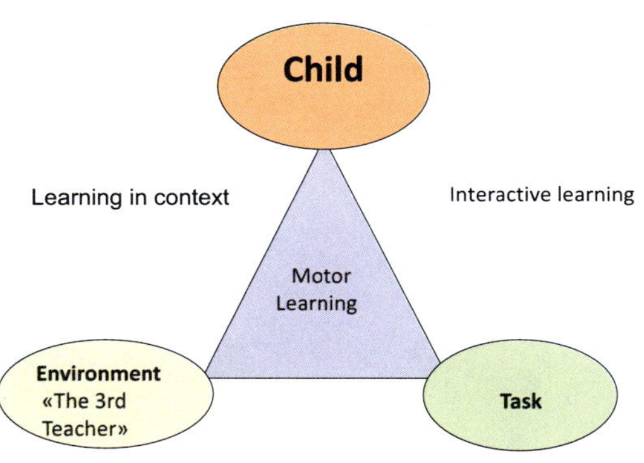

Fig. 6.1 Environment-behavior related contexts for learning. Interactive learning occurs in a contextual interaction between the child, the environment, and the task to be learned

6.2 Traditional Playgrounds

Playgrounds are designed to facilitate children's play and are aimed at enhancing children's physical, social, emotional, and cognitive development (Hart 1993). Different types of playgrounds, traditional, contemporary, adventure, and creative, were anticipated to promote different forms of play. The traditional playground is characterized as flat, barren often covered with asphalt, and equipped with climbing bars, a swing, a sandpit, a seesaw, and a slide. Usually the equipment is made of metal or plastic.

> While these "metal jungles" were of sturdy construction and required little maintenance, cold iron and steel structures mounted over brick or concrete surfaces were not aesthetically pleasing and above all were hazardous. Falls into these non-resilient surfaces did result in some serious injuries. These dangerous "metal jungles" still haunt us today in some public parks and schools (Hartle and Johnson 1993, p. 17).

Such playgrounds were found to be not very challenging and even the very young ones or those with motor behavior deficits did not have their potentials explored on these playgrounds (Frost 1992). A typical traditional playground can be characterized as minimum-sized, squared, and fenced in with childproofed gates and fences. It is flat without or with sparse vegetation and has five "compulsory" play apparatuses: a swing, a seesaw, a sandpit, a slide, and a climbing house. Such playgrounds are still the dominant play environments offered to children (Fig. 6.2).

Contemporary playgrounds are often designed by architects, are somewhat sculptured but containing the traditional play equipment, though constructed in different materials, such as wood and plastic. More creative play environments were facilitated in adventure playgrounds, creative playgrounds, and construction playgrounds. These were characterized by the availability of raw building materials and tools for constructions, a playhouse, and re-usable materials like tires (Hartle and Johnson 1993; Worthman and Frost 1990). Children spent more time on the playgrounds where the dominating play form was movement play, dramatic play, and more social- and cognitive behaviors (Frost and Campbell 1985). Comparative studies

Fig. 6.2 The traditional playground built using iron and plastic affording a minimum of play opportunities

of different play environments, however, showed that it was the specific features available and their context within the environment that enhanced more creative play behavior in children (Frost and Strickland 1985). Increased complexity seemed to increase the amount of play both quantitatively and qualitatively (Nielsen 2009). There is a common notion that outdoor play environments increase play behavior in children and play materials in physical environments seemed to have direct influences on the play opportunities for children (Nielsen 2009). Moore et al. (1992) suggested that a well-designed and well-managed play environment should provide children with developmental opportunities for motor skills and social development through the opportunities for playing and learning.

Children develop perceptual-motor skills through natural spontaneous interaction with the environment. They seek out stimulation and physically explore, discover, and evaluate the environment in relation to their needs. (Fjørtoft 2004). Such skills are generally named basic motor skills and comprise locomotion (rolling, crawling, climbing, jumping, hurdling, hopping, running, walking, pulling, pushing, throwing, etc.) and the context in which those skills are performed will imply movement qualities of co-ordination and balance, perception of body and space, rhythm and temporal awareness, rebound and airborne movements, projection and reception of movement. The perceptual and motor information the child establishes by performing such skills in complex environments enable the child to develop responses adaptable to challenging movement situations (Gallahue and Ozmun 2000). Play is considered as vitally important to the growth of the whole child, and facilitation of a learning environment may enhance the learning and development processes in children (Pellegrini and Smith 1998).

6.3 Outdoor Play Environments: Why Nature?

Play activities have proved to increase with the complexity of the environment and the opportunities for play (Frost and Strickland 1985; Nielsen 2009). It is also described how children's play become more vigorous outdoors than indoors and how play forms take another group- and gender constellations outdoors than indoors (Andersen et al. 2017; Pellegrini and Smith 1998). Options for choice and opportunities for play, and the possibility to construct and re-organize the play settings are irreplaceable values in children's play environments. Titman (1994) showed very clearly the children's preferences for outdoor play environments. Environmental qualities that the children appreciated the most were colors in nature, trees, woodland, shifting topography, shaded areas, meadows, places for climbing, and construction and challenging places for exploring and experience. This documents that children have a desire for more complex, challenging, and exciting play environments than the traditional playgrounds usually offered to them. From Titman's study it was also obvious that children appreciate nature:

> We go into the woods and build dens and swings. I like climbing, it's wicked. On a playground you know it's safe so it takes the fun out. When you're climbing a tree you can use your imagination more. Playgrounds just hinder you. . . . Climbing trees is good if you're bored, it makes you feel good, it's a nice feeling when your belly turns over (Titman 1994, p. 26, 36).

Rivkin (1990) mentioned some specific qualities to the outdoor room: The "realness "of physical attributes, children prefer real things over toys and sham. Furthermore, she mentioned children's propensity to symbolism and images that is part of dramatic play and sense of "placeness" demands a certain environment for magic. Children prefer corners and bushes to exposed space. The layering of the landscape affords looking through and gives the sense of depth and diversity. Moreover, she emphasized forms and shapes of landscapes and objects: open-ended spaces and forms often have associative qualities and give meaning to children's play and imagination. Likewise, lines and shapes in the landscape give the children a conception of space and forms. For example, the children prefer multifaceted forms to plain forms and they relate more to softened edges and curves in the landscape. Several other qualities such as experience of places, which engage the senses in textures, sounds, and fragrant smells are essential for nature connection. The novelty and unpredictability, the unusualness and incongruity, surprise and discovery are qualities that intrigue children.

Moore and Wong (1997) describing the project of turning a Yard from an asphalt square into naturalized settings, an Environmental Yard, with assistance from the school children. The children's perception of the Yard after the re-formation emphasized diversity, richness, a place to belong, caring for nature, and more friendly atmosphere. Interviews with the children 5 and 20 years later expressed the fascination of the yard, the complexity and memory of plants and animals. Not least were the landscape features that afforded play recalled: the little clearings, the bridge over the stream, stepping-stones in the pond, all the bushes, and the trees to climb. The children who spent time in the Environmental Yard expressed greater environmental awareness, they attended natural events, they were more innovative in their play. Fantasy play increased using objects that were readily available from the environment. The children also became more interactive with the natural environment outside school (Moore and Wong 1997, pp. 181–191).

A recent study by Ito et al. (2010, 2016) described the planning, design, and construction of a school biotope in Japan. "Process planning" and "Multi-functional landscape planning" (MFLP) were used to plan the school biotope. These design methods allowed changes of the natural conditions and children's usage with the time and creation of multi-functional play space. By overlapping different layers of functions such as vegetation, water, nature restoration, regional ecosystem conservation, playground, and ecological learning, a multi-functional space was designed. Unlike "zoning," MFLP does not divide a space into clear functional areas. The overlapping of layers creates multi-functional areas where, for example, children who are playing by the water can also learn about ecology at the same time. Thus, during the creation of a multi-functional play area, children are able to engage in "various activities" as its different layers are added on top of each other (Ito et al. 2016).

In collaboration with the schoolchildren, the planners worked out the total project of restoring a gravel parking lot into a green biotope. The flat landscape was turned into a place with slope topography covered with mud from paddy fields. Trees were planted and a pond with a trickling stream running through the landscape. Soon after the biotope was inhabited by water-bound insects, terrestrial insects, ducks, and herons. This green landscape afforded lots of activities to the children: exploring the water-biotopes, studying the ecology of plants and animals, and running around in the undulated landscape and it is used for studying the ecology of plants and animals in some school subjects (Fig. 6.3).

6.4 Learning Through Landscapes

The didactic approach to basic skills acquisition is that of "learning through landscapes" where the terrain is the facilitator for diverse movements that challenge motor behavior and where the tasks were conditioning to any individual qualification. The model below (Fig. 6.4) shows the connection between outdoor learning environments, physical activity, and motor development.

Fig. 6.3 The school biotope in Ikiminami primary school, Fukuoka, Japan. Source: Laboratory of Environmental Design (Keitaro Ito's Lab.), Kyushu Institute of Technology, 2017

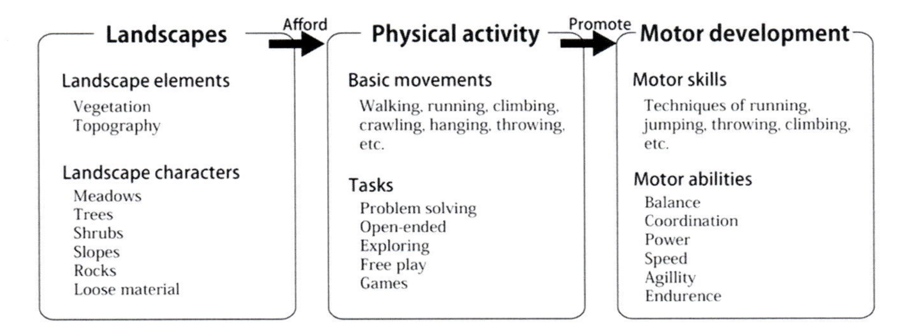

Fig. 6.4 A didactic approach to motor learning indicating the context between landscape elements and characters and the affordances for children's physical activity resulting in development of motor skills and abilities (Fjørtoft 2004)

Figure 6.4 illustrates how the context of landscape structures expressed as topography and vegetation may afford physical activity expressed as the basic movements in an open ended, explorative, and problem-solving approach. The results are indicated to be motor learning and development expressed as motor skills as improved techniques of the basic movements, running, jumping, climbing, throwing, etc., inclusive developing motor abilities like co-ordination, balance, strength, endurance, and flexibility. Through bodily experiments, children explore details and quality of movements such as speed, agility, power, and endurance. The link between the environment and the motor behavior realizes the affordances of the landscape. Affordances are the functions environmental objects can provide to an individual (Gibson 1979, 1986). Affordances are individual, according to the experience, the imagination, and the intentions of the individual. Learning environments afford physical activity, which in turn promote motor development. To promote motor development and achieve motor skills it is crucial that determinants for learning environments and physical activity are adequate for the given purposes. In outdoor environments, such determinants are described as topography and vegetation. The topography could be flat, sloping, hilly, steep, and/or rocky, and the vegetation is represented by diversity in plant units. Physiognomy of the vegetation represents the structure of different plants; for example, the branching of a tree and the density of shrubs. Loose materials are sticks, branches, logs, leaves, cones, stones, etc. The outdoor environment changes over time and with seasons. The winter season may afford a lot of challenging activities on snow and ice. Frost (1992) described children's playscapes as any landscape where children play. Children perceive environments as functions: functions to move, functions to build, functions to hide, to play, etc. Physical environments afford such challenges to children in different ways, and the more diverse, the more challenging. Children perceive and interpret such offers as functions to play and they operationalize the affordances into action. For example, a properly branched tree will be perceived by the child as "climbable," it affords climbing on, and the child will intuitively climb it (Heft 1988). A shrub vegetation may be perceived as a site for constructing a den, or suitable for role-play like a play house. Fjørtoft and Sageie (2000) describe children's play in a natural playscape whereby landscape characters and qualities correspond to children's use of landscape features. Children's play in complex natural environment showed significant effects on their motor development and fitness acquisition (Fjørtoft 2000).

6.5 The Children's Preferences

Several studies have documented children's preferences for play environments which differ quite fundamentally from the traditional and stereotype playgrounds more often provided. Based on research over time there is some evidence for phrasing children's preferences (Fjørtoft and Sageie 2000; Frost 1992; Gibson and Pick 2000; Moore and Wong 1997).

The children's preferences for playscapes could be listed as follows:

- Unstructured playscapes
- Open-ended tasks and options
- Open-ended space, lines, and shapes
- Manipulating objects, loose parts

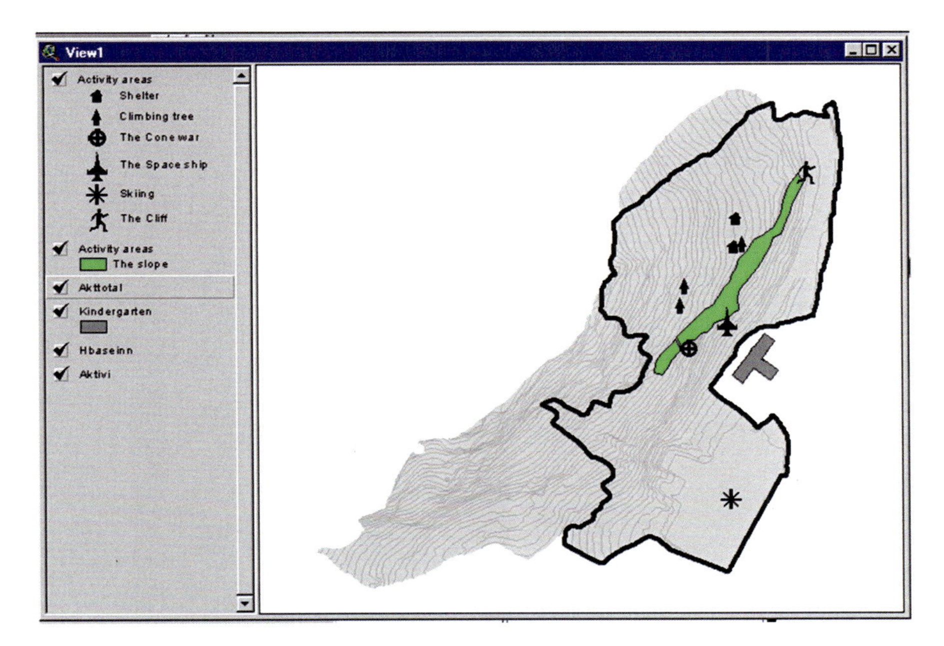

Fig. 6.5 Children's playscape and their favorite places: Shelters, climbing trees, the "cone war," the "space ship" the cliff and the skiing area. (Fjørtoft and Sageie 2000)

- Challenging playscapes
- Individualization
- Green areas: Lawns for tumbling, trees for climbing, bushes for hiding, flowers to pick, nice colors, placeness…

These preferences are seldom available at compulsory, traditional playgrounds. However, in natural settings such affordances may be more realistic.

In a natural forest playscape the children found their favorite places (Fjørtoft and Sageie 2000). There were some specific spots in the forest that the children named and used more frequently. Those were "the cone war," "the space ship," and "the steep slope." The names indicate the activities taking place there and it is an expression of the characteristics that free play takes when stimulated by the functional affordances of a natural environment. Cones are fit for throwing—whatever the target is, and appeal to functional play and battling. The "space ship" is a rock underneath the cliff, and affords functional and fantasy play. These places were located close to the kindergarten in a coniferous forest. The landscape structure and the function of the vegetation encourage the activities taking place there. The steep slope afforded sliding in the winter and nature studies in the summer as the steep slope turns into a hill of mosses (Fig. 6.5).

Natural environments afford many opportunities for free play and physical activity. The affordances of a landscape constitute landscape characters as determinants for children's play and physical activity. Children interpret the landscape

Fig. 6.6 The affordance of a tree: climbing and imaginative play: "the climbing cat"

structures as functions: the function to climb, to slide, to hide, to run, and throw. For example, the affordances of climbing are the trees and the rocks, which are to be climbed according to their structural constitution of branches, slope, and projections. Such structural determinants constitute functional affordances as they challenge activities that develop climbing skills in children. Figure 6.6 shows a girl using the affordances of the tree for climbing. It is the structure of the tree that promoted the development of this unique climbing technique in the girl pretending to be a cat. Climbing rocks demands other climbing techniques, more like mounting by clutching with hands and feet. It is the rock that affords the climbing and demands that particular climbing technique. For smaller kids a branched trunk of a tree invites challenging climbing and crawling (Fig. 6.7). The materiality of the environment affords challenges and experiences that promote motor learning. The children respond by exploring, discovering, and facing the challenges by mastering perceptual-motor skills in context with the environment (Fjørtoft et al. 2009).

6.6 Landscape Characteristics and Affordances for Play

Natural environments represent dynamic and rough playscapes that challenge motor activity in children. The topography, like slopes and rocks, afford natural obstacles that children have to cope with. The vegetation provides shelters and trees for climbing. The meadows are for running and tumbling. Description of physical

Fig. 6.7 A toddler exploring the affordances of the tree trunk: climbing and crawling

environments usually focuses almost exclusively on forms. Heft (1988) suggested an alternative approach to describe the environment, which focused on function rather than form. The functional approach corresponds better to the children's relations to their environment. Intuitively children use their environment for physical challenges and play; they perceive the functions of the landscape and use them for play. The central concept guiding children's examination of their environment is that of affordance and the way the children perceive their environments and correspond by action through movement behavior (Gibson and Pick 2000; Heft 1998). For example, if a rock is big enough to fit the hand, it might be perceived as an object to grasp or to throw, it affords grasping or throwing. A tree that is appropriately branched will likewise be perceived as climb-on-able, it affords climbing on. Table 6.1 illustrates the affordances of landscape characters and functional play activities. The characters of the landscape such as different types of forest afford play with loose materials in building dens, and other constructions. Trees and shrubs afford playing hide and seek or role and fantasy play. Open meadows invite running and games like tag, catch, and other games with running and high speed. Topography also plays a vital role for different play activities throughout the year, dependent on seasons, See Table 6.1.

Natural environments afford such possibilities and affordances. Waters (2017) explained affordances as the complementarity between the child and the environment. Natural features are important qualities of play environments, and the natural features allow a wide range of learning opportunities not available from other playground options (Ferreira et al. 2007). Figure 6.4 illustrates the connection between characteristics and qualities of a natural environment and the affordances

Table 6.1 Landscape characters and affordances for play activities (Fjørtoft and Sageie 2000)

Landscape characteristics	Characters	Play activities
Vegetation		
Trees	Deciduous, conifer, mixed	Climbing, construction play, building dens
Shrubs	Open Scattered Dense	Running, play tag, catch and seek, construction play, fantasy and role-play Hiding, hide and seek
Meadows	Open, flat, even	Running, play tag, catch and seek, acrobatics, skiing, building and playing with snow (winter)
Topography		
Slope	Slope <30 degrees	Rolling, crawling, sliding, downhill skiing, cross. Country skiing, ski-jump (winter)
Roughness	Rocks, cliffs, boulders	Climbing and bouldering

of different kinds of play and physical activity resulting in development of motor skills and abilities (Fjørtoft 2004).

In Scandinavia it has become popular for kindergartners to spend more time outdoors in the natural environment. Some kindergartens organized as outdoor schools, where the children, aged 3 to 6, spend all or most of the day outdoors in a natural environment. Playing in a natural environment seems to have positive effects on children; they become more creative in their play, and the play activities and play forms are increasing (Lysklett and Berger 2017).

6.7 Developmental Benefits from Playing in Outdoor and Natural Environments

Several studies have described the value of playing in natural environments, but rather few studies have documented developmental effects on children. An early study was conducted on how a natural environment in Norway provided a stimulating playscape for kindergarten children, and how different features in the landscape afforded play activities. A quasi-experimental study on 5–7 years old children was performed with one experimental group and one control group comparable in age and socio-economic background. The experimental group was playing outdoors in a natural environment for 2–3 h every day, while the control group played in a traditional playground over the same time period. The study lasted for 1 year and included different seasons and weather conditions. The impact of such outdoor activities on children's motor fitness was tested with the EUROFIT motor fitness test (Adam et al. 1988). The results showed a better improvement from pre- to post test in the experimental group compared to the reference group. Significant differences ($p < 0.01$) were found in balance and co-ordination abilities. The study indicated relations between all-round play in the natural environment and the effect on motor development in the children (Fjørtoft 2001, 2004).

A recent study by Tortella et al. (2016) confirmed the importance of multi-functional playgrounds and the effect on motor skill development in Italian children. However, the study showed a better developmental effect on the experimental group given instructions in their play activities. This may indicate that constructed playgrounds do not immediately invite the complementarity between the child and the environment, or the constructed equipment does not afford sufficient challenges for movement play and motor development in children.

Based on the WHO (2020) recommendations of daily physical activity for children and youth more studies have been focusing on the intensity of activity rather than functional motor development. Such studies indicate that children are more physically active outdoors than indoors. Generally, girls were found to be less active than boys, but as active being outdoors (Andersen et al. 2017). It is a general worry concerning children being more passive and that sedentary behavior is increasing (Jones et al. 2013).

6.8 More Green Play Environments in Urban Settings

Several studies over a considerable time span have focused on the importance of natural environments in children's development and well-being (Frost 1992; Hart 1979; Moore and Wong 1997; Rivkin 1990; Titman 1994, ao.). These studies also confirm the children's preferences for natural and unstructured playscapes (Frost 1998; Fjørtoft 2000, 2004; Fjørtoft and Sageie 2000, ao.). The study of the Japanese school biotope (Ito et al. 2014) provides an example of how it is possible to develop green playground settings in urban areas with contributing conservation of regional biodiversity. As many of these studies also have exemplified the possibility of establishing green playground settings in urban areas, this should be a challenge to planners, designers, and educators. Playground greening is positive for children's physical, explorative, and imaginative play and should be of high priority for future playgrounds and schoolyards. This will benefit children's playing and learning as well corresponding to The Convention on the Rights of the Child (UN 1990) and the WHO (2020) recommendations of daily physical activity. In addition, urban natural environment provides opportunities for variety of learning (Akoumianaki-Ioannidou et al. 2016; Shawket 2016; Wolsink 2016), which means that urban green spaces have potentials as educational resources. Natural environments must be adapted to meet these needs in order to facilitate children's assimilation of knowledge about the complex ecological relationships in various contexts where they live (Azlina and Zulkiflee 2012). More activities based on practice and experiences in nature must be included in the curricula, and applied environmental education must be offered. Outdoor classrooms should be brought into the agenda by creating suitable conditions for school gardens. Outdoor spaces in the city must be designed in such a way as to enable children to interact with the natural elements (Acar 2014). To design these spaces, "landscape-based" design approach using natural materials such as plant materials, land forms, and landscape elements offers different and varied types of development (Herrington and Studtmann 1998). Moreover, "Multi-functional landscape planning" (Ito et al. 2010, 2016) could put green playgrounds into ecological context, realized through naturalized playgrounds. This would play a crucial role for conservation of the regional ecosystem. As being the qualities of green environment, biodiversity supports ecosystem services generating the value of playground for children's development. A component of biodiversity can contribute to several ecosystem services (Aslaksen et al. 2015). Biodiversity has essential roles at all levels of ecosystem services: as regulator of ecosystem processes that underpin ecosystem services, as an ecosystem service in itself (e.g., biodiversity at the level of genes and species can contribute directly to goods and their values), and as a good in itself that is subject to valuation, economic, or otherwise (Mace et al. 2012). As urbanization and rapid growth of urban population are expected to expand rapidly in the present and future decades in cities across the world, it is important that the ecosystem services in urban areas and the ecosystems including the biodiversity that provide them are understood and valued by city planners and political decision makers (Bolund and Hunhammar 1999). The current issues of urban planning, and common focus of planning and growth management efforts is to establish green infrastructure to maximize urban biodiversity conservation (Hostetler et al. 2011). The landscape designers consider "landscape" as an "Omniscape" (Numata 1996; Arakawa and Fujii 1999) in which it is much more important to think of landscape planning as a learnscape, embracing not only the joy of seeing, but exciting a more holistic way of using body and senses for learning (Ito et al. 2016). Thus, green playground must be considered as an essential quality of green infrastructure. Urban planning should provide public open spaces capable of performing ecosystem services for children as their function. Future projects and implementation of creating and managing green playgrounds are needed through collaborative work of planners, designers, ecologists, educators, and children. This will support the quality of green playgrounds and contribute to children's development and well-being throughout their life span.

References

Acar H (2014) Learning environments for children in outdoor spaces. Procedia Soc Behav Sci 141:846–853

Adam C, Klissouras V, Ravazallo M, Renson R, Tuxworth W (1988) EUROFIT: European test if physical fitness. Council Europe, Committee for the Development of Sport, Rome

Akoumianaki-Ioannidou A, Paraskevopoulou AT, Tachou V (2016) Schoolgrounds as a resource of green space to increase child-plant contact. Urban For Urban Green 20:375–386

Andersen E, Borch-Jenssen J, Øvreås S, Ellingsen H, Jørgensen KA, Moser T (2017) Objectively measured physical activity level and sedentary behavior in Norwegian children during a week in preschool. Prev Med Rep 7:130–135

Andersen LB, Harro M, Sardinha LB, Froberg K, Ekelund U, Brage S, Anderssen SA (2006) Physical activity and clustered cardiovascular risk in children: a cross-sectional study (the European youth heart study). Lancet 368:299–304
Arakawa S, Fujii H (1999) Seimei no kenchiku [Life architecture]. Suiseisha, Tokyo
Aslaksen I, Nybø S, Framstad E, Garnåsjordet PA, Skarpaas O (2015) Biodiversity and ecosystem services: the nature index for Norway. Ecosyst Serv 12:108–116
Azlina W, Zulkiflee AS (2012) A pilot study: the impact of outdoor play spaces on kindergarten children. Procedia Soc Behav Sci 38:275–283
Biddle SJ, Gorely T, Stensel DJ (2004) Health-enhancing physical activity and sedentary behavior in children and adolescents. J Sports Sci 22:679–701
Bolund P, Hunhammar S (1999) Ecosystem services in urban areas. Ecol Econ 29(2):293–301
Clements R (2004) An investigation of the status of outdoor play. Contemp Issues Early Child 5 (1):68–80
Ferreira I, van der Horst K, Wendel-Woset W, Kremers S, van Lenthe FJ, Brug J (2007) Environmental correlates of physical activity in youth – a review and update. Obes Rev 8:129–154
Fjørtoft I (2000) Landscape as playscape. Learning effects from playing in a natural environment on motor development in children. Doctoral dissertation. Norwegian University of Sport and Physical Education, Oslo
Fjørtoft I (2001) The natural environment as a playground for children: the impact of outdoor play activities in pre-primary school children. Early Childhood Educ J 29(2):111–117
Fjørtoft I (2004) Landscape as playscape. The effect of natural environments on children's play and motor development. Child Youth Environ 14(2):21–44. http://www.colorado.edu/journals/cye/
Fjørtoft I, Gundersen KA (2007) Promoting motor learning in young children through landscapes. In: Liukkonen J et al (eds) Psychology for physical educators: student in focus. Human Kinetics Publishers
Fjørtoft, I., Kristoffersen, B. and Sageie, J (2009) Children in schoolyards: Tracking movement patterns and physical activity using global positioning system and heart rate monitoring. landscape and urban planning 93 (3-4):210-217
Fjørtoft I, Sageie J (2000) The natural environment as a playground for children. Landscape description and analyses of a natural playscape. Landsc Urban Plan 48:83–97
Fjørtoft I, Löfman O, Thorén KH (2010) Schoolyard physical activity in 14-year-ols adolescents assessed by mobile GPS and heart rate monitoring analysed by GIS. Scand J Public Health 38 (5):28–37
Frost JL (1992) Play and playscapes. Delmar Publishers Inc, New York
Frost JL (1998) Neuroscience, play, and child development. Paper presented at the meeting of the IPA/USA Triennial National Conference, Longmont, CO
Frost JL, Campbell SD (1985) Equipment choices of primary-age children on conventional and creative playgrounds. In: Frost L, Sunderlin S (eds) J, When children play: proceedings of the international conference on play and play environments. Association for Childhood Education International
Frost JL, Strickland E (1985) Equipment choices of young children during free play. In: Frost JL, Sunderlin S (eds) When children play: proceedings of the international conference on play and play environments. Association for Childhood Education International
Gallahue DL, Ozmun JC (2000) Understanding motor development: infants, children, adolescents, adults. McGraw-Hill College
Gibson E, Pick A (2000) An ecological approach to perceptual learning and development. Oxford University Press, New York, pp 24–25
Gibson J (1979) The ecological approach to visual perception. Houghton Mifflin Company, Boston
Gibson J (1986) The ecological approach to visual perception. Erlbaum, Hillsdale, p 129
Grahn P, Mårtensson F, Lindblad B, Nilsson P, Ekman A (1997) UTE PÅ DAGIS. Stad & Land nr 145. Movium, Sveriges Lantbruksuniversitet, Alnarp
Hart CH (ed) (1993) Children on playgrounds. Research perspectives and applications. State University of New York Press, Albany

Hart R (1979) Children's experience of place. Irvington, New York
Hartle L, Johnson JE (1993) Historical and contemporary influences of outdoor play environments. In: Hart CH (ed) Children on playgrounds. Research perspectives and applications. State University of New York Press, Albany
Heft H (1988) Affordances of children's environments: a functional approach to environmental description. Child Environ Q 5:29–37
Heft H (1998) Towards a functional ecology of behavior and development: the legacy of Joachim F. Wohlwill. In: Goerlitz D, Harloff HJ, Mey G, Valsiner J (eds) Children, cities, and psychological theories: developing relationships. Walter De Gruyter, Berlin, pp 85–110
Herrington S, Studtmann K (1998) Landscape interventions: new directions for the design of children's outdoor play environments. Landsc Urban Plan 42(2–4):191–205
Hostetler M, Allen W, Meurk C (2011) Conserving urban biodiversity? Creating green infrastructure is only the first step. Landsc Urban Plan 100:369–371
Ito K, Fjørtoft I, Manabe T, Masuda K, Kamada M, Fujuwara K (2010) Landscape design and children's participation in a Japanese primary school – planning process of school biotope for 5 years. In: Muller N, Werner P, Kelcey JG (eds) Urban biodiversity and design. Blackwell Academic Publishing, Oxford, pp 441–453
Ito K, Fjørtoft I, Manabe T, Kamada M (2014) Landscape design for urban biodiversity and ecological education in Japan: approach from process planning and multifunctional landscape planning. In: Nakagosi N, Mabuhay JA (eds) Designing low carbon societies in landscapes. Springer, Tokyo, pp 73–86
Ito K, Sudo T, Fjørtoft I (2016) Ecological design: collaborative landscape design with school children. In: Murnaghan AMF, Shillington LJ (eds) Children, nature, cities. Routledge, pp 95–209
Jones RA, Hinkley T, Okely AD, Salmon J (2013) Tracking physical activity and sedentary behavior in childhood: a systematic review. Am J Prev Med 44(6):651–658
Karsten L (2005) It all used to be better? Different generations on continuity and change in urban children's daily use as space. Am J Prev Med 23:15–25
Lysklett OB, Berger HW (2017) What are the characteristics of nature preschools in Norway, and how do they organize their daily activities? J Adventure Educ Outdoor Learn 17(2):95–107
Mace GM, Norris K, Fitter AH (2012) Biodiversity and ecosystem services: a multilayered relationship. Trends Ecol Evol 27(1):19–26
Moore R, Wong HH (1997) Natural learning. Creating environments for rediscovering nature's way of learning. The life history of an environmental schoolyard. MIG Communications, Berkley
Moore RC (1986) Childhood's domain: play and space in child development. Croom Helm, London
Moore RC, Goltsman SM, Iacofano DS (eds) (1992) Play for all guidelines: planning, design, and management of outdoor settings for all children, 2nd edn. MIG Communications, Berkley
Nielsen G (2009) School ground characteristics as a predictor of children's daily physical activity – some methods and results. Institute of Sport and Exercise Sciences, University of Copenhagen. Foredrag på Nordiskt metodseminarium kring barns utomhusmiljö. Sveriges Lantbruksuniversitet, Alnarp
Numata M (1996) Landscape ecology. Asakura Shoten, Tokyo
Olivieira AF, Moreira C, Abreu S, Santos R (2014) Environmental determinants of physical activity in children: a systematic review. Arch Exerc Health Dis 4(2):254–261
Pellegrini AD, Smith PK (1998) Physical activity play: the nature and function of a neglected aspect of play. Child Dev 69(3):577–598
Rivkin MS (1990) Outdoor play – what happens here?. In: Wortham S, Frost JL (eds) Playgrounds for young children. National survey and perspectives. A Project of the American Association for Leisure and Recreation. An Association of the American Alliance for Health, Physical Education, Recreation and Dance
Rivkin MS (1995) The great outdoors: restoring children's rights to play outside. National Association for the Education of Young Children. Washington, DC

Sallis JF, Glanz K (2006) The role of built environment in physical activity, eating, and obesity in childhood. Futur Child 16(1):89–108
Shawket IM (2016) Educational methods instruct outdoor design principles: contributing to a better environment. Procedia Environ Sci 34:222–232
Titman W (1994) Special places, special people. The hidden curriculum of school grounds. WWF UK (World Wide Fund For Nature)/Learning through Landscapes
Tortella P, Haga M, Sigmundsson H, Fumagalli G (2016) Motor skill development in Italian pre-school children induced by structured activities in a specific playground. PLoS One 11(7): e0160244
UN: The Convention on the Rights of the Child 2 September 1990
Waters J (2017) Affordance theory in outdoor play. In: Waller T et al (eds) The SAGE handbook of outdoor play and learning. SAGE Publ. Ltd, London
WHO (2020) WHO guidelines on physical activity and sedentary behaviour. World Health Organization, Geneva (Licence: CC BY-NC-SA 3.0 IGO)
Wolsink M (2016) Viewpoint "Sustainable City" requires "recognition"—the example of environmental education under pressure from the compact city. Land Use Policy 52:174–180
Worthman SC, Frost JL (1990) Playground for young children: National survey and perspectives. American Alliance for Health, Physical Education, Recreation and Dance, Reston, VA

Chapter 7
Greening School Grounds: Schools' Role in a Biodiversity Process

Ching-fen Yang

Abstract This paper reviews project cases to explore the role of schools in improving environmental biodiversity. In the past two decades, numerous projects have been implemented in Taiwan to promote environmental sustainability in schools. This process must be continued, and environmental education must be enacted concurrently to cultivate green concepts in people. Three projects are reviewed to describe local-based users' participatory processes for achieving environmental transformation and sustaining environmental concept cultivation. In addition, connecting individual school grounds can build a solid foundation for preparing a diverse biological environment.

Keywords Biodiversity process · School grounds · Environmental sustainability · Environmental education

Schools are vital public spaces in urban areas. If they can be transformed into open spaces with natural characteristics, they can provide biodiversity benefits for the surrounding area, especially in crowded cities, and transform school grounds into learning spaces that are more diversified by introducing corresponding curriculum. The International School Ground Association has gathered professionals and organizations from around the world who agree with promoting the concept of greening school grounds, which adds value to outdoor school spaces for learning about and promoting biologically diverse environments.

Taiwan has 4280 schools that comprise17,129 ha in total (refer to Table 7.1, Fig. 7.1); 11,699 ha of which is located in urban areas, accounting for 12.58% of the total public facility space. The building coverage rate is between 40% and 50% of total school sites. Converting at least half of outdoor school grounds into green spaces can bring significant benefits to surrounding areas.

C.-f. Yang (✉)
Building and Planning Research Foundation, National Taiwan University, Taipei, Taiwan
e-mail: ychingfen@ntu.edu.tw

K. Ito (ed.), *Urban Biodiversity and Ecological Design for Sustainable Cities*,
https://doi.org/10.1007/978-4-431-56856-8_7

Table 7.1 Number of schools at various levels in Taiwan as of 2013

	Primary	Junior high	Senior high	college	total
School number	2670	932	517	161	4280
School area(ha)	5277	2345	2901	6606	17,129
Average school area (ha/school)	2	3	6	41	4

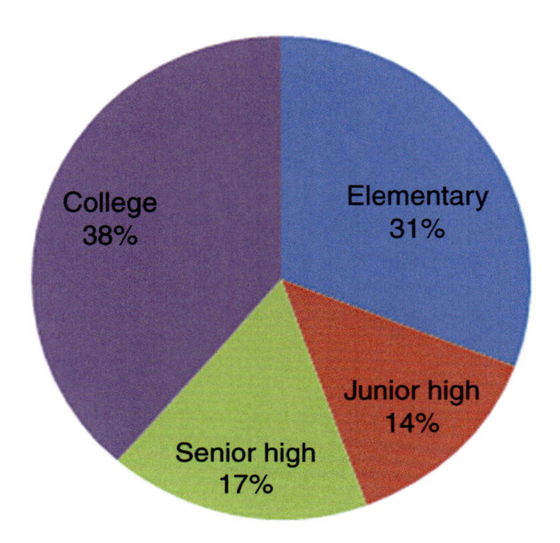

Fig. 7.1 Area ratio of schools at various levels in Taiwan

In the past, school buildings in Taiwan were constructed according to School Building Standards. All space except for that occupied by buildings is generally considered outdoor space, including playgrounds, tracks, playground equipment, landscaping elements, and other easy-to-manage and relatively safe spaces. In recent years, with the introduction of new school ground construction models and a focus on environmental education, green and energy conservation concepts have gradually been included in the design of school grounds. This shift in design practices has emphasized the benefits of school ground open spaces for cities, transformed the appearance of school grounds, and changed the role that school grounds play in their surrounding environments.

The Taiwan Sustainable Campus Project (hereafter referred to as the "Sustainable Campus"), initiated by the Ministry of Education, and the Verdant School Project (hereafter referred to as the "Verdant School"), organized by Taipei City, are two projects begun in the past two decades that are related to the ground greening of schools and improving their availability to the public. The former has focused on introducing the concept of environmental sustainability and corresponding environmental education to school grounds; the latter has focused on the role of schools in cities and communities, and, through participatory design, has increased openness and enhanced the ecological performance of schools. In addition, school-based curriculum reflects characteristics of its corresponding school. For example, the Evergreen Lily Elementary School, with its indigenous tribal characteristics, offers course on building a *tala-baliuw/tapau*[1] that pass on traditional indigenous wisdom regarding homelands. These three aforementioned projects increase the possibility of school grounds improving their biodiversity as well as both the natural and human environment. These plans are also designed to teach school members management concepts for sustainable environmental spaces and then to expand on them. This paper introduces these projects, reflects on their meaning, and proposes roles that schools can play in promoting a biologically diverse environment.

7.1 The Taiwan Sustainable Campus Project

7.1.1 Project Summary[2]

The Ministry of Education's Sustainable Campus Technical Expert Team began planning the Sustainable Campus in 2001 to continue the spirit of the New School Ground Movement, a collaboration between the Ministry of Education and civilian organizations to rebuild schools after the earthquake of September 21. The policy is based on the Executive Yuan's "Challenge 2008 National Development Plan: Water and Green Construction Plan" and the United Nation's Rio Declaration on Environment and Development regarding human sustainability development and survival.

The first phase of this project recruited 23 elementary and junior high schools. From 2003 to 2014, to fully promote this project, approximately 888 schools were approved to participate in the project. Among these schools, the majority were elementary schools, accounting for approximately 71% of the total. Official statistics of results from earlier participating schools showed that the green coverage rate increased from 23.4% to 35.8% and that the total water permeable surface coverage rate reached 12.8%.

This project used a small amount of subsidies[3] and tour assistance from specialized counseling teams to support each school grounds in transforming in a sustainable direction. First, each school proposed an improvement plan, applied for funds, and invited spatial professionals to manage the improvement project. The improvement focus in the policy was stated as follows: In addition to developing energy-saving, resource-saving, healthy, and comfortable school buildings as well as ecologically friendly and recyclable school ground environments, this project can establish applications of local Sustainable Campus technology and a reference for future evaluations. School grounds can be a starting point for promoting community

[1]*Tala-baliuw/tapau* meaning a type of small, temporary structure, and farmland in Rukai and Paiwan, respectively.

[2]Refer to Taiwan Sustainable Campus Project, https://www.esdtaiwan.edu.tw/English/intro.php (Taiwan Sustainable Campus Project n.d.).

[3]For the first 2 years, each school had received less than one million NTD. Beginning in 2004, projects that integrated peripheral schools received two million NTD at most of each.

rebuilding projects. School public spaces can serve as models through a participatory process to achieve the consensus of neighboring communities. These public spaces can highlight local characteristics, conform to local conditions, consolidate community awareness, and create an ecological education model that tightly links communities with school grounds. Thus, the subsidy items follow the four main themes: (1) the resource and power cycle, (2) site-corresponding sustainability, (3) the ecological cycle, and (4) healthy buildings.

7.1.2 Taiwan Green School Partnership Network[4]

Similar ideas were first proposed in 1999, 2 years before the start of the Sustainable Campus. The Taiwan Environmental Protection Department provided the Graduate Institute of Environmental Education of the National Taiwan Normal University (hereafter referred to as the "NTNU Environmental Education Institute") with funding and requested the institute design learning activities that provided students with hands-on experiences of environmental protection. The NTNU Environmental Education Institute developed the Taiwan Green School Partnership Network (hereafter referred to as "Greenschool") based on the concepts and experiences of Canada's SEEDS and the American GREEN, GLOBE Program[5]. Greenschool links the mechanism to the network and uses rewards to encourage idea sharing. Greenschool emphasizes the process of "ecological thinking, humanistic care, partnership, mobile learning, and resource exchange." This is designed to guide the teachers and students of participating schools to perform voluntary improvements based on their own situation as well as the involved school's abilities and resources.

The greatest result of Greenschool is its facilitation of teaching materials being shared in classes. One of the reasons for this result is that this project is combined with and promoted within the Grade 1–9 curriculum. Changes in teaching methods have significantly increased the demand for this type of cross-field integrated course.

This project is also closely linked with the Ministry of Education's Sustainable Campus. Schools that wish to apply for Sustainable Campus subsidies must be a registered Greenschool partner. Schools with more than six leafs[6] can use their

[4]Refer to Taiwan Green School Partnership Network, https://www.greenschool.moe.edu.tw/gs2/eng/ (Taiwan Green School Partnership Network n.d.).

[5]SEEDS stands for Society Environment Energy Development Studies. In January 2014, SEEDS merged with another charity, Connections Education Society, and changed its name to SEEDS Connections. GREEN stands for Global Rivers Environmental Education Network; GLOBE stands for Global Learning and Observations to Benefit the Environment.

[6]Schools that register as Greenschools (also called a "partner") can report green activity improvement at any time. After reporting, the central office responds and provides assistance. After the action improvement has been reported, the center issues a different number of leafs as a reward, depending on the characteristic of the change. Accumulated leafs become trees, and then a Greenschool.

reported contents to meet the application requirements. This method differs from the past method that separated education (software) and building (hardware) requirements, which is one of the main tenets of this project. The hardware concepts specific to Greenschool include Ecological Restoration and Maintenance as well as Sustainable Buildings, which create a different and diversified school ground environment by relating to their own surrounding area, and cultural, historical, and ecological characteristics. The Greenschool software was implemented in tandem with the Grade 1–9 curriculum. Each school's corresponding school ground changes provide teaching materials that conform to each school's characteristics. In the future, these schools can be linked to neighboring schools that have different education characteristics to form a diverse environmental education network.

7.1.3 The Objectives of the Sustainable Campus

A Sustainable Campus is not limited within a school. Its greater ambition is to combine schools and communities into a green hub, and to become an initiator for local sustainability development. The objectives of the Sustainable Campus include improving sustainability, ecology, environmental protection, and health themes. The project should be school building renovation accompanied by educational concepts. The ultimate goal of the Sustainable Campus is for a community and its schools to be combined into an initiative unit for local sustainability activity and then linked with other units to make Taiwan environmentally sustainable.

Based on this concept, the Sustainable Campus can be considered a simplified version of a "sustainable city." In order to create a Sustainable Campus that exhibits the following characteristics: (1) school grounds that perform biologically diverse and ecology with appropriately managed natural resources; (2) school grounds that demonstrate safety, comfort, high quality, openness, rich creativity, and local cultural learning as well as living; and (3) school grounds that demonstrate practical energy conserving technology and highly efficient energy use. Schools were encouraged to use adequate resource management to form a school ground space that incorporated the local culture and natural habitat; this dynamic process may also be used as a teaching tool.

7.1.4 Problems of Existing School Grounds Not Being Sustainable

However, as an improvement project, the Sustainable Campus must first respond to the sustainability flaws of existing school buildings. The exemplification of such a problem involves a lack of comprehensive school ground planning, the greening method not conforming to environmentally friendly practices (e.g., only laying down lawn that does not include multilayer or mixed greening), a low school ground water permeability rate (e.g., large areas of non-permeable polyurethane track), school ground allocation that lacks consideration of site characteristics and microclimate factors (e.g., a building's east–west orientation based on sun exposure), improper material use (e.g., materials produced through a CO_2 emission process and materials used that cannot be recycled or reused), a large amount of construction waste, a lack of a water saving system and storage facility, and entry of unprocessed school sewage into the urban system (Hsiao et al. 2001). Regarding resource management, current problems in school buildings are attributable to a reliance on unusable resources and a lack of waste material reuse. In social and cultural aspects, most school spaces and related activities are restricted to within school grounds, offering little opportunity to interact with society and culture outside the schools.

7.1.5 Recommending Improvement Items of the Sustainable Campus

The Ministry of Education has not responded to all the aforementioned problems. However, the Ministry has recommended improvement items corresponding to the project's four main themes (the resource and power cycle, site-corresponding sustainability, the ecological cycle, and healthy buildings). In addition to school space improvements, other items directly correspond to teaching requirements such as farms and ecological ponds. Although different schools have different items, most participating schools choose the recommended items that are appropriate for their schools.

For elementary and junior high schools, ecological ponds, farms, leaf litter and kitchen waste compost systems, and water permeable surface improvements are popular improvement items. These schools tend to choose demonstration facilities that can be used for teaching. These items are all within the subsidy scope of the Ministry of Education's funding. Other items such as conducting comprehensive school ground planning, as well as re-consideration of site characteristics and microclimate factors for adjusting building orientation, are not subsidized by this project.

7.1.6 Case 1. Shenkeng Elementary School Experiment

Shenkeng Elementary School, in Taipei County, was established in 1899. The school area is 2.7 ha, located in an urban area near the tourist destination of Shenkeng Old Street. On the school's south side is Jingmei River (Fig. 7.2), while a grassy field lays to the west. Shenkeng Junior High is east of the elementary school. Both schools are influential cultural and educational centers. The school grounds

Fig. 7.2 Jingmei River on the south side of Shenkeng Elementary School

have been rebuilt several times, but not according to any comprehensive plan. The old buildings are mostly one or two story buildings while the two new buildings are four and five stories. Except for the track (approximately 0.5 ha), the outdoor space is mostly gardens and plants. The greatest natural resources on the school grounds are several old trees.

Shenkeng Elementary School was registered as a Taiwan Greenschool in 2001, and has continually participated in the Sustainable Campus since 2002. In 2003, the elementary school was deemed a Taipei County environmentally sustainable education center and a school that followed the Ministry of Economic Affairs' energy education model. Shenkeng Elementary School can be considered a benchmark for the promotion of sustainability.

The Sustainable Campus was initially adopted at Shenkeng Elementary with an attempt to save a ring-cupped oak[7] whose roots had been sealed by cement (Fig. 7.3). The school later expanded its tree-planting practices. Other items it undertook included resource recycling and reuse (i.e., trash separation, resource recycling, and leaf litter composting), environmental management (i.e., increasing the school grounds' water permeable surfaces as well as performing environmental sound and

[7]This is not listed in the Ministry of Education's project because it was subsidized by the Shenkeng Township Office and Farmer's Association as well as the Taipei County Government in 2001 and 2002, respectively.

Fig. 7.3 The Ring-cupped Oak rescued from ground sealed off by cement

light improvements), natural resource reuse practices (i.e., the creation of a rainwater closure system and the use of solar energy), and finally, connections between classes and environmental activities, in which the implementation process was implemented to teach students about the school ground rebuilding project.

7.1.6.1 Creation of the Ecological Corridor in Northern Taiwan

Among Shenkeng Elementary School's improvement items, the creation of the Ecological Corridor in Northern Taiwan (Fig. 7.4) is a relatively large one. Initially, a wetland ecological area was created on the south side of the school grounds, near the Jingmei River. This area simulated a pond environment and was constructed from second-hand materials. In the second year, this ecological area was extended north. To demonstrate river ecology, the area was designed to simulate the Jingmei River's up-, mid-, and downstream ecological characteristics. Clay was used for waterproofing. The sides were constructed from concrete that resembled rocks to create a porous environment. Water is supplemented from the rainwater recycling system on top of the school buildings. Native water plants from the surrounding

Fig. 7.4 The Ecological Corridor in Northern Taiwan

Shenkeng area were selected for planting. Deadwood has been placed on the side of the ecological corridor pond to serve as a habitat for creatures. The length of the entire corridor is approximately 146 m.

This project achieved the following accomplishments: (1) it rearranged the circulation and divided the school grounds into dynamic and passive areas; (2) it increased biological diversification and attempted to link the Jingmei River into the ecological corridor; and (3) by removing the wall between the elementary school and Shenkeng Junior High, it connected the green areas between the two school grounds. The most exciting outcome of the establishment of the ecological area was that the migratory bird, red-necked Phalarope, had chosen to settle there.[8]

[8]However, owing to the leaking problem, the school terminated this corridor experiment around 2008.

7.1.6.2 Evaluation of Benefits from the Shenkeng Elementary School's Suitable School Grounds

A. Physical Space
The Shenkeng Elementary school reported that the rainwater recycling system and the Ecological Corridor cost a total of three million NTD, not including labor and materials required for follow-up maintenance of the ecological pond. However, regarding educational purpose, this system may be highly beneficial. In addition, the Ecological Corridor facilitated breaching the borders of the Elementary and Junior high schools. The benefits cannot be determined purely on the basis of revenue and expenditure. The school estimated that after it participated in the Sustainable Campus, the combined green coverage rate and water permeable surface area reached 50% of the total site area.
B. Cultivation of Space Management Professionals
The project initiators, the school principal and director, emphasized that the implementation process of this project included teacher, parent, and student opinions. Although there were also opposing opinions, the school's students and teachers had positive evaluations about the school ground transformation (Lin 2005). Many outdoor learning courses were conducted. Students noticed the changes in the school ground environment. Some students indicated that they had never used the maple path (Fig. 7.5) before, but after surface improvements, that they began to enjoy the cool air along the path. After the area surrounding the trees in the courtyard was changed to a water permeable surface and had wood platforms added, it became a popular location for the school. This allowed students to become more involved with the natural environment.

7.1.7 Case 2. National Taiwan University Experiment

National Taiwan University (NTU) was established in 1928. The main school grounds occupy approximately 114 ha, as of 2008. The average building coverage rate is 24% of the total area. The green area coverage is 39%, the water permeable area coverage is 69%, and the conservative estimate of green coverage rate is at least 51% of the total area. The school is located in an urban area and is an essential open space in southern Taipei City. Furthermore, the NTU farm on the southeast of the campus is a vital green area. The area is near Xindian Creek and the mountainous natural environment in southern Taipei City. Thus, this university is appropriately located for becoming a Sustainable Campus.

NTU participated in the Ministry of Education's Sustainable Campus Transformation Experimental Project in 2003 and has since continued its implementation of the Sustainable Campus. Items promoted in recent years have been based on the core goals of the Sustainable Campus, which include reducing water and electricity use, reducing trash production, valuing limited resources, lowering the environmental

Fig. 7.5 The Maple Path

burden, building sewage and sewage processing facilities on the main school grounds, fulfilling social responsibility as a citizen, maintaining school ground ecology and humanistic functions, strengthening environmental education, and sharing resources with the public. Of the university's numerous departments, related operations are managed by the Environmental Protection and Occupational Safety and Health Center.

7.1.7.1 Liugongjun Pool Setup

In addition to the aforementioned experiment, NTU began the Liugongjun Restoration Project in 2001. The project restored the original Liugongjun path that runs from the NTU farm to the Drunken Moon Lake along Palm Avenue. By using special funding from the Liugong Irrigation Association, ecological methods were used to restore the path to its original appearance. The project intended to provide at least three functions: turned the path into a landscape facility, offered ecological education site, and improved the microclimate of surrounding area.

Fig. 7.6 The reserved Liugongjun canal

Liugongjun was built in 1763 and was a crucial early irrigation canal system for farms in the Taipei basin. After Taipei City became urbanized, farmland gradually disappeared, and the canals lost their relevance. Sewers, parking lots and roads were constructed to replace the canals. The canal through the NTU main school grounds is the Daan branch canal, which has been converted into a drainage culvert in recent years. The banks of the Drunken Moon Lake, which originally contained rice paddies and ponds, were gradually covered by cement in accordance with area development. The pool on the west side of the campus was originally a lily pond that was fed with underground water. The Liugongjun Restoration Project changed the farm and surrounding areas into public space and turned the pond into a 1.08 ha pool (Figs. 7.6 and 7.7). The area of the pool is 2330 m^2 and the canal banks are approximately 4000 m^2 (National Taiwan University 2003).

The pool's planning originated from the transformation of NTU's teaching and experimental farm into a half human-constructed ecological farming environment that is open to the public and forms a leisure green belt. Because of its different levels of openness, the pool's transformation changed not only the landscape but also the ecological system, thus creating a potential threat to the original bird habitat.

Fig. 7.7 The Liugongjun pool with background of NTU experimental farm land

7.1.7.2 Design Concept of the Ecological Construction Method[9]

A. Design of the Pool
   The design of the pool includes the following:

   (a) Porous space to diversify habitats: For example, the original trenches of the canal were changed to mason stones with gaps reserved for insect habitats. Parts of the water-impermeable solid bottom of the old farm irrigation section were removed.
   (b) A multilayer open water area and micro terrain grass–lake environment: For example, irregular lake shores and concrete edges were created to increase the transitional area between water and land. Another example is the building of three environments of different water depths in the water area to provide for the requirements of different creatures.
   (c) Building a bird habitat: For example, providing birds with open water and shallow banks where they can forage and rest, establishing diversified lakeside vegetation environment for birds to hide and build nests with, building diversified grassy vegetation environments for birds, and building fences to prevent humans and other animals from disturbing and hunting

[9]Refer to National Taiwan University (2003).

birds. A riprap-based green island was placed in the middle of the water body for birds to conceal themselves in.

(d) Creating habitats for indicator species, which include fireflies and dragonflies.

(e) The presence of a water source: The water source was the neighboring Life Science Building basement, which originally had to be discharged of the groundwater it collected. This also serves to prevent floods and has a conservation function. If there is over flooding, the excessive water is discharged through the drainage. Since September, 2015, the water source is the creek water from nearby Hsindian River.

B. Planting Plan

Plants appropriate to the existing NTU environment were considered for planting. A basic growth requirement for plants is their humidity resistance. The spatial characteristics and requirements of functionality were also considered. These considerations were matched with patterns of plant growth, seasonal changes, flowering characteristics, texture, plant shapes, and the specific characteristics and experience of building a space. The project planted mostly native plants and nectar plants to attract birds and insects.

7.1.7.3 Evaluation of Benefits of NTU's Pool

A. Physical Space

The long-term records of biological species changes from the observations of the NTU Conserve Club throughout the pool project at the NTU farm showed that the NTU farm was once a substantial habitat for birds in the southern Taipei region. As the space opened up to the public and half of the area was lost because of construction, aggressive and unexpected contentious species entered the area. However, as the water area increased, the number of dragonfly species also increased. The NTU farm, which has been responsible for this project, indicated that a 1–3 year stabilization period is required for adjusting the growth of different species during different seasons to stabilize the ecological operation of the pool (Fig. 7.8).

B. Cultivation of Space Management Professionals

NTU is a prominent education institution in Taiwan. Therefore, it values the simultaneous advancements of environmental improvements and education. In addition to related departments conducting research and establishing courses on the NTU ecological environment, the NTU farm canal so assume public education responsibilities such as organizing activities during the initial stages after the pool is completed. These activities include inviting guests to release native species such as paradise fish, water scorpions, rose bitterling, and freshwater mussels into the pool to stabilize its water quality. This would serve as an example of a Sustainable Campus.

Fig. 7.8 A "Slow down for crossing animals" sign placed nearby

7.2 Taipei City Verdant School Project

Compared with the aforementioned national Sustainable Campus, the Taipei City government initiated the Verdant School in 2007 in accordance with an urban design perspective. Taipei City has 340 schools at different levels. The total area of these schools is 1636 ha, which is 6% of the entire land area of Taipei City (27,180 ha). Figure 7.9 shows the distribution of Taipei City schools. It illustrates how substantial an increase in city green spaces would occur if school grounds became public green spaces like parks.

7.2.1 Project Summary[10]

The project is named "The Love That You Can See: Verdant School." The term "verdant" refers to the expectation to increase the greenery of school borders, turn dark corners into bright green areas, and change school landscapes. This project is a part of the Taipei's Great Looks Series, which has objectives similar to those of the Sustainable Campus, where schools are used to promote city green ecological concepts. School and community members' participation in the transformation

[10]Refer to Department of Urban Development (2010).

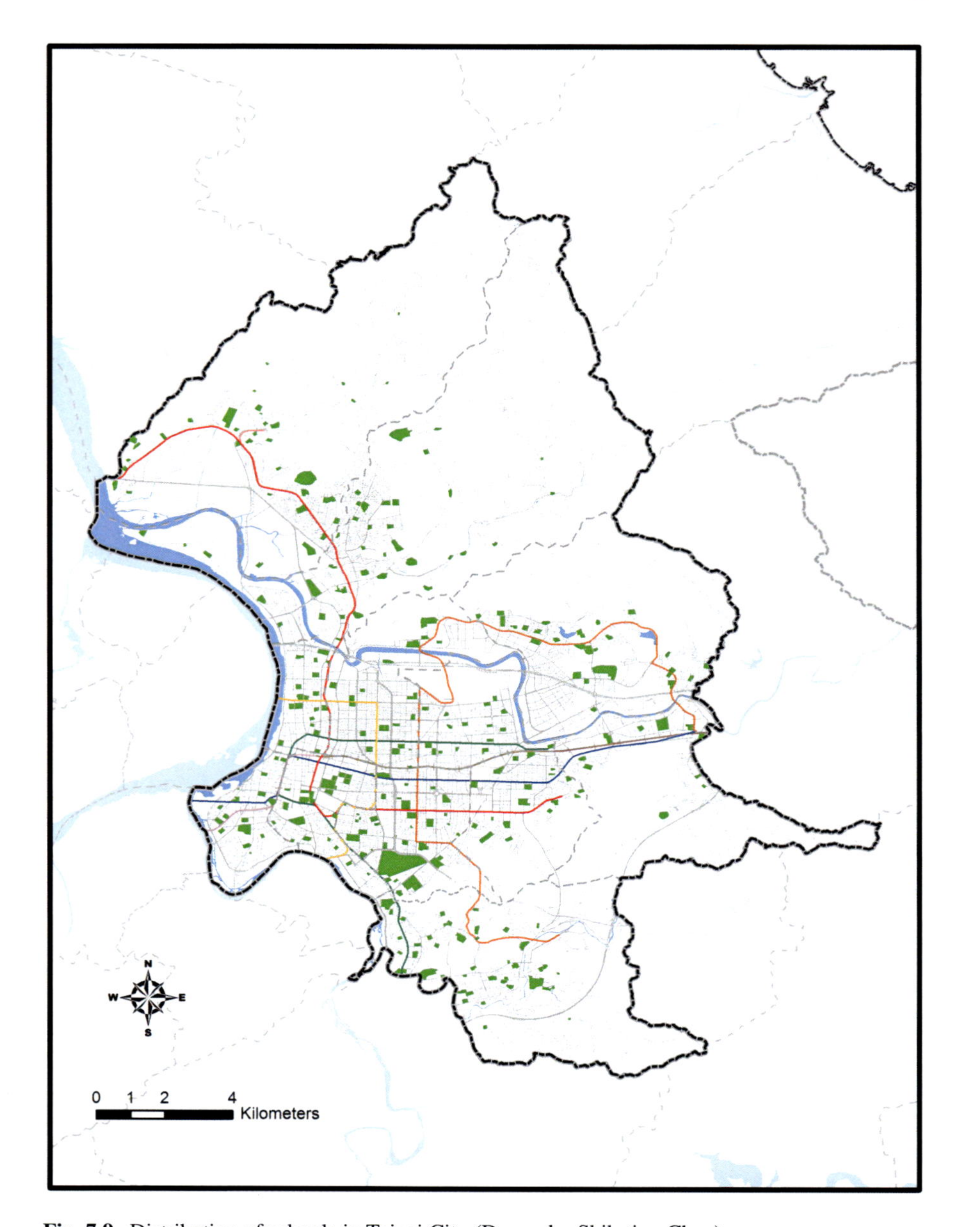

Fig. 7.9 Distribution of schools in Taipei City (Drawn by Shih-ting Chen)

process expands school greening, beautifies the community corridor landscape, and promotes the city's overall green image. Project objectives are as follows:

A. To improve the overall quality of the surrounding environment.
B. To lower CO_2 emissions and combat global warming trends.

C. To create more urban bird and wildlife habitats.
D. To increase open spaces and cultural diversity.
E. To encourage teachers, students, and the public to participate in urban greening work.
F. To restore the diversity of the urban landscape.
G. To strengthen the affinity between school grounds and their surrounding environments.

7.2.2 Space Renovation

The goal of space renovation is based on an ecological and favorable perspective of the surrounding environment. First, barriers around school borders are removed. In the past, schools used cement walls to demarcate the school area. In 1984, the Yilan county government was the first to use green fencing, low walls, and other soft designs to replace cement walls and connect school grounds with the community. However, although a few other schools had taken similar actions, most schools continued to use solid hard walls for security and management purposes until the initiation of this plan. In addition, school ground landscaping is not considered a simple amplification of plantings. This project applies the aesthetics of reduction to lower the heights and amounts of hard surfaces, and to retain a comfortable school ground environment.

Opening a school to the public is not only a change of physical space, but also a consensus of sharing the school resources with the neighboring community. This project also emphasizes a participatory process from school members, the planning and designing team, and neighboring residents to collaborate.

7.2.3 Evaluation of the Benefits of the Verdant School: Strengthening the Place Identity of Schools

A total of 18 schools participated in the first phase of this project, all of which were located in urban areas. The greening and making available of school grounds were highlighted as providing green spaces in the midst of an urban city. The total greening area in the first phase was approximately 6000 m^2 and the total length of altered wall was approximately 3000 m. In the following year, an additional 39 schools participated.

This project also strengthened the place identity of schools. The objective of the project was to increase school greenery and connect schools to other public trees, for example, those by streets, and neighborhood parks. However, because of the bottom up design process, schools performed differently. Although the schools may have only created small green areas, they also created more collective memories and

Fig. 7.10 Lower down the wall height to make the school border friendly. Beitou dolls are placed on the wall

developed their place identity[11]. These projects also became models for amicable connections between school grounds and their surrounding communities (Fig. 7.10).

Beitou Elementary School offers an apt example of developing a place identity. The school was established in 1902 and the school grounds occupy approximately 1.82 ha. The school is near the mountainous area of northern Taipei and has favorable ecological conditions; however, urbanization has degraded the site's conditions by affecting nearby plants (Fig. 7.11). For example, a famous century-old coral tree was negatively affected. Architects invited students and teachers to share ideas on how to remove the school wall while saving the old tree. Students noticed that the school wall separated two large trees that were originally planted on the school grounds, the *coral tree* and a *large-leaved banyan*. The *banyan* was originally inside the school wall, but was subsequently on the other side of the wall because the construction of a pedestrian path reduced the school area by 3 meters. This became a Romeo and Juliet story, for which students wrote a tale entitled *The*

[11]For example, the collaborative architect of Beitou Elementary School, Henry Bo-rui Xu depicted the characteristic of the place name—Beitou (*Patauw* in Pinpu, meaning of witch) and designed a Beitou doll for the school.

Fig. 7.11 The Banyan used to stand on the border wall

Story of Tree #73 based on this incident, eventually winning first prize in the "Little Director Big Dream" competition held by the Taiwan Public Television Service. The project team then negotiated with the neighboring Beitou police station and successfully reunited these two trees (Fig. 7.12).

This process not only greens physical space but also forms a collective memory for participants and enhances a school's identity. This is not a standardized environmental education course but rather a vehicle for instilling deep feelings toward environmental change, which can facilitate the cultivation of future spatial managers.

Fig. 7.12 The reunion Banyan and Coral trees

7.3 Evergreen Lily Elementary School's *Tala-baliuw/Tapau* Project

School ground greening and ecological improvements can not only promote environmental biodiversity but also strengthen the relationship between a school and its surrounding area. However, a school, as an education site or even a community center, should possess an even higher level of relevance.

Taiwan's school buildings have been transformed from the standardized forms to emphasizing local atmosphere and establishing school characteristic. This has been reflected in physical space and course design.

The following case did not occur in an urban area, but rather in a relocated village after a disaster. This school has an indigenous tribal theme, and implementation of school characteristic enabled the school to take on cultural inheritance responsibilities.

Table 7.2 The meanings of work hut, Tala-baliuw/Tapau, and home

	Type		
Meaning	Work hut	Tala-baliuw/Tapau	Home (Kabalhivane)
Spiritual sustenance		○	○
Residence		○	○
Farming	○	○	

7.3.1 Beginning of the Project

From September 2012 to January 2013, the NTU Building & Planning Research Foundation postdisaster reconstruction task force and Evergreen Lily Elementary School cooperated in arranging a series of cultural activities, building a *tala-baliuw/ tapau*, for sixth graders.

Once a week the children learned to build a work hut near the school basketball court and how to farm and manage the land. The aforementioned Sustainable Campus and Verdant School have also constructed structures or developed farms. However, the construction of physical space in this project had a cultural significance. This cross-generation practice was used to give children hands-on experience to learn traditional wisdom and teachings.

Typhoon Morakot in 2009 severely damaged indigenous villages in eastern and southern Taiwan. To recover from the typhoon damage, some villages had to relocated from their original living area (mostly mountain areas) to a government designated placement. The Rukai tribe's Kucapungane and Paiwan tribe's Davalan and Makazayazaya had to be relocated to Rinari. In addition to constructing residences and public spaces in Rinari, Evergreen Lily Elementary School of Rinari was established as a site for national education.

The original intention of the Evergreen Lily Elementary School's *tala-baliuw/ tapau* project was to facilitate passing on indigenous tribal culture, especially indigenous tribes' wisdom of mountain life. Even if the village had to be relocated and the residents live a lowland lifestyle, it was believed that they still ought to maintain their mountain wisdom. This wisdom had been accumulated over long periods, and involves respect for the environment and considers people's interaction with nature through an ecological perspective.

7.3.2 Main Concepts of a Tala-baliuw/Tapau[12]

In the Rukai language, a work hut, *tala-baliuw/tapau*, and home have different meanings (Table 7.2). A work hut is a simple temporary shelter from the rain and sun. A structure that is more intricately built and requires more thought is called a

[12]Refer to Kadresengan (2012), Building and Planning Research Foundation (2013).

tala-baliuw/tapau. This type of structure is a small, temporary home that can be used for long periods in all seasons.

Farmland can be placed around the *tala-baliuw/tapau*, and because this structure serves as a small home, tribal people do not require spending time going back and forth from their permanent home to the farmland. This place functions like a home in the village. In this situation, a *tala-baliuw/tapau* not only provides rest and communication functions, but also has a mental therapy function. If relevant events occur, then people still must return to their real homes (Kabalhivane). Thus, the *tala-baliuw/tapau* is another spiritual retreat with a role of both a work hut and a permanent home.

There is no standard process for constructing a *tala-baliuw/tapau*. Despite any initial plans, people must adjust to the site conditions. The building process reflects the requirements of living. Each construction step is designed to meet a requirement. In addition, the building process reflects the techniques and wisdom of indigenous culture. Sometimes, it also creates stories.

7.3.3 Main Work Items for Constructing a Tala-baliuw/Tapau

7.3.3.1 Process of Building a Work Hut

A. Practicing Environmental Ethics: Notifying the Environmental Spirits and the Owner of the Land (Figs. 7.13, 7.14, and 7.15).
   Establishing a work hut is a crucial part of building a *tala-baliuw/tapau*. The selection of the location, use of materials, the construction, and the use of the hut must be closely linked to the environment.
   Elderly people believe that people will eventually return to the soil, and that any environmental burdens should be avoided. Thus, before people can conduct any action in the environment, the "owner of the traditional territory" must be notified. For example, the Makazayazaya traditional territory was used for this course, and the land belongs to the chief of the Makazayazaya. Thus, the chief had to be notified of any activities on the land. This is an expression of land ethics, which cultivates respect for the land in the next generation as well as gratitude toward the land and gratitude to the creator.
B. Preparing the Land
   The site area should not be too wide. In addition to considering the amount of space required for daily life, there should be adequate space for bedding and activities.
C. Wisdom of Collecting Hut Materials and Building
   The selection of building materials is on the basis of traditional wisdom. The characteristics of wood during different periods must be understood. One example is the use of dead wood, which has been hardened by the weather of different seasons. The loose outer parts have been consumed by termites, leaving the

Fig. 7.13 Blessing and notification ceremony

strong and hard core behind. Use of this material can extend the life of the structure. The work hut was constructed from not only deadwood and bamboo but also thatch, which composed the roof and walls. Stone slate was used to stabilize the foundation. Vines replaced wires as the connection material. Environment resources must be used wisely in accordance with seasonal cycles and seasonal characteristics to achieve the best results.

The building orientation must consider lighting and wind direction to determine the opening for doors and windows. Next, the builder must analyze the building procedures to ensure that construction is possible for one person. The builder must be able to adequately use surrounding materials, analyze the steps accurately, and build the structure competently. The space inside the work hut is

Fig. 7.14 A tribe teacher demonstrated laying out thatched roof

Fig. 7.15 A work hut

built based on considerations of the family demands; that is, customized to the user's use and based on living experience and habits.

To adjust the microclimate around a structure, the surrounding area is planted with shade trees. In addition to providing shade, these trees also have other values. For example, their fruits can serve as food and their branches can be used as firewood, building material, or farm equipment. The trees also form a small ecological circle.

7.3.3.2 Building a Home (Part 2): Building a Farmland

Operation of a farm requires being in accordance with seasonal conditions and having operations performed efficiently. There are two tasks for building a farmland: land preparation and planting. In terms of land preparation, in practice, students must first understand the geography, loosen the soil, and differentiate the ridge. Soil should be loosened from its bottom to top and the terrain should be followed when stacking stones into ridges (Fig. 7.16). In terms of planting, because of the season, fall, the course taught the planting of sweet potatoes, one of the main tribal staple foods. Pumpkins could be planted next to the sweet potatoes and vines were used to define borders. After the sweet potatoes, taros were planted. Following the New Year, millet and red quinoa were planted (Fig. 7.17). For crop configuration, the edges could be planted with pigeon peas. Because of the height of pigeon peas, they would not interfere with plants in the middle of the farm field when planted on the edge. Large trees are not planted close to the farm field because the canopy would hinder the bottom plants from dew.

In addition to farm fieldwork, an ecology where animals and plants can coexist should be developed to make living comfortable for humans.

Finally, a life tree should be planted next to a *tala-baliuw/tapau* to represent and identify the owner, the owner's tribe, and the environment.

Fig. 7.16 Students learned to differentiate the ridge on farmland

Fig. 7.17 A student cultivated red quinoa—traditional indigenous food (Curtsey of Director Guo-Xing Tsai of Evergreen Lily Elementary School)

7.3.4 Evaluation of Tala-baliuw/Tapau Benefits: Passing Down Traditional Land Wisdom and Cultivating a Space Manager

Although this project was only planned for one semester, the *tala-baliuw/tapau* (Fig. 7.18) has already become an influential place for the school. Students from each year can now learn from the existing foundation. The *tala-baliuw/tapau* demonstrates the construction knowledge of the native people and adds to the school ground characteristics. Occasionally, the school has also planned work hut maintenance courses that enable students to review work hut building techniques with elderly people. The farm has also been planted with different crops according to the season. Now, each grade has their own farmlands on school grounds.

These tasks should have been learned by the younger generation from their elderly people at their traditional residence during daily activities as part of the development of conventional wisdom that trains them as space managers. However, environmental changes forced them to relocate to another place. This course was not merely an accumulation of traditional technologies. Rather, students also learned about the humble attitude that mountain indigenous tribes traditionally hold toward land and sustainable environments.

Fig. 7.18 A Tala-baliuw/Tapau with foreground of farmland and background of a work hut

7.4 Practical Reflection: A School's Role in Creating Biodiversity

Because of the educational nature of schools, they play a different role from other organizations in promoting environmental biodiversity. In practice, environmental biodiversity includes the transformation of school grounds and an environmental education that cultivates spatial managers by providing them with sustainability concepts.

The cases mentioned had different orientations. The primary objective of the Ministry of Education's Sustainable Campus and Taipei City's Verdant School was to transform physical space and introduce environmental education, as well as to alter the surrounding environment. In their practical implementation, these projects focused on changes of physical space and then conducted environmental education and participation under this basis. School teachers planned courses where students could have hands-on experience while introducing general environmental knowledge. The learning and improving process usually ended once the projects finished. In comparison, the *tala-baliuw/tapau* project was more focused on passing down traditional wisdom concerning the land and its use. Demonstration by elderly people and hands-on operations performed by young people reproduced traditional wisdom and lifestyles in the implementing school. This process developed spatial managers that possessed traditional mountain wisdom. In other words, concept formation was prior to the building process. Furthermore, as later maintenance according to seasonal changes was performed, it reflected a more dynamic feature of the project.

The aforementioned projects promoted different levels of biodiversity in a physical space and developed people's relationship with the environment. On school grounds, popular courses such as the environmental education or *tala-baliuw/tapau* courses taught a sustainability role people could play in the environment to cultivate environmental sustainability concepts in future space managers. Regarding building an interface between schools and communities, environmental activities were performed to promote social exchange between the two parties. One example of this was replacing school walls with green fences, thereby providing an open space where people could gather. This type of action had the possibility of promoting environmental sustainability.

Under these aforementioned conditions, the following topics are worthy of discussion to explore the role that schools play in the biodiversity process and in future challenges.

7.4.1 Transformations that Schools Require to Promote Biodiversity

A common question asked while implementing Sustainable Campus, "Does the project really require making school ground improvements, or do we need to enhance other aspects instead?" Based on the aforementioned results, the Sustainable Campus not only creates greener school grounds, but also focuses on promoting sustainability concepts through the improvement process. However, results show that during the project implementation period, the concepts are still awaiting promotion.

7.4.1.1 Sustainability Concepts Being Unpopular; Implementers Being Criticized

In items listed by the Ministry of Education, the performing of "multilayered ecological greening" most resembles the school ground greening promoted in the past. In the past, school ground landscaping emphasized aesthetics, whereas the ecological concerns are more related to greening quality and quantity (e.g., carbon fixation), and species diversification (e.g., including native plants and plants that attract butterflies and birds) as well. Multilayered greening refers to the practice of mixing different trees, trees of different heights, shrubs, flowers, and vines. Trees are allowed to grow freely and are trimmed only when necessary. The results of this method are different from the aesthetics the public is familiar with. This difference has led to a misunderstanding in people who do not understand ecological sustainability. For example, some have removed "weeds" to maintain neat and aesthetic school grounds.

According to the Sustainable Campus concept, maintaining a wilderness on school grounds has a greater priority than does school ground aesthetics. However, because of sustainability concepts not being popularized, their results have been misunderstood. Consequently, sustainability concepts must be continually promoted.

7.4.1.2 Conflicts Between Green Buildings and Learning Environment Require Smart Problem Solving

Overall, green buildings should create a comfortable and practical environment. However, conflicts may arise when considering what is practical and what is ideal for the learning space. For example, two-way hallways or wide hallways that extend classroom space were originally intended to facilitate diversifying teaching forms. However, their lighting depth is not beneficial to green building index calculations. For another example, if water conservation improvements and the increase of water permeable surfaces (such as lawns and plantings) do not consider space used by people, they may reduce the space available to students. These problems may undermine people's confidence in adopting sustainable improvements. Therefore, these items all require a balanced design to achieve the best for both priorities and are some of the greatest challenges for school ground buildings.

7.4.2 Review of Physical Space Construction: The Authentic Ecological Engineering Method

Many schools claimed that their Sustainable Campus has implemented ecological method. However, achieving ecological engineering method standards is not easy. An ecological method should promote a self-designed process at least.

Take the example of ecological ponds, most cases used natural or native materials to reduce the impact of artificial material on the natural environment. The natural environment was simulated to achieve biodiversity standards. However, to date, these ecological pond systems cannot be self-designed. Schools still have to invest labors and material costs to maintain the stability of the pond ecological system. This does not conform to the spirit of the ecological method. Ecology ponds may have demonstration benefits (e.g., facilitating recognition of native water plants) for elementary school teaching, but preserving the original ecological environment when constructing a school may be more meaningful than creating a human made ecological environment with the ecological engineering method afterwards.

7.4.3 Review of the Physical Space Building Process: Plan, Design, Construction, and Maintenance for Learning and Space Transformation

The spirit of the Sustainable Campus and Verdant School is considered the entire dynamic process as pedagogical tool. Since schools began this project, it has provided a learning milieu for teachers, students, and community members. The participatory process, including discussion of design and implementation methods with spatial professionals, and the subsequent maintenance after the construction is completed, challenges the school administration system and educational scheme. Therefore, the Sustainable Campus and Verdant School do not merely achieve physical transformation but also continue the implementation of environmental education. This process can eliminate the myth that natural space is only designed to fulfill traditional teaching requirements. A reassessment of which items really conform to the intention of sustainability on campus must also be performed.

7.4.3.1 Comprehensive Planning for Gradual Improvements

Previous sections describe problems that have accumulated on school grounds over long periods. School ground sustainability requires overall improvement. However, because these improvement items require substantial funding, they cannot be implemented in one project. This has significantly reduced the effects of implementing environmental sustainability on school grounds. Therefore, each school should conduct a comprehensive plan with sustainable concept and implement it in phases and areas.

7.4.3.2 Linking Resources and Constructing a Long-Term Partnership

During this process, a school must make external links with other schools, communities, departments, and professionals to form a partnership. Regarding physical improvements, schools require professional consultations to provide recommendations for long-term improvements. Regarding the role of a school, schools require interaction with different groups to elaborate on its characters. Thus, by linking with sufficient resources, schools can become the initiator for implementing sustainability concepts. Furthermore, the opening of school grounds and connecting of green spaces can develop a foundation for changing the overall environment.

7.4.4 Users' Participation in the Building Process Facilitates the Formation of Sustainability Concepts

The aforementioned projects provide users with different levels of participation. Most of the current environmental education in schools conduct courses after space transformation has been completed, which cover subjects such as observing animal and plant ecology and using plants to fabricate products. This is a more passive form of participation. However, the formation of sustainability concepts for spatial managers should assign priority to active participation and experience. Therefore, the process can develop a place identity as in the Verdant School example and can enable participants to implement and experience local traditional wisdom as in the *tala-baliuw/tapau* project.

In addition, the cultivation of spatial management must be linked to local culture. Concepts from everyday life practices must be understood. Thus, school ground transformations must include cultural considerations. In this type of learning environment the biologically diverse natural environment is locally produced, and thereby, can be sustained and continually maintained by local spatial managers.

7.4.5 Conclusion: A School's Key Role in Promoting a Biologically Diverse Environment

A school is a vital education organization in a community. The delivery of knowledge and implementation of these aforementioned projects on school grounds can effectively expand their benefits. The aforementioned Sustainable Campus, Verdant School, and *tala-baliuw/tapau* projects have not only improved local biodiversity, but have also used different methods to introduce environmental sustainability concepts. The provision of this concept facilitates maintaining school ground environmental sustainability. In the future, the recognition and connection of sustainability concepts can contribute toward developing a green network and can become a strong foundation for comprehensive environmental biodiversity.

References

Building and Planning Research Foundation (2013) Evergreen Lily Elementary School Tala-baliuw/Tapau activity records (Unpublished manuscript)

Department of Urban Development (2010) Taipei city government, love you can see – verdant school. Taipei great looks series 5 – school ground green fencing transformation results. Taipei City Government, Taipei

Hsiao C-p, Chen R-l, Lin H-t (2001) Design code of green building for primary school and secondary school. Architecture and Building Research Institute, Ministry of the Interior, ROC (Taiwan), Taipei

Kadresengan, Auvini (2012) Tala-baliuw/Tapau (Unpublished manuscript)
Lin, Chien-tzung (2005) Revitalization of a one hundred-year-old school: a case study of the innovative management in Sustainable Campus. Unpublished master's thesis
National Taiwan University (2003) The comprehensive planning for Liugongjun restoration. National Taiwan University, Taipei
Taiwan Green School Partnership Network (n.d.). https://www.greenschool.moe.edu.tw/eng/about.htm. Accessed 15 Nov 2015
Taiwan Sustainable Campus Project (n.d.). http://esdtaiwan.edu.tw/esdtaiwan_english/. Accessed 15 Nov 2015

Chapter 8
Natural Environment and Management for Children's Play and Learning in Kindergarten in an Urban Forest in Kyoto, Japan

Tomomi Sudo, Shwe Yee Lin, Hayato Hasegawa, Keitaro Ito, Taro Yamashita, and Ikuko Yamashita

Abstract There is a great and growing concern about the connection between children and nature in modern society. Conservation and management of natural areas is very important both for providing nature experiences for children and for wildlife. Therefore, improving quality of green spaces to preserve biodiversity as well as to promote children's direct experiences of nature is required in urban areas. The main aim of this research is to explore the quality and management of the existing urban natural environment for children from the ecological perspective through a case study of a kindergarten in an urban forest in Kyoto, Japan. The data were collected focusing on (1) educational characteristics of the kindergarten, (2) physical environment of playground, (3) play activities, and (4) teacher's awareness of nature through the environmental survey and interview. From the results, it was concluded that the substantial qualities that provide direct experience of nature for children are diverse affordances of nature, functions for play, and its ecosystems. This recognizes that children's play activities and its outcomes from nature can be called ecosystem services. Thus playgrounds need to employ an approach for ecosystem management, and utilizing the existing natural places as a playground have the potential to be a new way of urban ecosystem conservation and management.

Keywords Children · Natural playground · Affordance · Ecosystem services · Ecosystem management

---

T. Sudo (✉) · S. Y. Lin · H. Hasegawa
Faculty of Civil Engineering, Kyushu Institute of Technology, Kitakyushu, Fukuoka, Japan
e-mail: sudo.tomomi313@mail.kyutech.jp

K. Ito
Laboratory of Environmental Design, Faculty of Civil Engineering, Kyushu Institute of Technology, Kitakyushu-city, Fukuoka, Japan

T. Yamashita · I. Yamashita
Kitashirakawa kindergarten, Kyoto, Japan

K. Ito (ed.), *Urban Biodiversity and Ecological Design for Sustainable Cities*,
https://doi.org/10.1007/978-4-431-56856-8_8

8.1 Introduction

Urbanization and fragmentation of the ecological network have degraded natural environments and then caused biodiversity loss in Japan as well as the other countries. Consequently, there are great and growing concerns about the connection between children and nature in modern society.

Learning through play is not a new topic for early childhood education (Ciolan 2013), and the benefit of being in nature for children's development, the value of natural playgrounds, and importance of nature experiences in early childhood have been discussed (e.g., Fjørtoft and Sageie 2000; Arbogast et al. 2009; Refshauge et al. 2012; Jansson et al. 2014; Mårtensson et al. 2014; Mustapa et al. 2015). Direct experience of nature, which involves actual physical contact with natural settings and nonhuman species, is the basis for children's physical and mental development, and it is different from indirect experience in zoos, aquarium, and botanical garden, or symbolic experience through books, TVs, and computers (Kellert 2002).

Furthermore, MEXT (Ministry of Education, Culture, Sports, Science and Technology, Japan) revised the School Education Law of Japan in 2007, that specified the importance of developing children's interests in familiar environments such as society, nature, and organisms and learning appropriate understanding and attitude to the environments. Inoue (2009), however, pointed out the lack of natural elements and awareness of ecological worldview in the present curriculum; stereotypical nature-based practices, such as plant growing, animal raising, and outdoor play, have been given to children in all kindergartens and nursery schools since the nineteenth century. Besides, nature space for children to play daily has declined by 80–90% in the 2000s, and outside space to play in groups has fallen to 1% of what it was in the 1950s (Senda 2015). It is therefore necessary to consider ways to provide children with rich nature experiences.

Conservation and management of natural areas is very important for both providing nature experiences for children and wildlife habitat (Ito et al. 2010). Therefore, efforts for conserving natural habitat and improving quality of green spaces are required to preserve biodiversity as well as to promote children's nature experiences in urban areas. Accordingly, the research explores the quality and its management of existing urban natural environment for children's play from the ecological perspective through a case study of a kindergarten in an urban forest in Kyoto, Japan.

8.2 Study Site: Kitashirakawa Kindergarten in Kyoto City, Japan

8.2.1 Location and Surrounding Environment

Kyoto city, Kyoto prefecture, known as a historical place of Japan, is located in the middle of the country (Fig. 8.1). It has a population of 1,474,669 (2013, Kyoto city

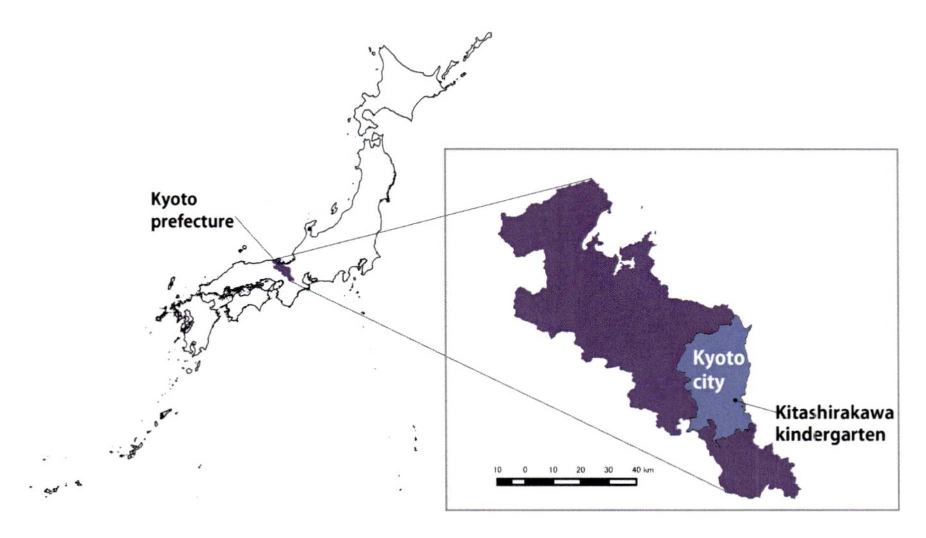

Fig. 8.1 Location of study site

statistic portal). The city area is 82,790 ha and 61,015 ha (73.7%) is covered by forest (Kyoto city 2013).

The study site, the Kitashirakawa kindergarten, is in the green area within the urban area of Kyoto city. The kindergarten, established in 1950, has been called "Oyama-no Yochien" which means "kindergarten on the hill." The kindergarten is in a forest, which connects to the mountain area surrounding Kyoto city (Fig. 8.2a). It has an elevation of about 100 m, higher than the surrounding urban area and varied topography (Fig. 8.2b).

8.2.2 Background of the Kitashirakawa Kindergarten

Eikichi Yamashita (EY) was the founder of the kindergarten. He worked as a writer and lived in one of the houses in the place with his family and his disciples before the kindergarten was set up. In the period after the end of World War II (1945), many people came back from the front. Because there was no kindergarten around the place, children had to travel far to reach a kindergarten. Huge demands to have a kindergarten were there. EY, therefore, decided to establish the kindergarten and started it in 1950. His wife, son (Ichiro Yamashita, IcY), and his disciples became teachers of the kindergarten. They reformed houses into classrooms and made a unique playground by using the special characteristics of the location and land features.

Fig. 8.2 (**a**) The aerial photograph; (**b**) The topographical map. Data Source: The Geospatial Information Authority of Japan

8.3 Research Methods

In this study, we focused on (1) educational characteristics of the kindergarten, (2) physical environment of playground, (3) play activities, and (4) teachers' awareness of nature. Investigation was conducted in July 2013 by field survey and interview with the current headmaster of the kindergarten Taro Yamashita (TY) and vice headmaster Ikuko Yamashita (IY). The interview was recorded and transcribed into text data.

8.3.1 Data Collection of Educational Characteristics of the Kindergarten

The Japanese early childhood education guidelines describe the importance of nature-based activities. However, practices are different in each kindergarten and depend on teachers (Inoue 2014). To know how natural environment is placed in the education in the study site, characteristics of the kindergarten were described by educational philosophy and curriculum. These data have been collected by interview and from a 50-year anniversary book written by former headmaster IcY (Yamashita 2000).

8.3.2 Physical Environment of Playground

The physical environment of the playground was described by its vegetation and physical character. The area of the playground was presented by interviewees. A vegetation map was made by referring to results of fieldwork investigation and geographic information of vegetation type provided from Biodiversity Center of Japan, Ministry of the Environment (http://gis.biodic.go.jp). The characteristics of the playground were observed by fieldwork investigation.

8.3.3 Play Activities

Data of play activities and insect species that children found while playing were extracted from interview text data and text data of *Ikuko diary* from September 2006 to December 2013. *Ikuko diary* written by IcY notes children's daily activities. How children play and what they play with were recorded. Finally, play activities were categorized by functional characters of environment features. Insect species were described with their characters of habitat.

To analyze the relationship between the children's play activities and natural environment, we use the theory of affordances (Gibson 1979). Application of this method for describing the value of natural environment to children has been explained (Heft 1988; Fjørtoft and Sageie 2000; Said 2012). The theory states that "perception of the environment inevitably leads to some course of action. Affordances in the environment that indicate possibilities for action, are perceived in a direct, immediate way with no sensory processing." Methods focusing on affordances are more appropriate for describing natural landscape as playscapes for children (Fjørtoft and Sageie 2000).

8.3.4 Teachers' Awareness of Nature

Data of the teachers' awareness of nature was extracted from interview text data. In order to analyze this data, the methodology of grounded theory was applied partly. Grounded theory is a form of qualitative research developed by Glaser and Strauss (1967). Regardless of type of data used, they are analyzed by means of a process termed *constant comparisons*. In doing constant comparisons, data are broken into manageable pieces with each piece compared for similarities and differences. Data that are similar in nature are grouped together under the same conceptual heading. (Cobin and Strauss 2015). Through further analysis, researchers form groups into categories and develop structure of the theory by core categories and other categories. However, we applied the first phase of grounded theory to reconstruct text data focusing on teachers' awareness of nature.

8.4 Results

8.4.1 Educational Characteristics of Kindergarten

8.4.1.1 Philosophies of Education

EY, the founder, and IY, the second headmaster, developed educational philosophies based on their experiences. TY, the third and current headmaster, has inherited their ideas. Three mottos have characterized the activities in the kindergarten; play is learning, walking up to the kindergarten every day, and nature is a friend.

Play is learning: Play is the most important activity for children to build up their minds and it will contribute to their future that must affect their whole life. In other words, play is real-world intelligence for living life. Children experience many kinds of situation through play and learn challenging, handling problems, creativity, cooperation with others and good feeling to share the time and emotion with friends.

Walking up to the kindergarten every day: The kindergarten is located on a hill and surrounded by forest. There is no access for cars and walking is the only way to

Table 8.1 The number of children and teachers (2013)

	3 years old group (male, female)	4 years old group (male, female)	5 years old group (male, female)
The number of children	44 (24, 20)	44 (23, 21)	38 (22, 16)
The number of teachers	3	2	2

reach the kindergarten. All children walk to the kindergarten every day on foot without their parents. Children are divided into several groups by their living area and they gather at each meeting point at the bottom of the mountain, then teachers lead them to the destination. They walk 1 km and climb about 200 stairs to reach the kindergarten every morning, even if it is rainy, windy, and snowy. This contributes to building their self-confidence and promoting their motor skills. Also it provides opportunities to feel changes of seasons.

Nature is a friend: Children feel the nature with five senses when they walk up to the kindergarten and play outside every day. They find bamboo shoots growing up in the spring, hear the cicadas' sounds in the summer, see the changing of leaf color in the autumn, feel the coldness and beauty of winter, and so on. Nature is always near the children in this kindergarten. Teachers do not call these practices "environmental education," because it is customary for children and teachers as a tradition of the kindergarten. Playing with nature is usual and not occasional experiences for them.

8.4.1.2 Curriculum

The children in this kindergarten are aged from 3 to 5 years old. They are divided into three groups by age, and each group has two or three classes (Table 8.1). The kindergarten starts at 9:00 and finishes at 14:00 usually (Table 8.2). When they arrive, it is free play time inside the kindergarten for about a half-hour. After morning activities and tidying up the room, there is free play time outside the kindergarten for less than an hour. And again there is free play time both inside and outside freely for an hour after lunch. Therefore, almost half of the school time is filled with free play activities both inside and outside. When a school day is finished, children go back home by walking down the mountain with teachers.

8.4.2 Physical Environment of Playground

Although the area owned by the kindergarten is about 2500 m^2 (playground A), the forest area surrounding the kindergarten is additionally used as a playground (B and C) (Fig. 8.3). Figure 8.3 shows vegetation and the playgrounds. The kindergarten is located along the edge of a bamboo forest, which had been maintained for food and materials for agricultural tools. Playgrounds B and C are in *Quercetum variabil—Q.*

Table 8.2 The daily time schedule

Time	Activity
8:10–8:20	Gather and start to walk to the kindergarten
9:00	Arrive at the kindergarten Inside free playing time
9:30	Tidy up the room
	Morning activity • Greetings • Sing songs • Some activity organized by teachers
	Outdoor free playing time
11:30	Lunch time[a]
12:30	Free playing time
13:30	Clean up and prepare to leave Read books or listen stories from teachers
14:00	Start to go back home[b]

[a]On Wednesday children leave kindergarten at 11:30. Art and craft class are arranged from 11:30 to 12:30 for applicants.

[b]Some playing activity are provided from 14:00 to 15:00 some days in a week. Children can apply it.

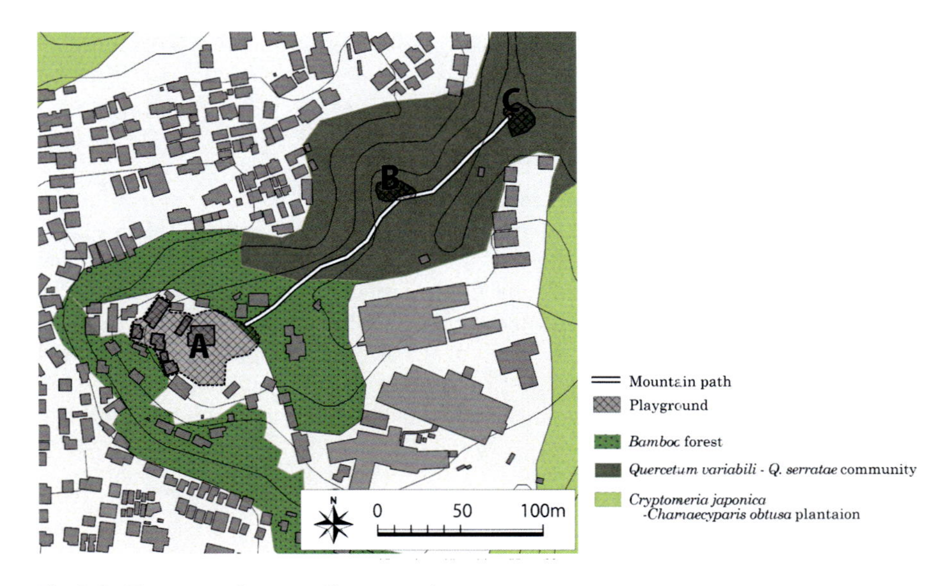

Fig. 8.3 Play area and surrounding vegetation

serrata community; a secondary forest that was maintained by periodical logging for human use (Suzuki 2001). *Cryptomeria japonica*, and *Chamaecyparis obtusa* plantation is artificial forest to produce timbers. It can be said that the kindergarten is in a

Satoyama landscape. *Satoyama* is a Japanese term for landscapes that comprise a mosaic of different ecosystem types including secondary forests, agricultural lands, irrigation ponds, and grasslands, along with human settlements. These landscapes have been formed and developed through prolonged interaction between humans and ecosystems, and are most often found in the rural and peri-urban areas of Japan (Duraiappah and Nakamura 2012).

8.4.2.1 Characteristic of Playground A

Playground A is the main area used daily in the free play time. It is about 2500 m^2 and is adjacent to bamboo forest, but separated by fences or buildings. The geological features make the playground unique and complicated (Fig. 8.4). The paths for connecting buildings and playgrounds are not flat but steep slopes or steps. Some play equipment, such as climbing frames and swings, are arranged on the flat ground area in front of the classroom 1. Below this flat playground, a small play area like a wide path is set. The boundary between the small play area and the upper flat area is covered by trees and shrubs. These two areas are connected by several types of slides and an overhead ladder and horizontal bars are arranged on it. The east part of the area, next to classroom 1, is covered by grass, and is called the *Secret garden*. This garden with a small pool and water biotope was constructed in 2013, and is maintained by teachers, children, and their parents. According to vice director IY, aquatic insects, small freshwater fish, and amphibians inhabit the biotope.

8.4.2.2 Characteristic of Playground B and C

Playgrounds B and C in the forest are connected by mountain paths. Depending on the condition of children and kindergarten programs, teachers normally take children to the areas B and C once a month on average. These playgrounds B and C are in deciduous oak forest, which is called the *Secret forest* (Fig. 8.5). Animal footprints are found on the path. Playground B is on the mountain path from the secret garden to playground C, and is a small open place surrounded by trees and stonework. Playground C is also a small open place but with denser trees than the B. The topography of playground B and C is considerably different.

8.4.3 Play Activities

8.4.3.1 Play Activities and Landscape Character

Table 8.3 shows play activities of children; environmental features are categorized into graspable manipulative/detached objects, attached objects, land features, and

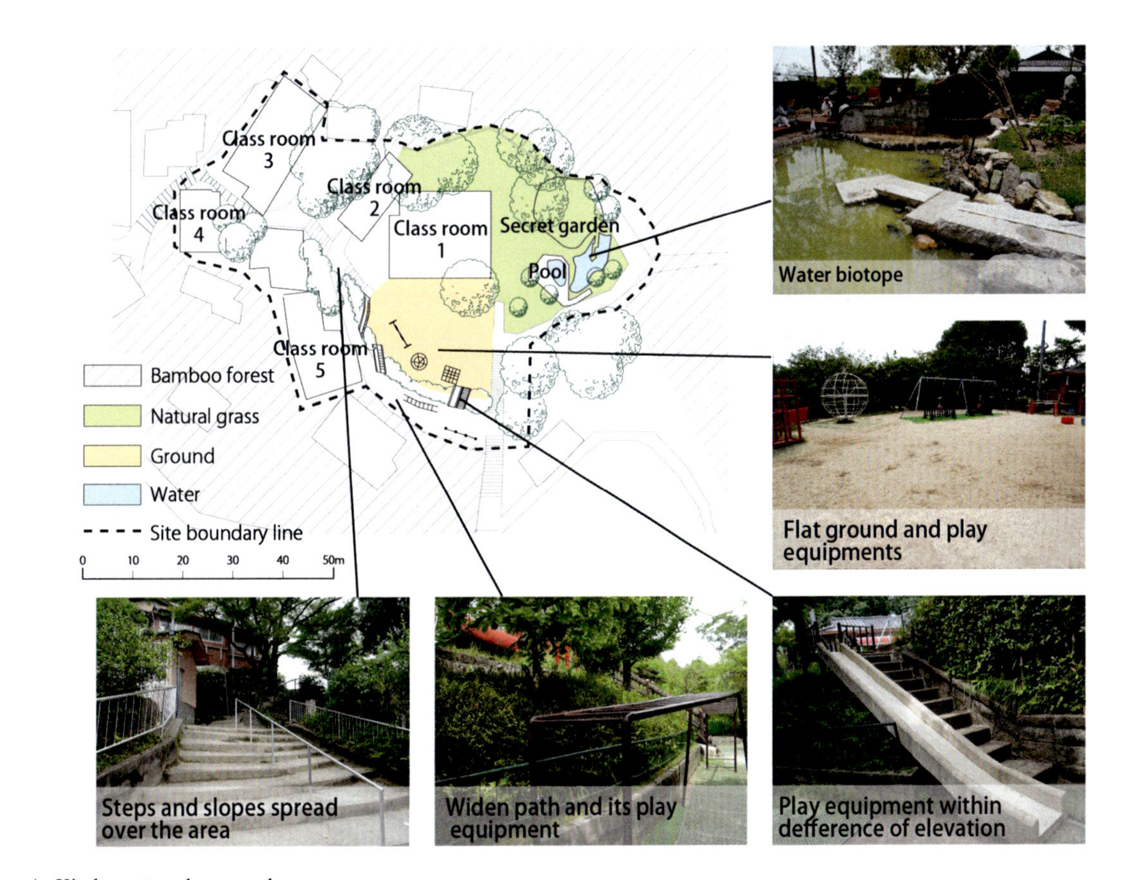

Fig. 8.4 Area A: Kindergarten playground

Fig. 8.5 (**a**) Playground B in the secret forest, (**b**) Playground C in the secret forest

seasons, and then the natural elements are subdivided into animals, plants, and the others.

Play activities are mostly categorized into graspable, manipulative/detached objects. The children perceive objects characterized as small things that they can keep in their hands such as animals like insects, snails, and small plants like nuts, seeds, and mushrooms.

Animals—animate, graspable, manipulative/detached object—offer various play activities for children. They are mainly the objects for observation, catching, and collecting, which can be caused by their spontaneous movement by which they attract children's attention. Moreover, play activities afforded by the animals will be different according to the species behavior at different places and times. They are also diverse in forms and colors. In imagination play, children make stories about insects which they found and imagine birds by birdsongs. In addition, listening is a characteristic of play activity caused by insects and birds.

Manipulative plants objects also afford kinds of play activities. The manipulative plants objects in Table 8.3 are marked as collectable which means that they are portable and multiple. Objects in this group are also flexible or have plasticity that enables children to deform objects that affords handicraft, picking up, and squashing. They are diverse in shapes, forms, colors, and changes in the time, season, and place as it is of animals. They also offer imagination play activities such as role play with likening nuts, branches, and leaves to kitchen tools. Additionally, eating is the characteristic activity offered by plants.

Attached plant objects mainly afford physical play activities with variety of body movement. The objects are stuck on the land and stable because heavy enough comparing children's body weight that can support children climbing on, hanging on, shaking and swinging on. They offer various kinds of options for physical activities because of variety of size, height, and form.

Terrain features, cliff and hill, also afford physical play activities. Different angles of inclination and surface material can offer different types of physical activity such as climbing, running down, and rolling. And snow afford sliding and rolling accompanied with land feature.

Table 8.3 Play activities related nature environment

Landscape character			Play activities	Number of the playing activity
Environment feature	Occurrences			
Graspable, manipulative/ detached object	Animals	Insects	Observation, observation the growth, collecting, catching, imagination, listening	79
		Birds	Observation, imagination, listening	3
		Amphibian	Observation, catching	3
		Reptile	Observation, catching	2
		Snail	Observation, collecting	2
		Crustacean	Observation, catching	2
		Annelid	Observation, collecting	1
		Fish	Observation, art and craft	1
	Plants	Fruits, nuts, seed, cone, propagule	Collecting, picking up, imagination, eating	24
		Leaves	Collecting, searching, imagination, game, art and craft, smelling, handicraft	19
		Flower	Observation, growing plants, collecting, imagination, handicraft, eating	11
		Mushroom	Picking up, collecting, imagination, squashing	6
		Branch, cupule, dry wood, cut bamboo	Collecting, imagination, handicraft	5
		Vegetable	Growing plants, picking up, art and craft	4
		Wild grasses	Eating	1
	The others	Stones, shell	Collecting	4
Attached objects	Plants	Trees	Climbing, jumping down, hanging, sitting, quivering	5
		Bamboo shoot	Digging up	1
		Vine	Swing	1
Terrain features	Cliff, hill		Climbing, running down, rolling	4
Seasons Substances	Snow		Sliding, rolling	2

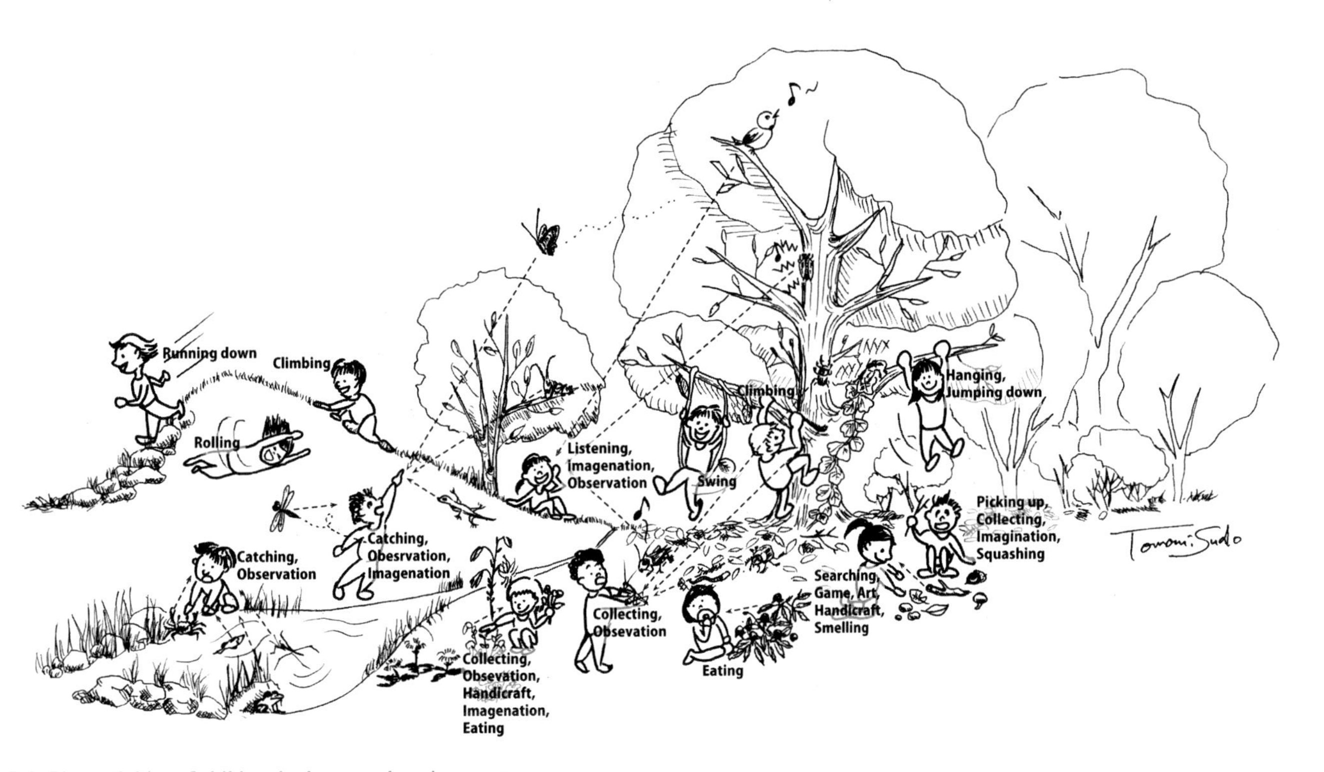

Fig. 8.6 Play activities of children in the natural environment

Figure 8.6 shows the abstraction of play activities of the children in natural environment according to Table 8.3. Objects and materials in natural landscape provide a variety of opportunities for children to play by using their physical abilities and five senses. A functional description of natural environments perceived by children, that is their affordances, offers a rich way of conceptualizing the ecological resources for children's learning and development.

8.4.3.2 Insect Species Which Have Appeared in Play Activities

Table 8.4 summarizes insect species, which are found by the children during their play activities and classification of their habitats. It consists of mostly Coleoptera, secondly Lepidoptera, Acridioidea, Mantodea, Hemiptera, Odonata, and Phasmida in that order. Half of the Coleoptera (the order of insect containing beetles) can be found in the broad-leaf forest, some in decayed tree, wood, a few in forest, forest floor, grassland, forest edge, and others. Most of the Lepidoptera found in forest have been described and the habitat for other species of insects is also mentioned (Table 8.4). This indicates that the playgrounds are in the wildlife habitats of species.

8.4.4 Teachers' Awareness of Nature

According to the results of the interview with the headmaster and vice headmaster, their awareness of nature was categorized as follows: (1) natural environment as a playground, (2) damage by Shika deer feeding, and (3) necessity of forest management.

8.4.4.1 Natural Environment as a Playground

What they think about nature in this kindergarten could be seen in dialogs;

[Headmaster] "It is natural behavior for us that children play in the nature thus it doesn't mean specific education. /Everything that children find in the nature can be play materials."

[Vice headmaster] "We are not in deep nature, there are roads and buildings close by, but we have trees which were planted by the first headmaster many years ago that attract insects now, and there is various vegetation in the forest. /Nature gives us gifts. We appreciate the gifts from nature. It is very good for them and us that they can have contact with living lives in nature."

They see the value of nature as children's playground, perceive it in ecosystems such as "trees attract insects" and appreciate what nature provides children; experience to contact living lives. Different text data also shows that teachers care very

Table 8.4 Insect species and their character of habitat based on text data of children's play

Order	Spices	Character of habitat						
		Forest	Broad-leaf forest	Forest floor	Forest edge	Decayed tree, wood	Grassland	Others
Coleoptera	*Carabus blaptoides*			○				
	Cicindela japonica				○			○
	Apoderus jekelii		○		○			
	Trypoxylus dichotomus		○					
	Anoplophora malasiaca		○					
	Mesosa longipennis		○			○		
	Batocera lineolata		○			○		
	Prionus insularis		○			○		
	Paraglenea fortunei							○
	Purpuricenus temminckii		○			○		
	Prismognathus angularis		○					
	Dorcus striatipennis		○					
	Figulus binodulus		○			○		
	Prosopocoilus inclinatus		○					
	Pterolophia zonata		○			○		
	Melolontha japonica		○					
	Anomala rufocuprea							○
	Maladera japonica				○		○	
	Rhomborrhina japonica		○					
	Tetrigus lewisi	○				○		
	Phelotrupes auratus			○				
	Phelotrupes laevistriatus			○	○			
	Sternuchopsis trifidus							○
	Curculio dentipes		○					
	Episomus turritus				○		○	
	Chrysochroa fulgidissima		○		○			
	Sastragala esakii		○					
	Psyrana japonica	○						
	Epilachna sp.						○	
	Acrothinium gaschkevitchii				○			
Lepidoptera	*Papilio xuthus*	○					○	
	Papilio protenor	○						
	Cephonodes hylas							
	Vanessa indica	○						
	Mycalesis francisca	○						
	Hestina japonica							
	Lithosia quadra	○						
	Rhagastis mongoliana				○			
	Saturnia jonasii							○
	Graphium sarpedon	○						
Acridioidea	*Loxoblemmus campestris*						○	
	Acrida cinerea						○	
Mantodea	*Tenodera aridiforia*				○			
Hemiptera	*Platypleura kaempferi*		○					
Odonata	*Anotogaster sieboldii*							○
Phasmida	*Neohirasea japonica*		○					

much when children meet insects, they are trying to share emotions with children such as being surprised and wondering about what they found in the nature.

8.4.4.2 Bark Stripping Damage by Shika Deer

Bark stripping damage by Shika deer has become a huge problem all over Japan recently (Kamada 2018), and it was mentioned in the interview investigation as well. The interviewees remarked about the impact of bark stripping damage caused by deer;

[Headmaster] "This year, some planted flowers were eaten by Shika deer."
[Vice headmaster] "And some trees have fallen because of the stripping damage. / Yes, they have been living around this area but they haven't come in the kindergarten before. But now they come beyond the boundary."

Damaged trees were found in playground in the forest during the fieldwork investigation. The interviewees also told us that since they planned a biotope in the kindergarten from 2012, they began to consider about surrounding forest condition seriously. Creating a biotope which means that they make a habitat in the forest is the opportunity for them to reconsider about forest ecology and relations between surrounding forest and of kindergarten. Additionally, deer coming in the children's play area causes further problems;

[Headmaster] "Some people say it (ticks) is not a serious problem but we should care for children. Their parents might be concerned as well."
[Vice headmaster] "Shika deer have ticks. If they come often into the garden, the risk of getting ticks would increase."

Severe fever with thrombocytopenia syndrome (SFTS) emerged as new tick-borne viral infectious diseases in Japan (Takashima 2013). A SFTS infected patient was first confirmed in 2013 in Japan, and it has been reported that 232 people have been infected from 2013 to 2017 in western Japan (National Institute of Infectious Diseases 2017).

These data show that they are conscious of surrounding forest condition. Bark stripping damage by Shika deer has noticeably been recognized recently. This problem is affecting the safety of the outdoor playground.

8.4.4.3 Necessity of Forest Management

They explained how they see the process of divested forest and necessity of management as follows:

[Headmaster] "Probably the consequences have been caused by us, humans, who have not done enough that are possible to do. Because we haven't done anything indeed deterioration of the natural environment has occurred. Now nature is showing us how it can be if we don't get involved."

[Vice headmaster] "The forest has been maintained for getting fire wood by people. It made the forest bright, then various plants grow and become habitat for insects, animals and so on. But it has been abandoned in recent years, overgrowth vegetation shields sunlight, making lower vegetation poor. We are trying to maintain it but it seems difficult because natural reproduction is more powerful than our efforts."

They mentioned *Satoyama* landscapes, the forest which has been used and maintained by people for living, and how it is changing in recent years.

Although they recognize the necessity of forest management, there are some obstacles such as land ownership and lack of knowledge for management. Human resource is also a huge problem because it is difficult for teachers alone. Nevertheless, they have some ideas for implementing forest management;

[Headmaster] "If we can work with children and parents, it will contribute to local society. This would provide a connection between local society and children, and we could build a community of children, parents, teachers and local residents through the forest management. It would be nice, we can share problems and roles, then work together."

[Vice headmaster] "It's a good idea that children participate in forest maintenance even if they can't do great work. Children play in the forest. They can care for the forest as well as they can care for toys after playing with them."

They find the value of the forest management from the social benefit in cooperating with children, parents, and teachers. It is reported that they started forest management with parents and they maintain the forest several times a year after interview investigation in 2013.

8.5 Discussion

8.5.1 Physical and Social Environment for Children

The educational philosophy and practices are supported by location and nature environment of the kindergarten. Kinoshita (1992) proposes that children's contact to nature is related to social and physical environment. Fjørtoft and Sageie (2000) noted that natural landscapes represent potential grounds for play and learning. The kindergarten is in the forest (Fig. 8.3) and it has different environments of playgrounds; A, B, and C (Figs. 8.4 and 8.5). Children spend approximately 2–3 h outside during play and on their round-trip walking route (Table 8.2). Also nature and play are centered in the education philosophy. These physical and social settings make it possible for children to play freely in the natural environment.

Young children learn how to respond to the world from people around them, therefore, how people react to children in nature is important for developing care for the environment (Chawla and Rivkin 2014). The kindergarten has developed the education philosophy utilizing the surrounding natural environment for children's learning through play since it was established in 1950. According to teachers' awareness of nature, it is described that they find the value of nature as foundation of their education, and they use natural environment with appreciations and respects. The natural environment is not perceived as just materials for play but in the context of the ecosystem. Teachers respond to children's discovering in nature by sharing emotions. These teachers' attitude to children and nature contributes to children's learning from nature. Furthermore, the value of nature as an educational resource, which has been developed traditionally, keeps teachers conscious of the surrounding natural environment.

8.5.2 *Quality of the Playground*

According to the studies on children in the outdoor environment, the functional significance of environmental features captures the awareness of the children rather than their structure (Heft 1988; Fjørtoft 2004), and children prefer a natural environment that can allow them to be active (Oka et al. 2004). Play activities are mostly categorized into graspable, manipulative/detached objects (Table 8.3). This result emphasizes that visual information of objects that are capable of being held in hands is perceived by early childhood children as functions for play.

Manipulative animals afford significant numbers of play activities (Table 8.3). Gibson (1979) states that "other animals are the most complex objects of perception that the environment presents to an observer. Animate objects are different from inanimate objects in a variety of ways but notably in the fact that they move spontaneously. . .other animals afford the observer not only behavior but also social interaction" (Gibson 1979). Animals described in this study are wild. They have specific habitat and live in interaction with each other and with environments, as was explained from insect species which children found in play (Table 8.4). This recognized that wild living things in ecosystem promote children's play activities. Senda (1998) noted that children's experiences of catching and playing with insect, which is known as a popular children's play activity in Japan, are not universal because the relationship between humans and animals is shaped by religious, cultural, and historical circumstance. It may indicate that the play activities with animals described in this study are particular children–nature relationship in Japan, however, we need more data from different cultural context for discussion.

Manipulative plant objects in Table 8.3 can be called *loose parts*. Loose parts are open-ended materials that can be used and manipulated in many ways (Nicholson 1971) and they stimulate children to consider a range of possible uses and meanings for the parts (Daly 2015). Louv (2008) stated that nature remains the richest source of loose parts. Loose parts in nature are produced in natural process. Therefore, these materials are provided repeatedly in accordance with natural life cycle as long as ecosystems sustain them.

The landscape features of nature can provide physical play activities which correspondingly stimulate the improvement of motor fitness in many different varieties of skills (Fjørtoft and Sageie 2000). In the natural environment, there are various structures, textures, features of natural elements and they change in shape, and color according to the season and weather. These qualities of natural environment as a playground for children are highlighted as learning effects on development of motoric and sensorial activities (Fjørtoft 2004).

The data of play activities was collected from text data but not from direct observation of children. This may limit detailed description of children's activities. In this study, however, diversity of affordances in the natural environment; animals, plants, and land features are substantial environment for children's play activities. This corresponds to the finding of Fjørtoft and Sageie (2000) that landscape diversity is related to different structures in the topography and the vegetation, which are important for children's spontaneous play and activities.

8.5.3 Benefits from Natural Playground

The varying experience in nature of children through play is perceived by means of imagination, listening, eating, smelling, and by making arts and crafts using their body and five senses (Table 8.3). The wealth of opportunities for experiencing nature by play raises the development of children. Study results reported by Kellert (2002) observed, "Children confront, in effect, nearly limitless contexts and opportunities in nature for developing and practicing the act of comprehension." He also indicated that the direct experience in the natural environment plays a significant, vital, and irreplaceable role in affective, cognitive, and evaluation development for children. It is also theorized that contact of nature during play in the natural environment has a positive impact on children's social and emotional development (Burdette and Whitaker 2005). Therefore, natural environment, also as a playground for children, is a unique space for holistic development of children.

As discussed previously, the natural playgrounds are sustained by the ecosystem. The character of ecosystem of the study site was defined as *Satoyama* landscapes, which involves diverse habitat and rich species (Saito and Shibata 2012). Put another way, the play activities in nature and their outcomes are ecosystem services from *Satoyama* landscapes. Ecosystem services are the benefits people obtain from ecosystems, which the Millennium Ecosystem Assessment (MA) describes as provisioning, regulating, cultural, and supporting services (The Millennium Ecosystem Assessment 2003). The play activities in nature can put into cultural ecosystem services. MA explains that cultural services are nonmaterial benefits people obtain from ecosystems through spiritual enrichment, cognitive development, reflection, recreation, and aesthetic experiences. Saito and Shibata (2012), and Kamada (2018) describe *Satoyama* ecosystem services and its changes. Changes in ecosystem services affect human well-being in many ways (The Millennium Ecosystem Assessment 2003). From this aspect, the degradation of *Satoyama* ecosystem affects children's play activities in natural environment and subsequently their development in Japan.

8.5.4 Forest Management and Problems

In *Satoyama* landscapes, people maintained natural landscapes to obtain a variety of goods and formed a system in which various land uses comprised a mosaic landscape (Takeuchi et al. 2016). However, in modernized association of people, the balance of *Satoyama* landscape and people's life has altered, which in turn caused abandonment of the management of forest and devastation of *Satoyama*. In addition, expansion of shika deer's population in both rural and urbanized areas is one of the problems which has caused serious damage to *Satoyama* landscapes in Japan (Kamada 2018). This is the result of extinction of predators of deer, global warming, deterioration of hunting culture, and transition of land use such as devastated *Satoyama* landscape. It was mentioned that the kindergarten also faces a problem of the bark stripping damage by Shika deer and fear of SFTS, and they are affecting the safety of the playground. It can be summarized that changes of human interaction in *Satoyama* landscapes affect quality of natural environment in the kindergarten, being children's playgrounds.

The teachers began to consider about the forest management because the problem caused by Shika deer had become prominent. The motivation of people who take environmental voluntary action is related to individual values (Asah et al. 2014). In the case of this kindergarten, the value of forest as playgrounds and the fact that playground is threatened raise their motivation of implementation of the management. The driving force is relevant that they obtain benefit from the forest through daily use.

Although they recognize the necessity of forest management, there are some obstacles such as land ownership, knowledge, and human resource for implementation of the forest management. Recent studies at *Urban-forest park at Takaragaike Park* in Kyoto show an example of urban forest management with a children centered community (Noda and Ogawa 2014). In that park, forest management for restoration of *Satoyama* landscape is implemented by combining children's play activities with management and developing a community including various stakeholders such as children, parents, NPOs, educational institutions, volunteers, city government, etc., for the purpose of children's free play in the forest (Noda 2013). Utilization of *Satoyama* landscapes as places for children's play and learning has the potential to be a new way of conservation of urban ecosystems.

8.6 Conclusion

Children's play activities and their outcomes are related to quality of playground and social environment. Direct experience of nature, which involves actual physical contact with natural setting and nonhuman species, is the basis for children's physical and mental development (Kellert 2002). Rich natural elements, diversity of species as well as topography, can provide opportunities for children to create many play activities (Fjørtoft and Sageie 2000; Louv 2008). Therefore, the substantial quality that provides direct experience of nature for children is diverse affordances of nature, functions for play, and its ecosystems. This implies that we need to see urban natural places not just as "green" but with its quality and multiple functions. That is landscape as "playscape" (Fjørtoft and Sageie 2000), "learnscape," and "omniscape" (Arakawa and Fujii 1999; Ito et al. 2014, 2016). The playground is the future arena where landscape architects have an opportunity to contribute to the quality of children's outdoor environment in a broad and collective way (Herrington and Studtman 1998). Furthermore, the individual's positive emotional affinity towards nature is not determined merely by the amount of natural environment but by the frequency of direct experiences with nature (Soga et al. 2016). Thus planners and landscape architects should realize that there are demands and possibilities for natural playgrounds in cities where children have convenient access.

In terms of interaction between human and natural landscapes, children's play activities and their outcomes from nature can be called ecosystem services. Thus playgrounds need to employ an approach for ecosystem management. Utilization of the secondary forest that people have used for their daily life as a place for children's play and learning has the potential to be a new way of urban ecosystem management. This point of view may help people to find value and potential of urban natural environment and that can drive motivation of involvement with the environment. Practical implementation for developing quality of natural playground needs the cooperation of the stakeholders from every field such as educators, planners, architectures, governors, and so on. In this study, we focused only on the benefit for children; however, the impact of children's activity on natural environment was not taken into consideration. Browning et al. (2013) demonstrate impacts of children's activities on natural landscapes so that sustainable management requires appropriate site selection, development, and maintenance. Future studies, thus, need to focus on sustainable ecosystem management, which support both preserving natural landscape and providing children's natural experiences, with understanding of children–nature interaction and its value.

Acknowledgements We would like to express our gratitude to the children and the teachers of Kitashirakawa kindergarten for cooperating in the study, and we thank for the insightful comments from Professor Mahito Kamada and English proofreading by Professor Ian Ruxton for writing the chapter. We also thank our colleagues from Laboratory of Environmental Design, Kyushu Institute of Technology and all the people who support the study. This study was supported by Kakenhi, Japan Society for Promotion of Science (JSPS), Grant-in-Aid for Scientific Research (C) (No. 23601013) in 2011–2013, and Grant-in-Aid for Scientific Research (B) (No. 15H02870) in 2015–2019.

References

Arakawa S, Fujii H (1999) Seimei-no-kenchiku (life architecture). Suiseisha, Tokyo. (in Japanese)

Arbogast LK, Kane CPB, Kriwan LJ, Hertel RB (2009) Vegetation and outdoor recess time at elementary schools: What are the connections? J Environ Psychol 29:450–456

Asah ST, Lenentine MM, Blahna DJ (2014) Benefits of urban landscape eco-volunteerism: mixed methods segmentation analysis and implications for volunteer retention. Landsc Urban Plann 123:108–113

Browning MHEM, Marion JL, Gregoire TG (2013) Sustainably connecting children with nature—an exploratory study of nature play area visitor impacts and their management. Landsc Urban Plann 119:104–112

Burdette HL, Whitaker RC (2005) Resurrecting free play in young children: looking beyond fitness and fatness to attention, affiliation and affect. JAMA Pediatric 159(1):46–50

Chawla L, Rivkin M (2014) Early childhood education for sustainability in the United States of America. In: Davis J, Elliott S (eds) Research in early childhood education for sustainability. Routledge, New York, pp 248–265

Ciolan L (2013) Play to learn, learn to play. Creating better opportunities for learning in early childhood. Soc Behav Sci 76:186–189

Cobin J, Strauss A (2015) Basics of qualitative research, techniques and procedures for developing grounded theory, 4th edn. SAGE, Thousand Oaks

Daly L (2015) Loose parts: inspiring play in young children. Redleaf Press, New York

Duraiappah KA, Nakamura K (2012) The Satoyama Satoumi assessment: objectives, focus and approach. In: Duraiapph KA, Nakamura K, Takeuchi K, Watanabe M, Nishi M (eds) Satoyama-Satoumi ecosystems and human well-being; socio-ecological production landscapes of Japan. United Nations of University Press, Tokyo, pp 1–16

Fjørtoft I (2004) Landscape as playscape: the effects of natural environments on children's play and motor development. Telemark Univ College J Child Environ Quarterly 14(2):21–44

Fjørtoft I, Sageie J (2000) The natural environment as a playground for children landscape description and analyses of a natural playscape. Landsc Urban Plann 48:83–97

Gibson JJ (1979) The ecological approach to visual perception: classic editions. Psychology Press, New York

Glaser B, Strauss A (1967) The discovery of grounded theory. Aldine, Chicago

Heft H (1988) Affordances of children's environments: a functional approach to environmental description. Denison Univ J Child Environ Quarterly 5(3):29–37

Herrington S, Studtman K (1998) Landscape interventions: new directions for the design of children's outdoor play environments. Landsc Urban Plann 42:191–205

Inoue M (2009) Review of the research on environmental education during early childhood in the past 20 years. Jpn J Environ Educ 19(1):95–108. (in Japanese with English abstract)

Inoue M (2014) Perspectives on early childhood environmental education in Japan; rethinking for a sustainable society. In: Davis J, Elliott S (eds) Research in early childhood education for sustainability; international perspectives and provocations. Routledge, New York, pp 79–96

Ito K, Fjørtoft I, Manabe T, Kamada M, Fujiwara K (2010) Landscape design and children's participation in a Japanese primary school-planning process of school biotope for 5 years. In: Muller N, Werner P, Kelcey JG (eds) Urban biodiversity and design. Blackwell, Oxford, pp 441–453

Ito K, Fjørtoft I, Manabe T, Kamada M (2014) Landscape design for urban biodiversity and ecological education in Japan: approach from process planning and multifunctional landscape planning. In: Nakagoshi N, Mabuhay JA (eds) Design low carbon societies in landscapes, ecological research monographs. Springer, Tokyo, pp 73–86

Ito K, Sudo T, Fjørtoft I (2016) Ecological design: collaborative landscape design with school children. In: Murnaghan A, Shillington L (eds) Children, nature, cities. Routledge, New York, pp 195–209

Jansson M, Gunnarsson A, Mårtensson F, Andersson S (2014) Children's perspectives on vegetation establishment: implications for school ground greening. Urban For Urban Green 13:166–174

Kamada M (2018) Satoyama landscape of Japan—past, present, and future. In: Hong S-K, Nakagoshi N (eds) Landscape ecology for sustainable society. Springer, Cham, pp 87–109. https://doi.org/10.1007/978-3-319-74328-8_6

Kellert SR (2002) Experiencing nature: affective, cognitive, and evaluative development in children. In: Kahn PH Jr, Kellert SR (eds) Children and nature: psychological, sociocultural, and evolutionary investigations. MIT Press, Cambridge, pp 116–151

Kinoshita I (1992) A study on children's contacts to nature in rural areas compared with urban areas: part 1 studies on aspects of functions for environmental study through children's play in rural spaces. J Architect Plann Environ Eng 431:107–118. (in Japanese with English abstract)

Kyoto City (2013) Kyoto city statistic portal. https://www2.city.kyoto.lg.jp/sogo/toukei/Population/index.html#maituki. Accessed 6 May 2017

Louv R (2008) Last child in the woods: saving our children from nature-deficit disorder, updated and expanded. Algonquin Books of Chapel Hill, Chapel Hill

Mårtensson F, Jansson M, Johansson M, Raustorp A, Kylin M, Boldemann C (2014) The role of greenery for physical activity play at school grounds. Urban For Urban Green 13:103–113

Mustapa N, Maliki N, Hamzah A (2015) Repositioning children's development needs in space planning: a review of connection to nature. Soc Behav Sci 170:330–339

National Institute of Infectious Diseases (2017) Severe fever with thrombocytopenia syndrome. https://www.niid.go.jp/niid/ja/diseases/sa/sfts.html. Accessed 6 May 2017

Nicholson S (1971) How not to cheat children, the theory of loose parts. Landsc Architect 62:30–34

Noda K (2013) Utilization and management of urban forest park: ecosystem conservation combining with playpark management in Takaragaike Park, Kyoto. Nature Osaka Study File 5. (in Japanese)

Noda K, Ogawa M (2014) Raising social community in urban forest: playpark work at Takaragaike Park, Kyoto. Parks Open Space 75(1):15–17. (in Japanese)

Oka T, Ito K, Yoshida S, Ikeda T, Imada M (2004) Study on the workshop with process planning for the environmental education in a forest park. Kyushu J For Res 57:158–162. (in Japanese)

Refshauge A, Stigsdotter KU, Cosco GN (2012) Adults' motivation for bringing their children to park playgrounds. Urban For Urban Green 11:396–405

Said I (2012) Affordances of nearby forest and orchard on children's performances. Soc Behav Sci 38:195–203

Saito O, Shibata H (2012) Satoyama and satoumi, and ecosystem services: a conceptual framework. In: Nakamura D, Watanabe T, Nishi (eds) Satoyama-satoumi ecosystems and human well-being: Socio-ecological production landscapes of Japan. United Nations University Press, Tokyo, pp 17–59

Senda M (1998) Bilingual: play space for children. Ichigaya, Tokyo

Senda M (2015) Safety in public spaces for children's play and learning. IATSS Res 38:103–115

Soga M, Gastiton KJ, Koayanagi TF, Kurisu K, Hanaki K (2016) Urban residents' perceptions of neighborhood nature; does the extinction of experience matter? Biol Conserv 203:143–150

Suzuki S (2001) A phytosociological classification system of the *Quercus serrata* forests in Japan. Veg Sci 18:61–74. (in Japanese with English abstract)

Takashima I (2013) Tick-borne viral infectious diseases. Med Entomol Zoology 64(2):61–66. (in Japanese with English abstract)

Takeuchi K, Ichikawa K, Elmqvist T (2016) Satoyama landscape as social–ecological system: historical changes and future perspective. Curr Opin Environ Sustain 19:30–39

The Millennium Ecosystem Assessment (2003) Ecosystems and human well-being: a framework for assessment. Island Press, Washington

Yamashita I (2000) Konomichi gojyu-nen (for fifty years). Uryu Bunko, Kyoto

Chapter 9
Ecological Evaluation of Landscape Components of the Tokushima Central Park Through Red-Clawed Crab (*Chiromantes haematocheir*)

Mahito Kamada and Sachiyo Inai

Abstract Red Claw Crab (*Chiromantes haematocheir*) requires a different habitat in each stage of their life cycle; estuary area at larval stage before metamorphosis, then forest during maturing and matured stages. Using this crab as an ecological indicator, suitability of landscape element types in urban park as wildlife habitat was evaluated. Results showed; (1) spaces in and along artificial stream were favored by the crab, which were located along a forest edge, while bare ground was disfavored, (2) in the forest, holes of stonewall were selectively used, (3) in the artificial stream, areas shaded by planted grasses and stones were preferred, and (4) at a time for larva spawning, a place with an old type stonewall, which has many large holes and vegetation, was selectively used. Water supply by an artificial stream and climate mitigation by the forest were essential functions for supporting the life of these crabs. Even in man-made structures for amenity, such as stream, stonewall, planted grasses and rocks, they successfully make use of such habitats. Therefore, it is evident that man-made structures can be designed multi-functionally for satisfying demands as habitat for wildlife and for the amenity of urban dwellers.

Keywords Urban green-park · Man-made structures · Amenity · *Chiromantes haematocheir* · Red claw crab · Life cycle

---

M. Kamada (✉)
Department of Civil and Environmental Engineering, Graduate School of Technology, Industrial and Social Sciences, Tokushima University, Tokushima, Japan
e-mail: kamada@ce.tokushima-u.ac.jp

S. Inai
Graduate School of Advanced Technology and Science, Tokushima University, Tokushima, Japan

Non-Profit Organization Tokushima Conservation Biological Society, Tokushima, Japan

K. Ito (ed.), *Urban Biodiversity and Ecological Design for Sustainable Cities*,
https://doi.org/10.1007/978-4-431-56856-8_9

9.1 Introduction

A historical place such as old garden in a city holds remnants of nature in the region and high-biodiversity as well as cultural monuments (Kümmerling and Müller 2008; Müller and Werner 2010). In Japan, the sites of castle in cities are usually open to people as park, and frequently hold well-developed forest in the area (Inai and Tahara 2006). Such places are very important for urban people to touch with nature in cities (Müller and Werner 2010).

Tokushima Central Park, which is situated at the center of Tokushima City with population of over 250,000 (Fig. 9.1), has been established at the site of castle. The castle used to be on and around a hill during Edo period, the era of a feudal society.

Fig. 9.1 Tokushima Central Park located in a center of Tokushima City along estuary area of Suketo River. A hill is covered by matured forest

The castle, except for stonewalls and moat, was removed in 1875 due to end of the period, and the area has been opened to public as a park from 1906 to present day (Nonaka 2015). The area has been continuously owned and conserved by the local government of Tokushima City, and was designated as historic property by Japanese government in 2006.

Conservative management by the local government has led to preserving the area from development, and it has allowed trees to grow on hill-slope. As a result, matured forest, which is mainly composed with *Elaeocarpus sylvestris* var. *ellipticus*, *Cinnamomum camphora*, *Celtis sinensis* var. *japonica*, *Aphananthe aspera*, has established on the hill (Morimoto et al. 1977; Kamada 2000, Fig. 9.1), and it has been designated as a natural monument by the local government of Tokushima City from 1963. The hill is now called as *Shiro-yama*, meaning *Castle-hill*, and familiar to people of Tokushima.

The Tokushima Central Park has spectrum of components from highly natural to highly artificial one. Hill-forest is adjacent to estuary area of Suketo River (Fig. 9.1), and both ecosystems provide wilderness to urban people. People's awareness of environmental issues is influenced crucially by their experiences of nature in their everyday surroundings (Savard et al. 2000; Müller and Werner 2010), and the circumstance of the Park is extremely advantageous to raise the environmental consciousness of urban people in Tokushima.

The local government has made several facilities as amenities for the people, such as trail for walking/jogging/cycling, tennis court, artificial stream and pond, rose garden, etc. People enjoy cherry blossoms every April (Fig. 9.2). According to answer of 3625 people on questionnaire (NPO Institute for Conservation Biology of Tokushima 2008), 83% of the people in Tokushima City have visited the Central Park, and main purposes were cherry blossom viewing (54%), walking (38%), refreshment (26%), visiting museum in the park (24%), rose garden (21%), and just passing (11%). People who visited there purposing use of natural forest were few; 9% were for forest therapy and 4% for nature study such as bird watch. Only 13% people know that the forest has been designated as a natural monument and 8% know that the dominant tree in the forest is *Elaeocarpus sylvestris* var. *ellipticus*.

Two challenges are there for use and management of the nature in the park. One is that only few portion of people use the nature in the park. Devices to increase chances for people to touch with the nature and to raise awareness should be established. Another is that planner, designer, and manager of the park have little ideas/methods to make facilities for supporting persistence of natural capital and for assisting people to get interest from the nature. From viewpoints of ecological planning and design, artificial facilities have to be made to support persistence of natural capital in the city and assist people to touch the wilderness easily (Fig. 9.3). It is necessary to develop thinking way to link ecological knowledge with design qualities and principles of landscape architecture (Ignatieva 2010).

Using red-clawed crab, which species inhabits in the Tokushima Central Park, we evaluate ecological condition of components of park landscape, and get implication for ecological design and education.

Fig. 9.2 Several facilities have been installed in the Tokushima Central Park, as amenities for the people

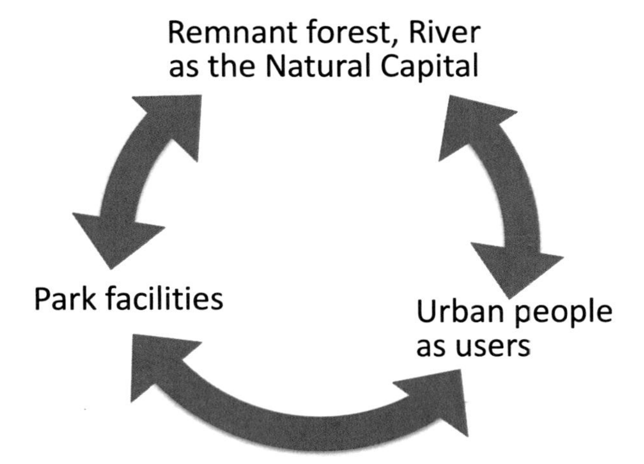

Fig. 9.3 Artificial facilities should be installed to support persistence of natural capital in the city and assist people to touch the wilderness easily

9.2 Red-Clawed Crab as an Ecological Indicator

Red-clawed crab, *Chiromantes haematocheir*, is Crustacea species, which occurs in eastern China, Taiwan, Korea Peninsula, and Japan. The crabs inhabit on land, forested area in particular, during maturing and matured period. Matured female come to riverside of estuary at night of flood tide in July and/or August for releasing zoea, the first stage of larva. Released larva floats in estuary and grows. Three or four weeks later from the release, larva of Megalopa stage land on tidal flat of riverside and grow to immature crab. Then immature crabs move to inland, and grow to be matured (Ono 1959; Suzuki 1981).

In order to complete life cycle of red-clawed crab, connectivity between land and estuary areas is important for larva to be released and to land. Physical structure of ecotone, such as riverbank and tidal flat, would affect success of releasing and landing. Heavy traffic may disturb crab's movement. Because the crab breathes through its gills, environmental conditions such as air humidity and water supply are essential for its life (Hashiguchi and Miyake 1967), as well as feed availability. Shelters to escape from enemies are also necessitated. Forest would keep air humidity and artificial stream provides water. Stonewall has holes and it would function as shelter. Thus components of park landscape affect survivorship of the crab. The red-clawed crab, therefore, is adequate to evaluate landscape components of the Tokushima Central Park from an ecological point of view (Fig. 9.4).

By finding environmental electivity of the crab, we evaluate ecological value of (a) forest and stonewall inside it, (b) parts of artificial stream, and (c) ecotone between hill-forest and river (Fig. 9.5). Results of the survey conducted in 2009 are shown in the following sections by referring to Inai et al. (2014).

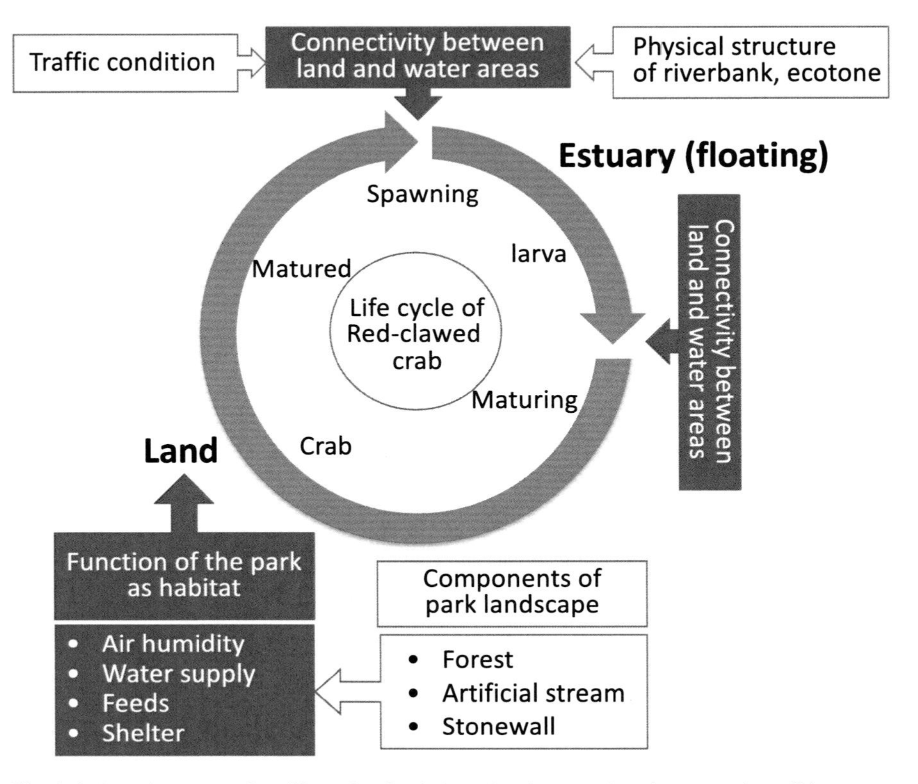

Fig. 9.4 In order to complete life cycle of red-clawed crab, several environmental conditions must be complete; connectivity between land and estuary areas, physical structure of ecotone, shelters, water supply, and air humidity. Thus the crab is adequate to evaluate ecological condition of landscape components in the Park

9.3 Ecological Evaluations of Components of the Park Landscape

9.3.1 Forest and Stonewall

Survey sites were set with regular intervals from the artificial stream (Fig. 9.6), and the number of the crabs was compared between inside and outside forest in a week of September. In the forest, the number was compared between sites with stonewall and without stonewall at the same elevation area.

No crabs were found at outside forest. The density per 10 m^2 in artificial stream was 0.6, and that in forest was 7.2, as shown in Fig. 9.7. Humidity was higher inside forest than outside, at the site of 0 m elevation nearest to artificial stream in particular (Fig. 9.8). For *Sesamops intermedium* that has similar life cycle with red-clawed crab, activities are inhibited when humidity becomes below 80% (Yoshida 1961). Period that humidity exceeded 80% was from June to October at a place of 0 m

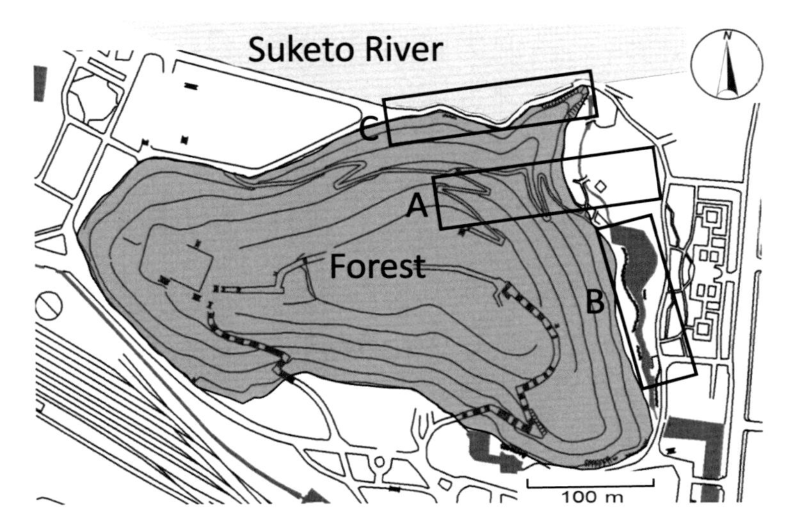

Fig. 9.5 Ecological value was evaluated at (**a**) forest and stonewall inside the forest, (**b**) artificial stream, (**c**) ecotone between hill-forest and river

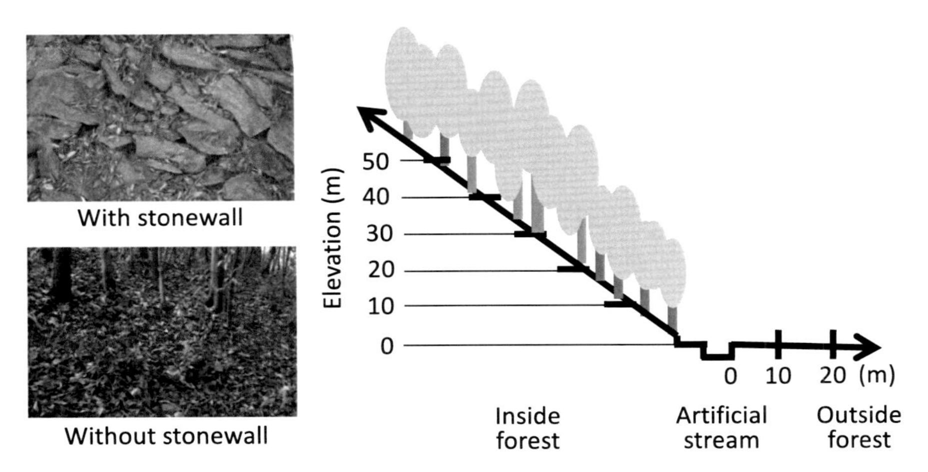

Fig. 9.6 Number of the crab was compared between inside and outside forest. In the forest, the number was compared between sites with stonewall and without stonewall at the same elevation

elevation in the forest, and from June to August at 20 m and 40 m elevation in the forest. At outside forest, however, humidity exceeded 80% only in July. The forest provides an essential space to the crab.

In the forest, almost crabs were found in holes of stonewall, and the density decreased exponentially with an elevation from the artificial stream (Fig. 9.9). The distribution pattern of the crabs is results of competition among crabs for water and shelter; the crabs, probably, favor to stay at a place adjacent to artificial stream

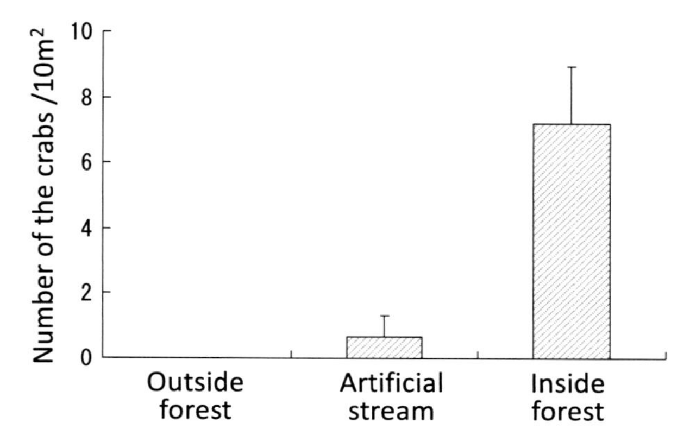

Fig. 9.7 Comparison of crab density at outside forest, in artificial stream, and inside the forest

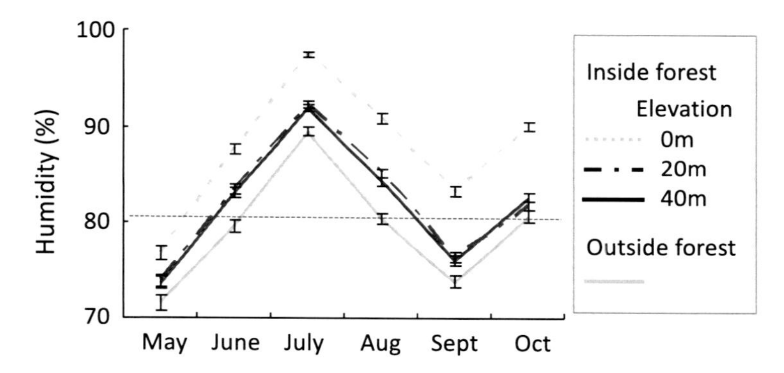

Fig. 9.8 Comparison of air humidity at different elevations inside the forest and outside the forest

because of water availability, but reluctantly move to stonewall to hide from predators. Thus the stonewall nearest to artificial stream becomes the best site to stay for the crabs. Number of holes at stonewall, however, are limited and weaker crabs expelled to places far from the stream. This would be allowed by existence of the forest, which functions to keep humidity.

9.3.2 Artificial Stream

Artificial stream, with the width of 10 m in maximum and 4 m in minimum (Figs. 9.5b, 9.10a), has two shore types (Fig. 9.10b, c). One is wet masonry type, which shore is covered by stones vertically without gaps. Another is that the grasses, such as *Ophiopogon jaburan*, are planted along shore and overhung the stream. Huge rocks are laid as ornaments in some part of the stream, and make a space between the bottom edge of the rock and stream surface (Fig. 9.10d).

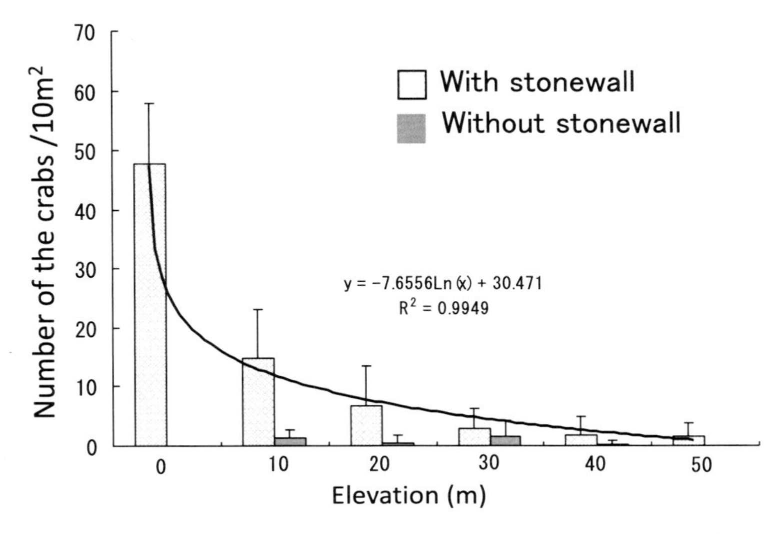

Fig. 9.9 Almost crabs were found in holes of stonewall in the forest. The density decreased exponentially with an elevation from the artificial stream

Fig. 9.10 Crab density in the artificial stream was compared at different physical conditions. (**a**) artificial stream, (**b**) stream bank of wet masonry, (**c**) stream bank with vegetation, (**d**) huge stone in the stream

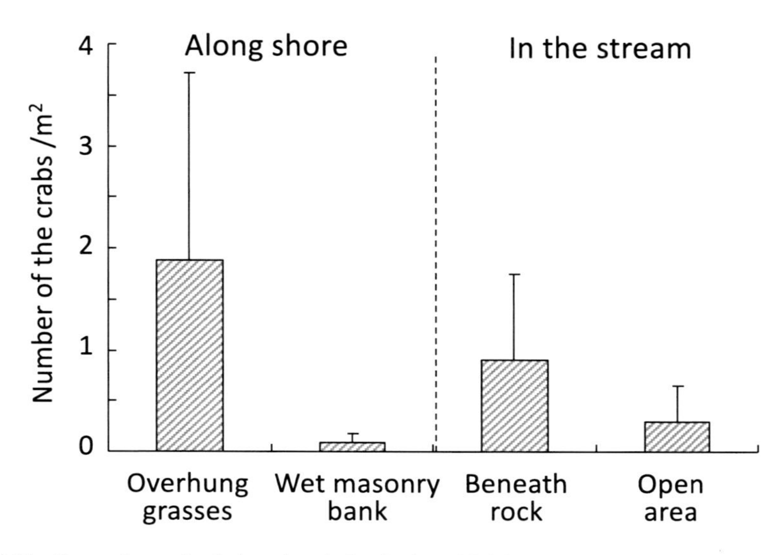

Fig. 9.11 Comparison of red-clawed crab density in artificial stream

Spatial use of the red-clawed crab was observed during 3 days in September. Comparing the density of the crab in relation to shore types (Fig. 9.11), it was much higher along shore with overhanging grasses (1.9/m^2) than along the shore with wet masonry bank (0.05/m^2). In the stream (Fig. 9.11), a space beneath the rock was much preferred by the crabs (0.9/m^2) than open area (0.25/m^2).

These results indicate that spatial heterogeneity, which provides space for the crabs to hide, is extremely important in designing components of park landscape. The cases, such as overhanging plants, rocks, and stream itself could provide examples/ideas for designing to provide ornamental satisfaction to people and lifeline to the crabs. Although reinforcement of stream bank by wet masonry is useful for landscaping, it is inadequate to sustain crab's life.

9.3.3 Ecotone Between Hill-Forest and River

At evening of flood tide, a day of full moon or new moon from middle of July to beginning of August, the red-clawed crabs of the Tokushima Central Park descend from the hill-forest to a shore of Suketo River to release larvae. Before reaching the river, the crabs have to cross a trail and climb down an embankment (Fig. 9.12).

Almost part of the embankment was renewed around 2000 and covered by wet masonry. Low water revetment of the new type is constructed of gabions, and set flatly under 0.7 m from water surface of flood tide level. Old type of embankment, which was made of dry masonry in 1928, was remained in only small area. Tidal flat

Fig. 9.12 Red-clawed crabs descend from the hill-forest to a shore of Suketo River to release larvae. (**a**) Water front, (**b**) crabs crossing trail, (**c**) time of crab's movement overlaps with that of human activities

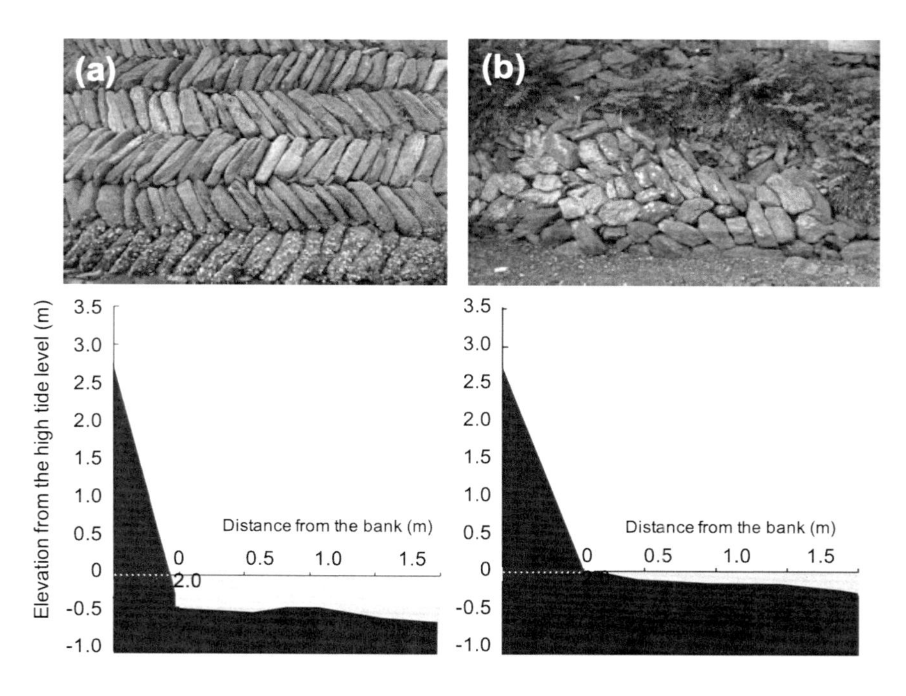

Fig. 9.13 Types of embankment. (**a**) New type of embankment is covered by wet masonry. Low water revetment is constructed of gabions, and set flatly under 0.7 m from water surface of flood tide level. (**b**) Old type of embankment is made of dry masonry. Tidal flat with gentle slope is composed of sands and small stones

with gentle slope, which is composed of sands and small stones, has remained in connecting with the bottom of the embankment (Fig. 9.13).

Mean depth of 81 holes between stones covering new type embankment is 10.8 cm, while the mean of 73 holes of old type is 30.9 cm (Fig. 9.14a). Vegetation cover on the old type embankment is 65% and that on the new type is 25% (Fig. 9.14b).

Number of crabs was observed at new and old embankments in the evenings with flood tide, from 19 to 21 o'clock in 22 July and in 6 August, the time of full tide was 19:00 and 19:02, respectively. The density of crabs is 35/m^2 at an old type embankment and 8/m^2 at new type (Fig. 9.14c). Red-clawed crabs begin rereleasing larvae when the tide begins to be out. The crabs, which reached to river shore before the beginning of ebb, have to wait at the embankment until the time. The old type embankment made of dry masonry is favored by the crabs, because it can serve much larger space to the crabs to hide from enemies than the new type made of wet masonry. Crab density, therefore, becomes much larger at the old type embankment.

Not only the structural quality of the new type embankment is low, but also the low water revetment that is laid flatly at deep from flood tide level is unsuitable for the crab to release her larvae. At a tidal flat with gentle slope (Fig. 9.13b), land area

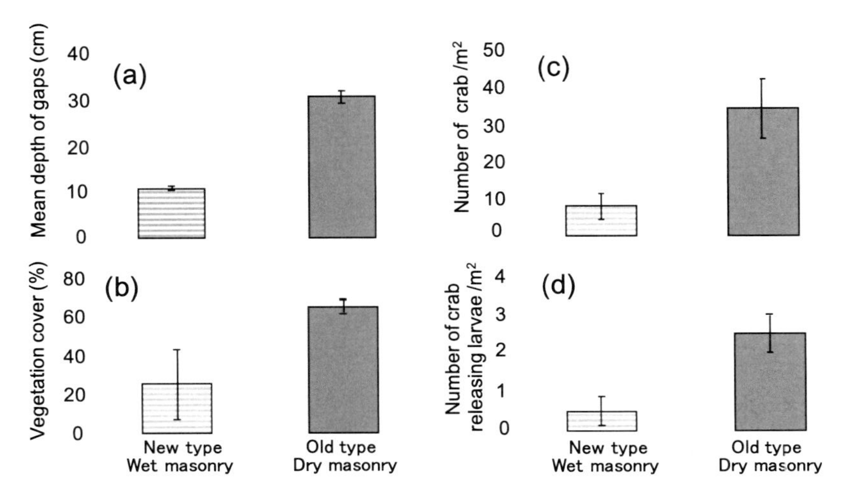

Fig. 9.14 Comparison of (**a**) hole depth, (**b**) vegetation coverage, (**c**) number of crabs, and (**d**) number of crabs releasing larvae between wet and dry masonry

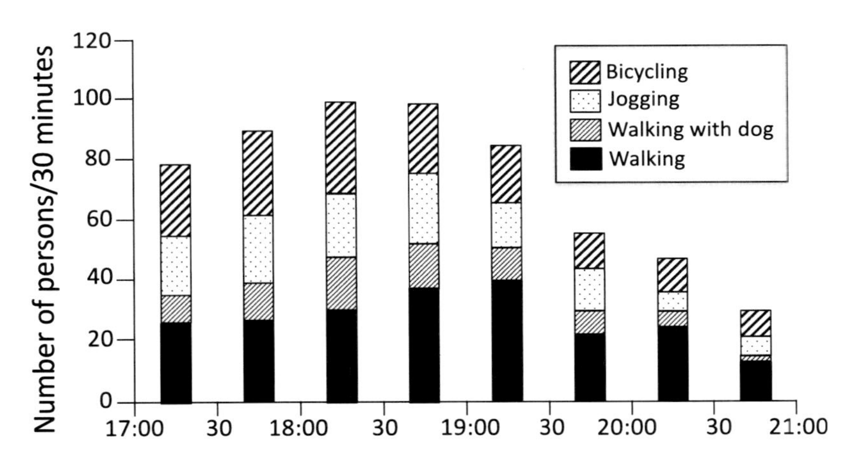

Fig. 9.15 Traffic at the trail

gradually appears along wth ebb tide, and thus the crab can spend longer time for releasing larvae than a tidal flat without slope (Fig. 9.13a). In fact, number of crabs releasing larvae was 2.5/m^2 at tidal flat in front of old embankment and that at new embankment was only 0.5/m^2 (Fig. 9.14d).

Red-clawed crabs hove to move to river shore by 19:00 of full tide. The crabs, therefore, have to cross a trail in the evening (Fig. 9.12c). It is the time becoming cool and suitable for people to make activities outside. Observation on traffic at the trail from 17:00 to 21:00 in 7 and 22 July and 6 August showed the average number of people who were walking, walking with dog, jogging, and bicycling increased in the evening and reached 100 in 18:00s (Fig. 9.15). Sunset time of the observed days was 19:16, 19:10, and 18:58, respectively. During observation, it was seen that crabs crossing the trail were frequently interrupted by the heavy traffic and sometimes squashed by feet and bicycles. Thus the condition is heavily disadvantage for the crab.

9.4 Conclusion

9.4.1 Recognition of Ecological Functions Through Red-Clawed Crabs

Based on the research results, we developed a program for environmental education and delivered to students of elementary school (Fig. 9.16). After a guidance (Fig. 9.16a), the children searched crabs at several places in the park (Fig. 9.16b, c) and found out (Fig. 9.16d). Then we leaded children to discuss each other about the places and their characteristics where the crabs could be found. They shared their experiences that the crabs could be found only in forest, and many were in streams

Fig. 9.16 Environmental education program was developed and delivered to students of elementary school

Fig. 9.17 NPO installed environmental education program in the event and delivered to parents and children

and holes of stonewalls, and they could understand that the forest could keep humidity and the holes of stone wall acted as a shelter for the crabs.

Results of a trial of educational program show that the crab can be an assisting device for people to recognize ecological functions of landscape components of the park. And hence another program matching with various generations was developed by a NPO, and delivered to the public, particularly for parents and children (Fig. 9.17).

9.4.2 Implications for Designing Park Landscape

Red-clawed crab is a very important wildlife for the city of Tokushima, because it can be functioned as a connector to link people with nature in the city. Measures for maintaining the crab population should be considered and carried out.

Importance of the forest has been recognized by the local government and designated as a natural monument. Water quality of Suketo River, habitat for larva of red-clawed crab, has been improved as a result of efforts in several decades. Measures for maintaining and improving habitat quality have been already installed for natural area in and adjacent to the Central Park. Values for the public of natural

Fig. 9.18 Sign to call attention to red-clawed crab crossing road, which has been set at Yuki-town as well as Naruto-city, Tokushima Prefecture (http://park7.wakwak.com/~ishikawaya/etc/news20020928.html)

areas remaining in a city seem easy to be recognized and conserved by the government.

In city parks, many artificial items have been installed for assisting activities of urban people. In a case of the Tokushima Central Park, artificial stream set for esthetic value provides essential value to the crab through providing water and keeping humidity stable. Rocks in the stream and vegetation along stream shore as esthetic objects give shelter to the crab. Although all these items have been installed for people, they function ecologically and support the crab to survive. The fact represents that esthetic items can install as ecological items without conflict. This is a very important viewpoint when we consider conservation of biodiversity in city (Rosenzweig 2003; Dearborn and Kark 2009). Changing the bank structure form wet masonry to dry masonry is essential to make better habitat for red-clawed crab and for other small animals, and probably it is beneficial for esthetic view at historic place such as the Tokushima Central Park with old castle.

Challenge is how to avoid human interference when crabs cross road to move to river shore for releasing larva. Setting sign to call people's attentions may be an idea, which has been installed such as in Yuki-town and Naruto-city of Tokushima Prefecture (Fig. 9.18). Fulfilling ethical responsibilities (Dearborn and Kark 2009) is the essential thing necessitating to urban dwellers.

Acknowledgments We grateful to Prof. Ito K of Kyusyu Institute of Technology for giving us the opportunity to summarize the study. We also thank to Dr. Ishimatsu K of Chugoku Regional Research Center for his great assistance in the editing process.

References

Dearborn DC, Kark S (2009) Motivations for conserving urban biodiversity. Conserv Biol 22:432–440

Hashiguchi Y, Miyake S (1967) Ecological studies of marsh crabs, *Sesarma* spp. II, habitats, copulation and egg-bearing season. Sci Bull Faculty Agric Kyusyu Univ 23(2):81–89. (in Japanese with English summary)

Ignatieva M (2010) Design and future of urban biodiversity. In: Müller N, Werner P, Kelcey JG (eds) Urban biodiversity and design. Wiley-Blackwell, Hoboken, pp 118–144

Inai S, Tahara N (2006) Role of forest developed at the site of a castle in a city. Planta 177:49–54. (in Japanese)

Inai S, Shinomiya R, Kawaguchi Y, Kamada M (2014) Functional evaluation of landscape elements of urban green park using red claws crab (*Chiromantes haematocheir*) as environmental indicator. Landsc Ecol Manage 19:57–68. (in Japanese with English abstract)

Kamada M (2000) Vegetation in Tokushima prefecture, Shikoku, Japan. In: Fujiwara K, Box EO (eds) A guide book of post-symposium excursion in Japan. IAVS Post-Symposium Excursion Committee, Yokohama, pp 127–137

Kümmerling M, Müller N (2008) 'Park an der Ilm'-Weimar (UNESCO world heritage site)—historical landscape gardens in Central Europe as early heritages for the development of ecological designed parks. Bfn-Skripten 229-2:27–44

Morimoto M, Ishii H, Konishi K, Miyai A (1977) Vegetation of Shiroyama. The Society for Nature Conservation of Tokushima, Tokushima City, pp 27–47. (in Japanese)

Müller N, Werner P (2010) Urban biodiversity and the case for implementing the conservation on biological diversity in towns and cities. In: Müller N, Werner P, Kelcey JG (eds) Urban biodiversity and design. Wiley-Blackwell, Hoboken, pp 3–33

Nonaka K (2015) Background and progress of the plan to turn the site of Tokushima Castle into a public park. J City Plan Inst Jpn 50(1):69–80. (in Japanese with English abstract)

NPO Institute for Conservation Biology of Tokushima (2008) Report on people's awareness on conservation of natural forest at Shiroyama, p 6 (in Japanese)

Ono Y (1959) The ecological studies on Brachyura in the estuary. Bull Mar biol Sta Asamushi, Tohoku Univ 9:145–148

Rosenzweig ML (2003) Win-win ecology: how earth's species can survive in the midst of human enterprise. Oxford University Press, New York

Savard J-PL, Clergeau P, Mennechez G (2000) Biodiversity concepts and urban ecosystems. Landsc Urban Plan 48:131–142

Suzuki S (1981) The life history of *Sesarma* (*Holometopus*) *haematocheir* (H. Milne Edwards) in the Miura peninsula. Res Crustasea 11:51–65

Yoshida M (1961) Daily rhythmic activity of a land-crab *Sesarma intermedia* (De Haan). Jpn J Ecol 11:160–162. (in Japanese with English abstract)

Chapter 10
Developing Urban Green Spaces and Effective Use of Rooftop Spaces for Cooling and Urban Biodiversity

Kazuhito Ishimatsu, Keitaro Ito, and Yasunori Mitani

Abstract Urban green spaces, which are important for people and wildlife, have been deteriorated due to urbanisation. As a result, our living environment is threatened by Unban Heat Island (UHI), poor urban biodiversity, and so on. It is, therefore, an urgent work to restore urban green spaces, despite that there are not enough open spaces. This is why rooftops of buildings, which had not previously been regarded as spaces for planting for vegetation, have been utilised as a type of open space, and so green roofing has become one of the gradually developing fields of urban ecological engineering. In Japan, the number of green roofs has increased for the past decade, especially extensive green roofs for cooling. However, our experiment shows that plants on extensive roofs can hardly affect the cooling effect because water evaporation on the soil surface is the most important for it. At the same time, green roofs are asked to be habitats for urban wildlife. Due to this, we explored brown/biodiverse roofing in UK, which is a relatively new type of extensive roofing used to provide brownfield wildlife with mimic brownfields. Many of the well-known urban environmental issues are caused by loss of biodiversity and natural habitats, mainly as a result of surface sealing through construction measures, increased UHI, and emission of green gases. Those issues can be partially mitigated by altering buildings' surface properties. The first step to avoid these issues is to increase the amount of open spaces and permeable surfaces as much as possible.

Keywords Urbanisation · Extensive green roof · Brown/biodiverse roof · Watered-soil covered roof · Urban heat island · Urban biodiversity · Green space · Habitat

K. Ishimatsu (✉)
Department of Civil Engineering, National Institute of Technology, Akashi College, Akashi-shi, Hyogo, Japan
e-mail: k.ishimatsu@akashi.ac.jp

K. Ito
Laboratory of Environmental Design, Faculty of Civil Engineering, Kyushu Institute of Technology, Kitakyushu-city, Fukuoka, Japan

Y. Mitani
Department of Electrical Engineering and Electronics, Kyushu Institute of Technology, Kitakyushu-shi, Fukuoka, Japan

K. Ito (ed.), *Urban Biodiversity and Ecological Design for Sustainable Cities*,
https://doi.org/10.1007/978-4-431-56856-8_10

10.1 Introduction

Urban green spaces can be important elements contributing to urban sustainability (Esbah et al. 2009) and ecosystem services, because they can contribute to the reduction of various types of pollution, the improvement of microclimate conditions the absorption of stormwater, and the prevention of flooding (Georgi and Dimitriou 2010). In particular, in case where urban green space includes not only grass but also trees, the contributions to the above functions of urban green spaces increase. For example, in terms of the improvement of the microclimate in urban spaces, deciduous trees offer shade during the summer and as Georgi and Dimitriou (2010) note, the suitable selection of the right species can enhance cooling through evapotranspiration reducing the temperature by up to 3.1 °C. Evapotranspiration creates pockets of lower temperature in an urban environment, known as the 'phenomenon of oases'. Conversely, it permits the sun to shine though the branches during the winter.

Furthermore, the spaces are an effective way of protecting significant features and scenic views, as well as providing buffers for differing land uses and valuable spaces for urban wildlife (Esbah et al. 2009). They shelter the native flora and fauna and help maintain native biodiversity (Esbah et al. 2009), and offer important harbours for remnant urban wildlife (Kong et al. 2010). Some species cope with, and survive in, urban areas better than others (Wood and Pullin 2002); for example, Sattler et al. (2010) note that urban areas host many arthropod species and cannot be regarded as species-poor environments. Green spaces can obviously provide vegetation and 'unsealed' surfaces, and ameliorate the detrimental effects of urbanisation on species assemblages by preserving or creating habitat, or by maintaining corridors for movement through the urban matrix (González-García et al. 2009; Smith et al. 2006). Although urban areas are among the most modified and complex of landscape, they still maintain a significant diversity of wildlife (Attwell 2000).

In the past decades, landscape ecology has become one of the most rapidly developing ecological fields worldwide (Kong et al. 2007), because increased urbanisation has had and continues to have a negative impact on urban green spaces (Kong and Nakagoshi 2006), and affects the urban microclimate (Georgi and Dimitriou 2010) as represented by 'Urban Heat Island (UHI)' phenomena. Unfortunately, urban green spaces, which are important for human and wildlife in the urban area, have largely been changed into man-made spaces. In other words, urban green spaces are increasingly heavily managed and used by people for recreation, such that the potential for biodiversity is reduced. In addition, open spaces are increasingly covered with hard surfaces rather than left green spaces. Thus, the absence of green spaces is characteristic of most contemporary cities globally (Georgi and Dimitriou 2010). In addition, loss and isolation of habitats due to urbanisation threaten biodiversity and warrant limits on development (Kong et al. 2010). If these areas become separated from one another by barriers such as large expanses of buildings and other human structures, species extinction may occur (Esbah et al. 2009), because the majority of the remnant urban wildlife is located in small fragments of indigenous vegetation that have been set aside during development (Rudd et al. 2002). These fragments are commonly subject to high levels of disturbance, due to the nature of human activities undertaken within or adjacent to these areas (Fernandez-Juricic 2000; Marzluff and Ewing 2001). Therefore the networks, which offer habitats and corridors that help conserve biodiversity (Kong et al. 2010), are also important to maintain the ecological components of a sustainable urban landscape (Sandström et al. 2006), as well as total green area. Because of this, rapid urbanisation makes people more aware of urban green space, and there is an increasing realisation that it is difficult to live without some contact with nature (Kong and Nakagoshi 2006).

However, there are rarely enough open spaces due to urban densification. This is why rooftops of buildings, which had not previously been regarded as spaces for planting for vegetation, have been utilised as a type of open space, and so green roofing has become one of the gradually developing fields of urban ecological engineering. A green roof is comprised of the following components: a roof structure; a waterproof membrane or vapour control layer, as insulation, a root barrier to protect the membrane, a drainage system, a filter cloth, a growing medium (soil) consisting of inorganic matter, organic material (straw, peat, wood, grass, sawdust) and air, and plants (Wilkinson and Reed 2009). Green roofs are mainly divided into two types: intensive and extensive (Molineux et al. 2009; Nagase and Dunnett 2010; Schrader and Böning 2006). Intensive green roofs are characterised by a thick layer (more than 200 mm) of growing medium or substrate, in which a wide range of plants and vegetation can be grown, particularly if irrigation is available. However, the relatively heavy weight of the substrate requires additional structural support by the building, and therefore only a limited range of buildings can be used for installing intensive green roofing. On the other hand, extensive green roofs are generally substrate based with a vegetated layer or sedum mat, either on its own with a sponge membrane for moisture retention or with a substrate base, offering between 25 and 100 mm deep root zones due to restrictions on weight loading on the building's structure. However, because of the thin substrate layer, the extensive roof environment is a harsh one for plant growth; limited water availability, wide temperature fluctuations, and high exposure to wind and solar radiation create high stress. As a result, a relatively small range of plant species is normally used for extensive green roofs. Sedum is a common and very suitable plant for use on extensive green roofing (Castleton et al. 2010; Oberndorfer et al. 2007).

According to Japanese Ministry of Land Infrastructure, Transport and Tourism (2009), the total green roof area in Japan has been increasing steadily during the past decade (Fig. 10.1). Almost all of these green roofs aim to improve thermal environment of midsummer (Iijima 2008), and about half of these roofs are extensive roofs with lawn grass or sedum (Japanese Ministry of Land Infrastructure, Transport and Tourism 2009). However, regarding sedum it can be killed by Japanese hot and humid summer, and disturb Japanese biodiversity because it is not a Japanese native plant. As well, it has been reported that evapotranspiration velocity from soil area without any plants is similar to one from lawn grass or sedum areas, therefore in

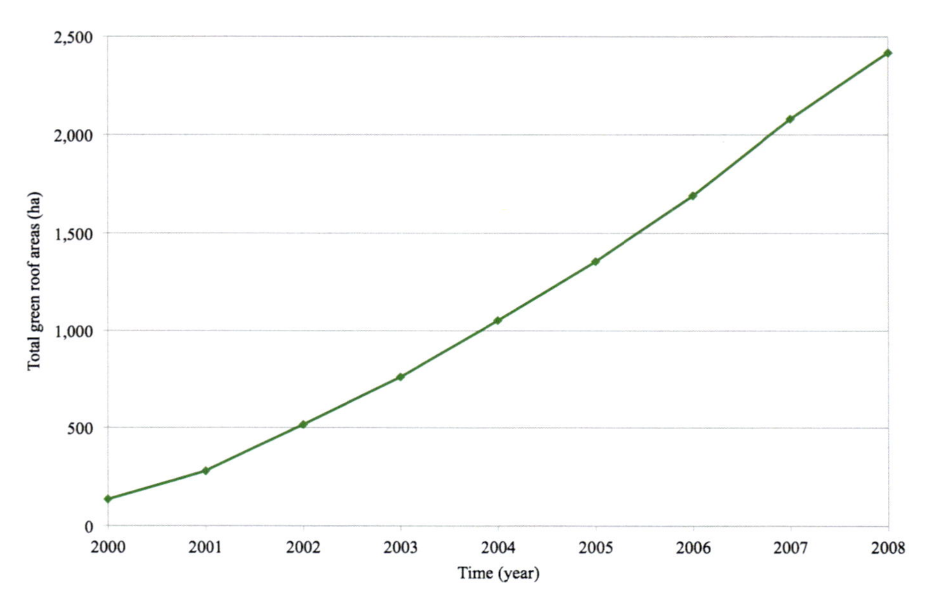

Fig. 10.1 Process of total green roof areas in Japan (Ministry of Land, Infrastructure and Tourism 2009)

terms of mitigating UHI watered-soil may be key, regardless of plants (Ferrante and Mihalakakou 2001; Ohno et al. 2006).

Furthermore, regarding not only sedum but also all plants, the process of evapotranspiration takes places in the plants' leaves rather than in the soil, drawing energy from the environment and not directly from the soil-covered building (Pearlmutter and Rosenfeld 2008). In contrast, the despite watered-soil layer on a roof can indeed be used as an effective means of evaporative cooling and has the particular benefit of producing this cooling during the hottest hours of the day (Al-Turki et al. 1995; Pearlmutter and Rosenfeld 2008). In addition, plants in green roofs prevent watered-soil from evaporating for cooling. In terms of cooling purposes, the role of plants on roofs is for shading (Ferrante and Mihalakakou 2001; Palomo and Barrio 1998). Therefore, it has been suggested that cover-materials other than plants could provide the wetted-soil surface with substantial shading, without the potential for cooling by evaporation, that is to say, the possible ways of reducing this energetic liability is to 'eliminate' the exposed roof by shading the roof area with some structures like tight weave mesh or in some way sheltering it with a substantially thick layer of soil (Pearlmutter and Rosenfeld 2008). The reduction in energy usage and external surface temperatures of roofs can also lead to a reduction in the UHI of the city centres (Wilkinson and Reed 2009).

Although several tests for substantiating the effect of evaporative roof for cooling has been implemented so far (Al-Turki and Zaki 1991; Al-Turki et al. 1995; Pearlmutter and Rosenfeld 2008), comparative experiment of alternative strategies for converting the theory into residential environment of human scale has never progressed. This chapter firstly aims to monitor how indoor thermal environment is affected by watered-soil covered roof by means of real residential models, and via an outside configuration of building in the Japanese urban area to explore the achievement of thermal comfort by encompassing natural methods and engineering technology. Secondly, in this chapter, we outline brown/biodiverse roofing in the UK, which is a relatively new type of extensive roofing for provision of mimic brownfields for brownfield wildlife, benefitting from techniques that offer diverse habitats under the severe condition resulting from the thin substrate layer. Finally, we discuss future prospects for developing urban green spaces.

10.2 Effect of Watered-Soil Covered Roof for Cooling

10.2.1 Preparatory Experiment

The preparatory experiment aimed to monitor the effect that plants on roofs have on cooling inside buildings and research possibilities for Japanese native plants for green roofs, which has taken place at the rooftop of civil engineering building at Tobata Campus of Kyushu Institute of Technology, Kitakyushu City, Japan, from 24 Aug 2007 to 2 Sep 2007.

10.2.1.1 Plants for Covering Roof

Plantain (*Plantago asiatica* L.) was selected as a Japanese native plant for green roofs, because the plant can endure severe condition on rooftop to be alive at cracks between asphalts or concretes, and efficiently cover soil with leaves due to perennial and rosette plant. In order to compare it with representative plants of existing Japanese extensive green roofs, Lawn (*Zoysia tenuifolia* Wild.) and Sedum (*Sedum mexicaum* Britt.) were adopted.

10.2.1.2 Experimental Setup

Four samples were composed of the following components (sorted by lowest to highest): a Styrofoam box, 150 × 250 × 150 mm; a drainage layer (FD DRAIN LN; produced by Tajima Roofing Ltd.), 500 × 500 mm; a root barrier membrane (FD Filter; produced by Tajima Roofing Ltd.), 500 × 500 mm; and light soil (FD Soil; produced by Tajima Roofing Ltd.), 50 mm. In addition, fifth sample was exposed without a set of miniature green roof (see Fig. 10.2). Temperature sensors (Ech2o probe; produced by Decagon Devices, Inc.) were set in each Styrofoam box, a meteorological sensor (Vantage Pro2; produced by Davis Instrument) was set close to the cells. Except the fifth cell, the other were given 700 ml irrigation to soil twice a day, 5:30 and 17:30. Table 10.1 shows the conditions of miniature green roofs.

Fig. 10.2 Implementation of preparatory experiment at the rooftop of civil engineering building at Tobata Campus of Kyushu Institute of Technology, Japan

Table 10.1 Conditions of miniature green roofs for the preparatory experiment

	Type of roofing
A1	Watered-soil 50 mm and plantain
A2	Watered-soil 50 mm and sedum
A3	Watered-soil 50 mm and lawn
A4	Watered-soil 50 mm
A5	Bare

10.2.1.3 Result and Discussion

Figure 10.3 shows hourly measured temperatures and solar radiation flux on 24 Aug 2007, from 6:00 to 18:00. Unfortunately, other days during this experimental period were not sunny all the day. We, therefore, selected the date because continual sunshine was important for the experiment.

Firstly, it can be seen that while A5 temperature rapidly rose with increasing solar radiation flux, the others mildly rose like the outside temperature. That A5 temperature rose as high as over 41 °C was due to the direct sunlight.

Secondly, compared A1, A2, A3, and A4, it is shown that the temperature in A2 is slightly lower than the others. The cause is thought that sedum effectively protected the soil surface from direct sunlight and made insulations of air pockets like mulching material, because sedum's three-dimensional volume was more massive than plantain and lawn. However, the effect of A2's advantage for cooling was only around 1° compared to A1, A3, and A4, therefore it can be safely said that the type of plants in extensive green roofs hardly affect the function for cooling.

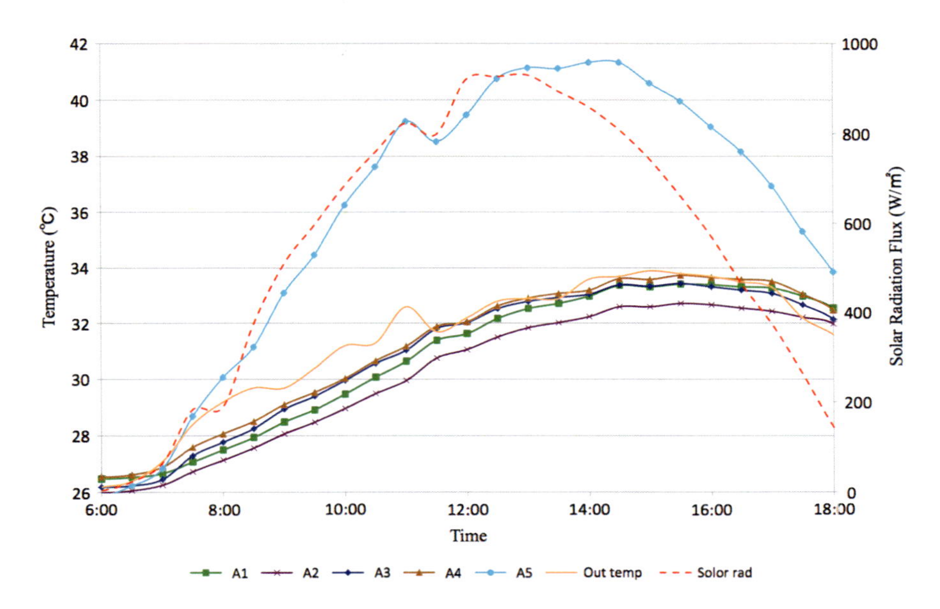

Fig. 10.3 Measured temperatures and solar radiation flux vs. time—24 Aug 2007 (after Ishimatsu et al. 2010)

It is often argued that given the fact that sedum can disturb Japanese urban ecosystem, to adopt it for green roofs is not beneficial. In addition, in Japan the law for controlling invasive alien species has been in force since in Jun 2005 and updated year on year. On 1 Feb 2010, 12 types of alien plants have been restricted (Ministry of the Environment 2010). Sedum has not been restricted yet, however, Ministry of the Environment has continued to collect more information on other alien species, and the number of restricted species has increased year on year. Thus green roofs covered by nonnative plants like sedum may not be a sustainable way. In terms of cooling, to cover roofs with watered-soil may be the most beneficial way because of lower cost for installing and maintenance than any green roofs, with only slightly lower performance. According to Palomo and Barrio (1998), to select light soil can reduce the thermal conductivity as well as weight, because a less dense soil has more air pockets and is hence a better insulator (Castleton et al. 2010). In case of needing plants for soil covered roof, vernacular native plants should be adopted as s covering plant instead of any nonnative plants.

10.2.2 Main Experiment

The main experiment aimed to monitor how indoor thermal environment is affected by watered-soil covered roof by means of real residential models, which has taken place at the Green-energy trial unit 'Green Cube' at Tobata Campus of Kyushu Institute of Technology, Japan, from 11 Aug 2009 to 5 Sep 2009.

10.2.2.1 Green Cube

Each Green Cube (see Fig. 10.4), 2500 × 2500 × 2500 mm, is composed of a concrete floor slab, steel frames, glass walls, and roof slab, and a Styrofoam thermal insulator and thin steel plate on the roof slab with an air conditioner, electricity consumption metre (Tibt-R; produced by Enegate Co., Ltd.), and a temperature sensor (Hygrochron; produced by KN Laboratories, Inc.). These cubes were totally sealed for the experimental period.

10.2.2.2 Plant for Covering Wall

Loofah (*Luffa cylindrica* (L.) Roem.) was selected for covering the glass walls, because the plant can grow up easily and blind the walls with large leaves. Each seeding was planted in each flowerpot, afterword, set along each wall which has a support net for helping Loofah to climb the wall.

10.2.2.3 Experimental Setup

The effect of watered-soil covered roof for cooling inside temperature was monitored by means of combining watered-soil covered roof, sunlight reflection coating roof and green wall. The watered-soil covered roofs were composed of a drainage layer (FD Drain LN; produced by Tajima Roofing Ltd.), 2500 × 2500 mm; and light soil (Vivasoil; produced by Toho-Reo Co., Ltd.), 50 mm. A meteorological sensor (Vantage Pro2; produced by Davis Instrument) was set on the field of the experimental equipment. The watered-soil covered roofs were irrigated with 10,000 ml

Fig. 10.4 Eco-energy trial unit 'Green Cube'

Table 10.2 Conditions of green cubes for the main experiment

	Roof	Wall
Cube 1	Watered-soil 50 mm	Loofah
Cube 2	Untreated	Loofah
Cube 3	Sunlight reflection	Untreated
Cube 4	Sunlight reflection	Loofah
Cube 5	Untreated	Untreated
Cube 6	Watered-soil 50 mm	Untreated

once a day, at 10:00. The flowerpots planted seedlings of Loofah were given 750 ml irrigation to the soil twice a day, at 10:00 and 16:00. Table 10.2 shows the conditions of green cubes.

10.2.2.4 Electricity Consumption

Power consumption data (Standard RS-485) measured by the electricity consumption metre was sent to IP network converter via a wireless terminal and onto the campus LAN. We monitored electricity consumption of each cube operating each air conditioner with 27 °C setting from 10:00 to 18:00 for sunny days only during the experiment.

10.2.2.5 Result and Discussion

Indoor Temperature

Figure 10.5 shows the measured mean indoor temperatures without running air conditioner for six sunny days in the daytime (11:00 to 16:00).

Firstly, a comparison between Cube 3 and Cube 6 shows that watered-soil covered roof is more effective than sunlight reflection roof regarding the cooling effect. Furthermore, the former is also superior to the latter in terms of providing insects and plants habitats. However, the way to paint roofs for sunlight reflection is so easy, and roof with steep slope cannot be covered by soil. Thus, it is, in general, sensible to consider soil covering for flat or gently sloping roofs, with other roofs painted for sunlight reflection.

Secondly, in three cubes with green walls, Cube 4 (with sunlight reflection roof and loofah walls) was the lowest, because Cube 4's ratio of green coverage at walls was the highest (Table 10.3). Therefore, it can be anticipated that Cube 4 would have been similar to Cube 1 and Cube 2 if those cubes were under exactly same conditions. To sum up the result of Cube1, Cube 2, and Cube 4, it is important for improving indoor thermal conditions to block building from direct sunlight.

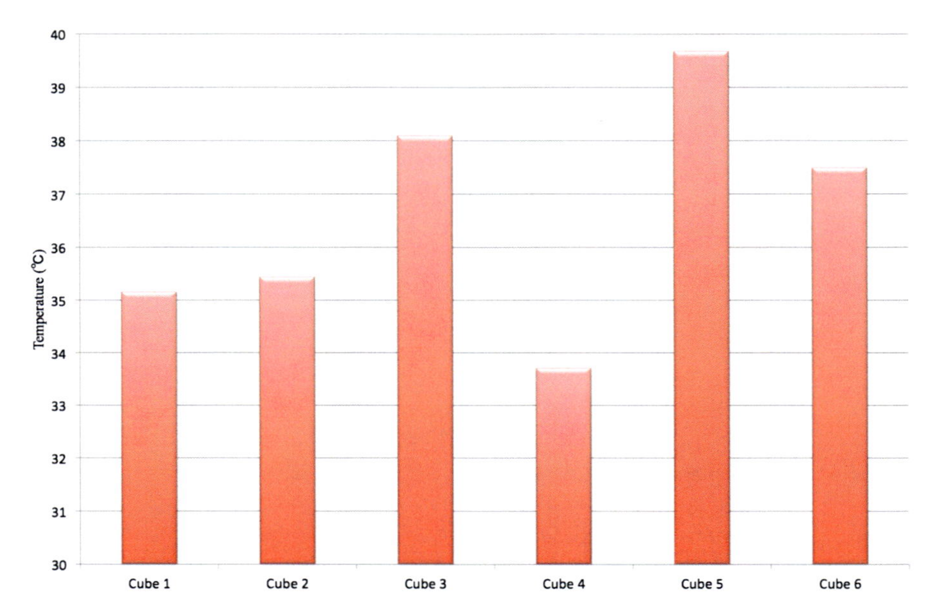

Fig. 10.5 Measured mean indoor temperatures without running air conditioner for six sunny days in the daytime (11:00 to 16:00) (after Ishimatsu et al. 2010)

Table 10.3 Ratio of green coverage at walls on 17 Aug 2009

	Green coverage (%)
Cube 1	63.0
Cube 2	68.2
Cube 4	76.5

Electricity Consumption

Figure 10.6 shows the measured mean indoor temperatures and electricity consumption with running air conditioner for eight sunny days in the daytime (11:00 to 16:00). Despite some influence from inherent variation in performance of individual air conditioner units, the results followed those of mean temperatures without running air conditioner closely (see Fig. 10.5). In addition, it was proven that watered-soil covered roof can reduce electricity consumption for cooling as well as sunlight reflection painted-roof. According to Castleton et al. (2010), the thicker the soil substrate on the roof, the better it reduces heat gain/loss into/out the building. Furthermore, green walls can enhance the effect for cooling. It is argued that in case it is difficult to cover walls with plants, another ways to blind walls should be used instead of plants, because main role of green walls in cooling was to protect from direct sunlight like curtain.

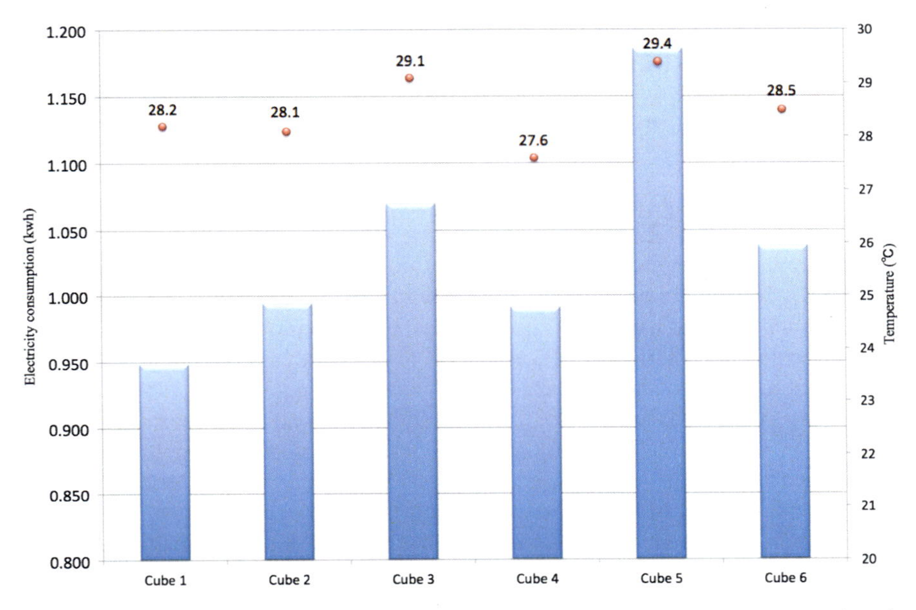

Fig. 10.6 Measured mean indoor temperatures and electricity consumption with running air conditioner for eight sunny days in the daytime (11:00 to 16:00) (after Ishimatsu et al. 2010)

10.2.3 What Is the Purpose of Green Roof?

The introduction of electro-mechanical air conditioning systems into the buildings, with their great energy expenditure, has become the standard alternative used to natural cooling (Palomo and Barrio 1998). Meanwhile, a climatic conscious design of outdoor spaces and the appropriate use of natural components are key elements to reduce the outcome of unsound evolution of urban areas where impermeable surfaces and denuded landscapes determine undesirable climatic effects and unhealthy life condition (Ferrante and Mihalakakou 2001). However, the ways to achieve a consensus for this purpose are never simple because this matter involves the complicated ecosystem components. The first step to overcome the challenge is to increase the amount of open space and permeable surfaces as much as possible (Ferrante and Mihalakakou 2001).

Although the environmental functions of green roofs (e.g., mitigation of rainwater runoff or UHI) are often promoted, the cost-effectiveness is not superior to alternatives due to the fact that installation and maintenance costs are not inexpensive. There are several alternative ways to achieve those environmental goals. It should be noted that green roof becomes essential only in the case of enhancement of urban biodiversity under the condition that open space in urban area is usually small.

10.3 A Method to Utilise Rooftop Spaces for Enhancement of Urban Biodiversity in the UK

10.3.1 What Are Brownfields?

In urban areas, the origin of brownfield (Fig. 10.7) is principally demolished buildings (houses and factories), although the definition also includes landfills, sand or gravel pits, industrial dumps, former collieries, and railway lands (Small et al. 2003). To sum up, brownfield refers to land that was previously developed for housing or industry but has, to differing degrees, been abandoned and recolonised by different ecological assemblages (Lorimer 2008; Schadek et al. 2009).

How did brownfield sites appear? Firstly, many buildings destroyed by bombing raids during World War II were not immediately rebuilt, and these vacant sites were colonised by wildlife (Grant 2006). Secondly, the process of industrial change has resulted in the creation of brownfields across Europe, particularly in urban areas (Grimski and Ferber 2001). As London's industry and docks declined, other sites were cleared and subsequently colonised by diverse vegetation (Grant 2006). While parks and gardens come to mind as obvious refuges for nature, plants, and animals are often more adventurous with regard to the places they colonise and use (Kadas

Fig. 10.7 Brownfiled in the City of London. These areas are quite important for urban wildlife (after Ishimatsu and Ito 2013)

2006). As our world becomes increasingly developed, many species of wildlife adapt in unpredictable ways (Brack Jr. 2006).

While brownfield is typically considered to have no or negative economic value, recent research suggests that there are many ecosystem services provided by such habitats (Robinson and Lundholm 2012). Brownfields provide habitat conditions similar to more natural habitats, and they may help maintain populations of some rare species (Eyre et al. 2003). Furthermore, compared with lawns and urban forest, the brownfield site showed higher levels of ecosystem service provision for indicators of habitat provision, both plant species and invertebrate diversity (Robinson and Lundholm 2012). In the UK, brownfields include some of the most species-diverse habitats left (Kadas 2006), and are thought to support a minimum of 12–15% of Britain's nationally rare invertebrate (Small et al. 2003). Eyre et al. (2003) surveyed a total of 78 brownfields for beetles between 1991 and 2001 throughout England, as a result generating a total of 182 records of 46 nationally rare species (16 ground, 10 rove, and 20 phytophagous species). A number of these species are more usually associated with other, more natural habitats such as riverine sediments, sandy heaths, and chalk grasslands. They note, brownfields are important habitats for beetles, and there is evidence that the situation is similar for other invertebrate groups. Wasteland habitats associated with urban brownfields are of intrinsic importance, relying on the codification of these habitats as distinctive habitats characterised by suits of species and abiotic conditions that fulfil a range of scientific criteria (Harrison and Davies 2002).

From the 1980s to the present day, however, with UK government policy encouraging reuse of abandoned sites such as brownfields, these sanctuaries for nature have been increasingly redeveloped (Grant 2006). The UK government has set a target of building 60% of new dwellings on previously developed land (Lorimer 2008). Redevelopment of brownfields is widely acknowledged as one of the major tools to achieve sustainable development (Grimski and Ferber 2001). Because one of the reasons for emergence of brownfields is economic structural change and the decline of traditional industries, they are frequently coupled with severe loss of jobs and, as a direct consequence, decline of the neighbourhoods around derelict sites or even of whole cities. Although new parks and green spaces have occasionally been created within redeveloped sites, these are, unfortunately, nearly always ecologically impoverished, lacking the diversity provided by the original vacant sites (Grant 2006).

Due to this, the amount of brownfields with nature conservation value in Britain is set to decrease dramatically under current home-building and regeneration policies (Harrison and Davies 2002; Small et al. 2003). Huge swathes of industrial brownfield along the Thames Estuary are slated for redevelopment, and this will have an immense impact on wildlife (Kadas 2006). A challenge that faced brownfield conservationist in East London was to persuade local residents and policy makers that valuable species and ecological assemblages could be found inhabiting brownfield sites in the city (Lorimer 2008). Angold et al. (2006) suggest that planners can have a positive impact on urban biodiversity by slowing the pace of redevelopment and by not hurrying to tidy up and redevelop brownfields.

One of the most successful strategies that has been employed by the third constituency in its efforts to campaign for urban biodiversity and brownfield conservation has been to compromise with developers of brownfields and to persuade them to install wildlife-friendly mitigation technologies on roofs of buildings (Lorimer 2008). This is why rooftops of buildings have been regarded as installment sites of threatened brownfields.

10.3.2 Brown/Biodiverse Roof

This is brown/biodiverse roof (Fig. 10.8), which is usually constructed for habitat mitigation in the UK, especially in London, as the only mitigation obliging constructors to install brown/biodiverse roofs comes from conservation of a rare bird species, the black redstart (Molineux et al. 2009). The black redstart is listed as a priority species for the London Biodiversity Action Plan sponsored by the London Biodiversity Partnership, as a Bird of conservation concern, and a Red Data Book species (London Wildlife Trust 2001). They are reliant on old vacant lots and brownfields and are thus now under threat from regeneration of much their breeding ground (Gedge 2003; London Wildlife Trust 2001).

Fig. 10.8 Brown/biodiverse roof at Royal Holloway University of London (after Ishimatsu and Ito 2013)

The roof membrane for each section is generally made of butyl rubber and protected by a nonwoven polypropylene geotextile fleece supported by a plywood deck (Grant 2006). To promote biodiversity, a variety of substrates are used, including a chalk and subsoil mixture, loamy topsoil, and gravel. In addition, crushed brick favours ruderal vegetation and can thus be used to replicate the brownfield biodiversity that was in place before development began (Lorimer 2008). The brown/biodiverse roof aims to provide vegetal and animal species with habitats while somehow managing to grow vegetation without any irrigation system or fertiliser.

10.3.3 Techniques to Enhance Biodiversity on Extensive Roofs

Firstly, from associated brownfields and other valuable vegetated areas, the top 150 mm of substrate must be carefully removed and appropriately stored so that some of the existing vegetation, seed bank, and soil organisms can be conserved (if suitable) for subsequent use on extensive roofs (Brenneisen 2006); for example, white and biting stonecrops have some of the most spectacular flowering displays and are very attractive to bees, butterflies, and other insects (Natural England 2007). In addition, adaptation of spider and beetle fauna to natural soil and other substrates such as sand and gravel from riverbanks seemed to be a factor for successful colonisation (Brenneisen).

Secondly, small logs laid across the substrate will not only provide shelter for insects but also create nesting sites for many small bees and wasps that burrow into dead timber (Natural England 2007; Robinson and Lundholm 2012).

Thirdly, designing brown/biodiverse roofs so that they have varying substrate depths and drainage regimes create a mosaic of microhabitats on and below the soil surface and can facilitate colonisation by more diverse flora and fauna (Brenneisen 2006). There is increasing use of locally derived lightweight granular waste materials as sustainable sources for roof substrates (Oberndorfer et al. 2007).

Lastly, Fig. 10.9 is a diagram of a typical cross-section of a brown/biodiverse roof, showing use of diverse soil surfaces and substrates to create a mosaic of wildlife habitats for colonisation by a more diverse flora and fauna.

In fact, it was reported that spiders, beetles, bees, wasps, ants, and so on, which can be seen at brownfields, were found on Laban Dance Centre (brown/biodiverse) roof in London (Kadas 2006). As spiders are predatory, they occupy a mid-trophic level in the food chain. The presence of spiders would suggest a varied invertebrate fauna present at the survey site. At Royal Holloway, University of London, moss forests, which provide cover for thousands of microscopic animals and habitats for other invertebrates (Natural England 2007), can be seen on brown/biodiverse roofing (Fig. 10.10). In addition, it was reported that ground-nesting birds utilised brown/

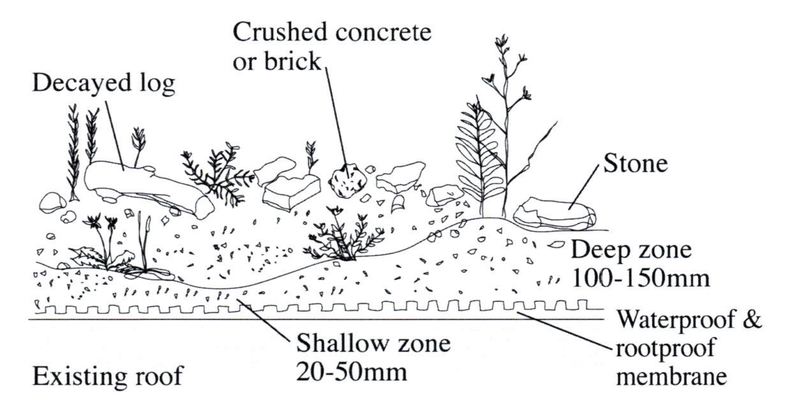

Fig. 10.9 Typical cross-section of a brown/biodiverse roof (after Ishimatsu and Ito 2013)

Fig. 10.10 Unplanted mosses at brown/biodiverse roof at Royal Holloway, University of London (after Ishimatsu and Ito 2013)

biodiverse roofs as a nesting location (Brenneisen 2006), though this example is from Switzerland.

Unfortunately, however, there is still a lack of information about wildlife on brown/biodiverse roofs due to the fact that they do not have a very long history. Brown/biodiverse roofing needs more time to be investigated because of its dependence on successions. In the presence of soil, rainfall, and sunlight, whether on rooftops or not, successions suitable for each environment will take place over time.

10.3.4 Adaptation and Limitation of Habitats on Rooftops

It should be noted that green roof industries also consider the concept of brown/ biodiverse roofs. Table 10.4 shows the requirements of each roof type in terms of frequency of maintenance, and construction and maintenance costs based on Fujita (2003). Although no data on costs of brown/biodiverse roof are available, it is clear that the construction cost of brown/biodiverse roof is lower than that of extensive roof with lawn, and the maintenance cost of brown/biodiverse roof is also lower than that of extensive roof with sedum, because they do not need any artificial vegetation layer but rather depend on succession. It is not always true that a soil layer without any plants has no ecological value; for example, there are some invertebrates which favour areas beneath stones or logs as their habitats.

In terms of ecological function, roof surfaces are assumed to be arranged in the following descending order: intensive roof, brown/biodiverse roof, extensive roof such as law or sedum, impermeable (Fig. 10.11). If budgets allocated for creating extensive roofs with lawn and sedum can be used for brown/biodiverse roofs, there will be much more permeable areas in urban area. In urban environments, vegetation has largely been replaced by dark and impervious surfaces (Oberndorfer et al. 2007). Many of the well-known urban environmental problems are caused by loss of biodiversity and natural habitats, mainly as a result of surface sealing through

Table 10.4 General maintenance frequency and costs of construction and maintenance required by roof type (assuming 200 m^2) (Fujita 2003)

	Extensive roof with lawn	Extensive roof with sedum	Intensive roof with shrub	Intensive roof with tree and lawn
Construction cost (JPY/m^2)	15,000 ~ 20,000	20,000 ~ 30,000	30,000 ~ 40,000	50,000 ~ 70,000
Yearly maintenance cost (maintenance breakdown) (JPY/m^2)	1800	650	2000	4,50
Overall check	3 times per year	3 times per year	3 times per year	12 times per year
Drainage cleaning	3 times per year	3 times per year	3 times per year	12 times per year
Pruning	–	–	2 times per year	2 times per year
Lawn mowing	3 times per year	–	–	3 times per year
Fertilisation	2 times per year	–	Once a year	2 times per year
Weeding	2 times per year	Once a year	2 times per year	2 times per year
Adjusting pole supporting tree	–	–	–	4 times per year
Pest control	–	–	Once a year	Every time pests occur
Irrigation system check	4 times per year	–	4 times per year	12 times per year

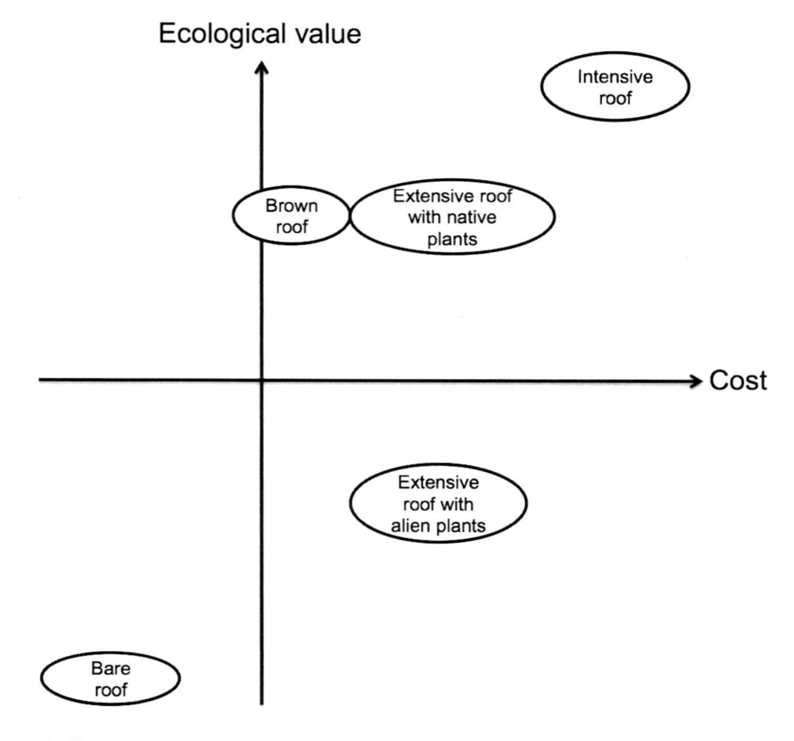

Fig. 10.11 Correlation diagram among various types of roofs (after Ishimatsu and Ito 2013)

construction measures, increased loads of heavy metals and organics, and emission of greenhouse gases (Schrader and Böning 2006). Those problems can be partially mitigated by altering buildings' surfaces properties (Oberndorfer et al. 2007). The first step to avoid these problems is to increase the amount of open spaces and permeable surfaces as much as possible (Ferrante and Mihalakakou 2001).

On the other hand, intensive roofs should be efficiently installed at equal distances to provide 'stepping stones' in urban areas not having enough open space for green areas. They can support more complicated biodiversity than brown/biodiverse roofs by offering valuable wildlife sanctuaries and providing better connectivity between existing habitats (Kim 2004). Some water areas should also be simultaneously installed, coupled with them if possible, because the combination of water and vegetation provides greater habitat diversity (Hunter and Hunter 2008). Despite the fact that intensive roofs are more beneficial for the urban environment, cost problems still remain, unfortunately (Table 3.4). Because of this, it is quite important to combine brown/biodiverse roofs with intensive roofs.

However, some animals cannot reach rooftops areas due to their restricted mobility, and earthworms are unable to survive on extensive roofs due to the limited substrate depth; they perish during high temperatures in summer because they cannot migrate to deeper and cooler regions of the soil (Brenneisen 2006). Thus, brown/

biodiverse roofs or intensive roofs should never be considered a justification for destroying natural or semi-natural habitats on the outskirts of cities and beyond (Schrader and Böning 2006). Those roofs are only a method to delay the weakening of urban biodiversity. It should be noted that to restore urban green spaces on the ground is controversially the most effective method for enhancing urban biodiversity.

10.4 Future Prospects for Developing Urban Green Spaces

Land and conservation management is increasingly concerned with regional-scale habitat analyses (Bunn et al. 2000) due to the fact that high-density development and rapid urban sprawl have affected the urban vegetation's composition and biodiversity (Kong et al. 2010). As a result urban area, unfortunately, provide an excellent opportunity to study the effects of habitat fragmentation, because urban green areas are typically surrounded by completely hostile man-made matrix, even though gardens in suburban areas may provide additional resources for some organisms (Öckinger et al. 2009). For example, Sandström et al. (2006) note there is a clear increase in the number of bird species as well as individuals from centre to the surroundings of the city in south-central Sweden. This fact visibly indicates that urban forests are an important component of the urban landscape in terms of at least bird species diversity.

Habitat fragmentation caused by urbanisation can be extreme within urban ecosystems, and fragments of natural vegetation may be too small or even too isolated to support some species (Savard et al. 2000). If they become isolated, small populations can lose genetic variation though inbreeding and genetic drift, and will become increasingly prone to extinction (Rouquette and Thompson 2007). Thus, seed dispersal and wildlife movements have a potentially profound effect on the population dynamics and constitute an essential survival process in patchy habitats (Angelibert and Giani 2003; Keller et al. 2010; Rouquette and Thompson 2007; Rudd et al. 2002).

Dispersal of wildlife is usually risky, and there will always be a balance of risk between living longer in an already occupied habitat and risking resources in an act of colonisation (Angelibert and Giani 2003). Dispersal will be efficient if the benefits of reaching a better site exceed the cost from the risk of death during dispersal (Angelibert and Giani 2003). Furthermore, whether or not patches can be recolonised depends on the availability of dispersing individuals and the ease with which these individuals can move about within the landscape (Kindlmann and Burel 2008). Due to this, new habitat should be created between the existing sites to reconnect the extant populations, and connectivity should be a key component of all management planning (Rouquette and Thompson 2007). In particular, it is often said that greenways (corridors) provide an opportunity to reduce the impacts of habitat fragmentation (Linehan et al. 1995). Given that the primary function of greenways is to provide linkages, they represent one of the most effective tools in preventing fragmentation and perhaps species loss at the regional level. However, it is difficult to implement it, as of today, because a lot of buildings have occupied urban areas. That means there is no open space. Therefore, rooftop spaces on buildings can support a development of green areas and ecological networks in urban areas.

Yet, landscape connectivity changes with the choice of measures. For example, connectivity measures based on distances may be appropriate for bird as the matrix and corridors may not be of great importance in this case. Measures based on the amount of corridors in the landscape may be appropriate for small mammals (e.g., carabid beetles) whose movement is affected by matrix permeability. Furthermore, road and railway lines are considered to be barriers against the movement of animals moving on the ground (Kamada 2005). Evidently, each of these measures will give us a different connectivity for the same landscape (Kindlmann and Burel 2008). Another very important component of network planning is the consideration of private and unprotected areas (e.g., brownfield). Those habitats can be an invaluable food and habitat source for a wide range of urban species and are essential in developing the matrix that supports the large numbers of corridors required for connectivity. Public education on gardening with native plants and providing proper habitats is another tool to enhance the connectivity of the region and improve the viability of the corridors. Conservation planners should be concerned with how to invest resources wisely to realise the greatest return, in terms of protected or enhanced biodiversity (Kadoya 2009).

This chapter suggests that rooftop space can contribute to enhance urban biodiversity. If possible, however, to create habitats on the ground is more preferable than on the rooftop due to accessibility. Moreover, if impervious spaces (e.g., concrete and asphalt) are replaced with pervious spaces (e.g., green area and brownfield), it is significant for not only urban biodiversity but also UHI. Even so, the character of habitat in urban areas desired for biodiversity is not completely compatible with the needs for a secure human environment (e.g., juvenile crime). As a consequence, the directives in management plans aim to remove potentially dangerous place (e.g., shrub on urban green spaces). Another conflict between biodiversity and people is the risk of falling trees or branches from large old trees, and accordingly insurance implications that can force the removal of standing dead wood. To deal with the planning dilemma between social security and biodiversity maintenance, one solution may be small-scale zoning planning in parks to create vegetated zones with different characteristics in urban green spaces (Sandström et al. 2006). Understanding the ecological interactions between built and natural areas within urban areas can help us manage and plan urban environments to promote diversity and ecological function (Newbound et al. 2010).

In conclusion, an important aspect in dealing with biodiversity is that not all species are equal. Species vary in size, shape, abundance, distribution, trophic position, ecological function feeding habitats, and desirability (Savard et al. 2000). Some species may play important roles in the community, so their absence would significantly affect several other species (Savard et al. 2000). For example, insect pollination is a vital ecosystem function in terrestrial systems (Robinson and

Lundholm 2012). At the same time, it can be predicted that it will be quite important to pay attention to brownfield wildlife in the near future, in countries facing severe population decline, especially Japan and Republic of Korea, where there is a possibility that brownfields will increase (Ishimatsu and Ito 2013). Conservation planners, therefore, have to consider the quality of urban green spaces with their aesthetic value. Öckinger et al. (2009) found that traditional parks had the lowest number of butterfly species because the parks usually lack most of the features that constitute suitable butterfly habitat. Typically, they have short grass turf that is cut regularly, low numbers of native plant species, and a low structural diversity. On the other hand, it was also found that brownfields had a high value for urban biodiversity and it may be more beneficial to leave brownfields unmanaged than to try to manage urban and suburban parks for biodiversity. But, what kind of green space do urban dwellers prefer? Mathey and Rink (2010) recognise brownfields as a new form of urban nature, and recommend that urban dwellers be informed about the biodiversity value of brownfield therewith learn to accept new aesthetical paradigms, under the condition of shrinking cities. From now on, this approach from both conservation biodiversity and landscape design will be more significant (Ishimatsu et al. 2012). Finally, it might be easy for some invasive alien plants to colonise there because brownfields develop due to the abandonment. It will be also necessary to monitor colonisaitons of invasive alien plants at brownfields. Brownfields might have a possibility to be ideal spaces for them.

References

Al-Turki AM, Zaki GM (1991) Energy saving through intermittent evaporative roof cooling. Energ Buildings 17:35–42

Al-Turki AM, Gari HN, Zaki GM (1995) Comparative study on reduction of cooling loads by roof gravel cover. Energ Buildings 25:1–5

Angelibert S, Giani N (2003) Dragonfly characteristics of three odonate species in a patchy habitat. Ecography 26:13–20

Angold PG, Sadler JP, Hill MO, Pullin A, Rushton S, Austin K, Small E, Wood B, Wadsworth R, Sanderson R, Thompson K (2006) Biodiversity in urban habitat patches. Sci Total Environ 360:196–204

Attwell K (2000) Urban land resources and urban planting: case studies from Denmark. Landsc Urban Plan 75:125–142

Brack V Jr (2006) Short-tailed shrews (Blarina brevicauda) exhibit unusual behavior in an urban environment. Urban Habitat 4(1):127–132

Brenneisen S (2006) Space for urban wildlife: designing green roofs as habitats in Switzerland. Urban Habitat 4(1):27–36

Bunn AG, Urban DL, Keitt TH (2000) Landscape connectivity: a conservation application of graph theory. J Environ Manage 59:265–278

Castleton HF, Stovin V, Beck SBM, Davison JB (2010) Green roofs: building energy savings and the potential for retrofit. Energ Buildings 42:1582–1591

Esbah H, Cook EA, Ewan J (2009) Effects of increasing urbanization on the ecological integrity of open space preserves. Environ Manag 43:846–862

Eyre MD, Luff ML, Woodward JC (2003) Beetles (Coleoptera) on brownfield sites in England: an important conservation resource? J Insect Conserv 7:223–231
Fernandez-Juricic E (2000) Avifaunal use of wooded streets in an urban landscape. Conserv Biol 14:513–521
Ferrante A, Mihalakakou G (2001) The influence of water, green and selected passive techniques on the rehabilitation of historical industrial building in urban areas. Sol Energy 70:245–253
Fujita S (2003) Shokubutsu no Sentei to Kanri. Nikkei architecture. In: Jitsurei ni Manabu Okujoryokuka. Nikkei Publications, Tokyo, pp 26–32. (in Japanese)
Gedge D (2003) From rubble to redstarts. http://www.laneroofing.co.uk/marketingbox/documents/generic/682732553_10Gedge.pdf. Accessed 8 Dec 2010
Georgi JN, Dimitriou D (2010) The contribution of urban green spaces to the improvement of environment in cities: case study of Chania. Greese Build Environ 45:1401–1414
González-García A, Belliure J, Gómez-Sal A, Dávila P (2009) The role of urban green spaces in fauna conservation: the case of the iguana (Ctenosaura similis) in the 'patios' of León city, Nicaragua. Biodivers Conserv 18:1909–1920
Grant G (2006) Extensive green roofs in London. Urban Habitat 4(1):51–65
Grimski D, Ferber U (2001) Urban brownfields in Europe. Land Contam Reclam 9(1):143–148
Harrison C, Davies G (2002) Conserving biodiversity that matters: practitioner's perspectives on brownfield development and urban nature conservation in London. J Environ Manage 65:95–108
Hunter MR, Hunter MD (2008) Designing for conservation of insects in the built environment. Insect Conserv Diversity 1:189–196
Iijima K (2008) Okujoryokukakukan no tayosei to donyushokubutsu. Jpn Soc Reveg Technol 34:338–343. (in Japanese)
Ishimatsu K, Ito K (2013) Brown/biodiverse roofs: a conservation action for threatened brownfields to support urban biodiversity. Landsc Ecol Eng 9(2):299–304
Ishimatsu K, Matthew L, Ito K, Yasunori M (2010) Green roofs for urban areas: effect of irrigated-soil covered roofs for cooling. Paper presented at the 7th biennial international workshop on advances in energy studies, Universitat Autònoma de Barcelona, Barcelona, 19–21 October 2010
Ishimatsu K, Ito K, Mitani Y (2012) Developing urban green spaces for biodiversity: a review. Landsc Ecol Manage 17(2):31–41
Kadas G (2006) Rare invertebrates colonizing green roofs in London. Urban Habitat 4(1):66–86
Kadoya T (2009) Assessing functional connectivity using empirical data. Popul Ecol 51:5–15
Kamada M (2005) Hierarchically structured approach for restoring natural forest: trial in Tokushima prefecture, Shikoku, Japan. Landsc Ecol Eng 1:61–70
Keller D, Brodbeck S, Flöss I, Vonwil G, Holderegger R (2010) Ecological and genetic measurements of dispersal in a threatened dragonfly. Biol Conserv 143:2658–2663
Kim KG (2004) The application of the biosphere reserve concept to urban areas: the case of green rooftops for habitat network in Seoul. Ann N Y Acad Sci 1023:187–214
Kindlmann P, Burel F (2008) Connectivity measures: a review. Landsc Ecol 23:879–890
Kong F, Nakagoshi N (2006) Spatial-temporal gradient analysis of urban green spaces in Jinan, China. Landsc Urban Plan 78:147–164
Kong F, Yin H, Nakagoshi N (2007) Using GIS and landscape metrics in the hedonic price modeling of the amenity value green space: a case study in Jinan City, China. Landsc Urban Plan 79:240–252
Kong F, Yin H, Nakagoshi N, Zong Y (2010) Urban green space network development for biodiversity conservation: identification based on graph theory and gravity modeling. Landsc Urban Plan 95:16–27
Linehan J, Gross M, Finn J (1995) Greenway developing a landscape ecological network approach. Landsc Urban Plan 33:179–193

London Wildlife Trust (2001) Advice for the Black Redstarts conservation in London. http://www.lbp.org.uk/downloads/Publications/Management/black_redstart_advice_note.pdf. Accessed 7 Sep 2010
Lorimer J (2008) Living roofs and brownfield wildlife: toward a fluid biogeography of UK nature conservation. Environ Plann 40:2042–2060
Marzluff JM, Ewing K (2001) Restoration of fragmented landscape for the conservation of birds: a general framework and specific recommendations for urbanizing landscapes. Restor Ecol 9:280–292
Mathey J, Rink D (2010) Urban wastelands: a chance for biodiversity in cities? Ecological aspects, social perceptions and acceptance of wilderness by residents. In: Muller N, Werner P, Kelcey JG (eds) Urban biodiversity and design. Willey-Blackwell, Oxford, pp 406–424
Ministry of Land, Infrastructure, Transport and Tourism (2009) Okujo/hekimen-ryokukakukan ha aratani donoteido sosyutsu saretanoka. http://www.mlit.go.jp/common/000045480.pdf. Accessed 4 Aug 2010 (in Japanese)
Ministry of the Environment (2010) Gairai-seibutsu-ho. http://www.env.go.jp/nature/intro/index.html. Accessed 10 Sep 2010 (in Japanese)
Molineux CJ, Fentiman CH, Gange AC (2009) Characterising alternative recycled waste materials for use as green roof growing media in the U.K. Ecol Eng 35:1507–1513
Nagase A, Dunnett N (2010) Drought tolerance in different vegetation types for extensive green roofs: effect of watering and diversity. Landsc Urban Plan 97:318–327
Natural England (2007) Living roofs. ISBN978-1-84754-016-4
Newbound M, Mccarthy MA, Lebel T (2010) Fungi and the urban environment: a review. Landsc Urban Plan 96:138–145
Oberndorfer E, Lundholm J, Bass B, Coffman RR, Doshi H, Dunnett N, Gaffin S, Köhler M, Liu KKY, Rowe B (2007) Green roofs as urban ecosystems: ecological structure, functions, and services. Bioscience 57(10):823–833
Öckinger E, Dannestam Å, Smith HG (2009) The importance of fragmentation and habitat quality of urban grasslands for butterfly diversity. Landsc Urban Plan 93:31–37
Ohno T, Yamamoto S, Maenaka H (2006) Evapotranspiration from sedum mat and Cynodon turf. Jpn Soc Reveg Technol 69:431–436. (in Japanese and with English summary)
Palomo E, Barrio D (1998) Analysis of the green roofs cooling potential in buildings. Energy Build 27:179–193
Pearlmutter D, Rosenfeld S (2008) Performance analysis of a simple roof cooling system with irrigated soil and two shading alternatives. Energ Buildings 40:855–864
Robinson SL, Lundholm JT (2012) Ecosystem services provide by urban spontaneous vegetation. Urban Ecosyst 15:3. https://doi.org/10.1007/s11252-012-0225-8
Rouquette JR, Thompson DJ (2007) Patterns of movement and dispersal in an endangered damselfly and the consequences for its management. J Appl Ecol 44:692–701
Rudd H, Vala J, Schaefer V (2002) Importance of backyard habitat in a comprehensive biodiversity conservation strategy. Restor Ecol 10:368–375
Sandström UG, Angelstam P, Mikusiński G (2006) Ecological diversity of birds in relation to the structure of urban green space. Landsc Urban Plan 77:39–53
Sattler T, Duelli P, Obrist KM, Arlettaz R, Moretti M (2010) Response of arthropod species richness and functional groups to human habitat structure and management. Landsc Ecol 25:941–954
Savard J-PL, Clergeau P, Mennechez G (2000) Biodiversity concepts and urban ecosystems. Landsc Urban Plan 48:131–142
Schadek U, Strauss B, Biederman R, Kleyer M (2009) Plant species richness, vegetation structure and soil resources of urban brownfield sites linked to successional age. Urban Ecosyst 12:115–126
Schrader S, Böning M (2006) Soil formation on green roofs and its contribution to urban biodiversity with emphasis on collembolans. Pedobiologia 50:347–356

Small EC, Sadler JP, Telfer MG (2003) Carabid beetle assemblages on urban derelict sites in Birmingham. UK J Insect Conserv 6:233–246

Smith RM, Warren PH, Thompson K, Gaston KJ (2006) Urban domestic gardens (VI): environmental correlates of invertebrate species richness. Biodivers Conserv 15:2415–2438

Wilkinson SJ, Reed R (2009) Green roof retrofit potential in the central business district. Prop Manag 27:284–301

Wood BC, Pullin AS (2002) Persistence of species in a fragmented urban landscape: the importance of dispersal ability and habitat availability for grassland butterflies. Biodivers Conserv 11:1451–1468

Part III
Towards Ecological Landscape Ecology and Planning for Future Cities

Chapter 11
Synergies in Urban Environmental Policy: Ecosystem Services and Biodiversity Co-benefits in São Paulo City, Brazil

Raquel Moreno-Peñaranda

Abstract While cities consolidate as centers of socio-economic development across the globe, two crucial yet interrelated urban environmental challenges remain largely unsolved. First, how to make cities more sustainable and hence overturn urban drivers of global environmental change. Second, how to increase well-being for city dwellers by promoting fair access to the benefits of local ecosystems and biodiversity. This is a tremendous challenge requiring innovative approaches to urban environmental policy that connect multiple goals at different scales to improve effectiveness and engagement. The urban co-benefits approach aligns local environmental goals city governments are particularly sensitive to (e.g. air pollution reduction) to global ones not easily prioritized in the urban agenda (e.g. climate change mitigation). Including ecosystem services and biodiversity co-benefits could lead to more comprehensive and effective urban policy actions, since (peri) urban ecosystems and their biodiversity can provide multiple well-being *and* sustainability benefits—for example, carbon management, microclimate regulation, biodiversity protection, recreation, food production, or community participation. This chapter explores four different initiatives relevant to ecosystems and biodiversity co-benefits in São Paulo city, Brazil, in terms of their potential for fostering synergies in urban environmental policy conducive to improving both well-being and sustainability. The study reveals that initiatives combining a multi-scale, multi-stakeholder, and multi-sectoral approach with a strong social inclusion component have greater co-benefit potential. Notwithstanding, obstacles for materializing such win–win approaches exist, from the planning to the implementation level, and include financial, regulatory, and managerial aspects. The chapter draws lessons from the city of São Paulo, which can be relevant to other (mega)cities confronting similar challenges.

R. Moreno-Peñaranda (✉)
Center for Research and Development of Higher Education, The University of Tokyo, Tokyo, Japan
e-mail: rmp@g.ecc.u-tokyo.ac.jp

K. Ito (ed.), *Urban Biodiversity and Ecological Design for Sustainable Cities*,
https://doi.org/10.1007/978-4-431-56856-8_11

Keywords Urban development · Climate change · Sustainability · Human well-being · Green house gas (GHG) emissions inventories · Protected areas · Urban agriculture

11.1 Introduction

The world is becoming inexorably urban. Over half of the world population lives nowadays in urban areas, from 34% in 1960 (UNDESA 2015). Cities across the globe face two crucial, interrelated environmental challenges: how to become more sustainable (that is, how to reduce their ecological footprints) *and* how to ensure the well-being of their inhabitants (from clean air and safe water to fair access to local ecosystems services and biodiversity). Tackling greenhouse gas emissions related to urban sectors (and their associated impacts on urban air pollution), together with environmental degradation and biodiversity loss linked not only to urban expansion but to urban consumption, are perhaps among the most paramount challenges for the future of our rapidly urbanizing world. Understanding the role of urban ecosystems and biodiversity for the provisioning of ecosystem services leading to decreasing the environmental impacts of cities inside and outside their boundaries is crucial to effectively address these challenges.

The Sustainability-Well-Being Nexus in Urban Development
Urban areas benefit from a myriad of different types of ecosystem services inside and outside their boundaries. Ecosystem services refer to the variety of goods and services provided by ecosystems that are essential for human life, including provisioning, regulating, supporting, and cultural services (Millennium Ecosystem Assessment (MA) 2005; TEEB 2010). From a global perspective, the sustainability of an increasingly urban world ultimately depends on reducing cities' ecological footprints—that is, the use of ecological goods and services outside city boundaries necessary for the functioning of urban areas (Rees and Wackernagel 2008). Urbanization processes contribute directly to environmental loss by appropriating and/or degrading natural ecosystems for development (McKinney 2002), which in turn has implications for biodiversity, upon which the functionality of ecosystems ultimately depends (Elmqvist et al. 2003). Biodiversity is especially affected by urban expansion: 29 out of 825 world eco-regions (home to 213 endemic terrestrial vertebrates) are more than 1/3 urbanized (Mcdonald et al. 2008). Urban consumption poses further sustainability challenges. Urban consumption patterns are among the leading causes of appropriation of ecosystem services (MA 2005). While cities occupy about 2% of the Earth's surface, they consume over 75% of its resources (UNDESA 2015). It has long been established that, as cities become more affluent, they tend to have bigger ecological footprints, thus posing risks of overexploitation and unsustainable use of natural resources and biodiversity (Folke et al. 1997). Around 80% of worldwide energy consumption and green house gas (GHG) emissions have been associated with urban activities (Hoornweg et al. 2011).

Besides sustainability concerns related to urbanization and urban consumption, which have a more global (or regional) dimension, another key environmental challenge for cities occurs at the local level, namely their ability to ensure sufficient and equitable access to the environmental goods and services necessary for the well-being of their inhabitants. Clean air, safe water, fresh foods, protection against heatwaves and flooding, proper spaces for recreation and spiritual fulfillment are, to name just a few, central issues in the urban sustainability-well-being nexus. Air quality has become a major urban environmental challenge in the last decades, especially in cities of developing countries, but also for industrialized ones (Mayer 2009; Baró et al. 2014): air pollution has been linked to severe health impacts among urban residents, involving premature deaths (including infant deaths), chronic bronchitis cases, and person days of work loss. Proper actions taken at the city level can significantly reduce the impacts of urban air pollution (Cifuentes et al. 2001). Likewise, although in a more indirect way, urban biodiversity is also of great importance for securing human well-being; for many urban residents, enjoyment and appreciation of nature will only come from urban spaces (Muller and Werner 2010).

From Climate Co-benefits to Broader Synergies in Urban Environmental Policy

Linking global and local dimensions in urban environmental policy can generate synergies conducive to sustainability *and* well-being outcomes. While urban environmental challenges have traditionally had a predominantly local policy dimension (e.g. air pollution, waste management) cities are currently confronted with global environmental problems—such as climate change—while addressing urban developmental challenges (McCormik et al. 2013). Coupling greenhouse gas mitigation strategies in cities with urban air pollution reduction can generate sustainability and well-being co-benefits deriving from improved environmental quality and development opportunities (Puppim de Oliveira 2013). Win–win climate co-benefits situations can emerge in direct ways-for example, GHG mitigation can lower fossil fuel combustion conducive to reducing air pollution and thus improve health outcomes (Darby and Kinney 2007). Co-benefits can also result through more complex processes mediated through behavioral and cultural factors. For instance, health improvements in urban populations can result from changes in transportation policies leading to increased physical activity which were initially designed for mitigation purposes (Woodcock et al. 2009) or from reduced meat consumption resulting from innovative food policies tackling urban land use change (Wilkinson et al. 2009). The climate co-benefits literature in cities has predominantly looked at the transport, energy, or waste sectors, while the role of land use (Shima et al. 2011) or, more generally, urban ecosystems (Baró et al. 2014) or biodiversity (Raymond et al. 2017) has not been yet fully explored.

The Role of Urban Ecosystems in Sustainability and Well-being: Beyond GHG Mitigation

Early studies have shown that urban ecosystems could be key for aligning the sustainability and well-being dimensions of urban environmental challenges related to climate change and biodiversity loss. If properly designed and managed, urban ecosystems can contribute to climate change mitigation (Nowak and Crane 2002; Baró et al. 2014). Urban vegetation, mainly trees and shrubs, directly sequesters and stores atmospheric carbon through photosynthesis, and urban soils store organic carbon from litterfall (until it is returned to the atmosphere by decomposition). The capacity of urban green space (vegetation and soils) to offsetting carbon emissions is relatively small when compared to other urban sectors (e.g. energy, transport) but not negligible. For example, in the city of Sacramento, California, data showed that urban forests contained 1.8% of regional carbon (McPherson 1998), or 0.5% in the city of Seoul (Jo 2002). Approximately 800 million tons of carbon were estimated to be stored in U.S. urban forests, corresponding to $22 billion equivalent in control costs (Nowak and Crane 2002). The importance of enhancing carbon storage capacity of urban areas through terrestrial carbon sequestration (bio-sequestration by forests and soils) has been addressed by the international climate change policy agenda for over a decade (IPCC 2007). Urban soils can sequester large amounts of carbon, despite regional variations and land-cover distributions (Pouyat et al. 2006). There is potential for urban tree plantings to be cost-effective in carbon credit markets (McHale et al. 2007). Moreover, effects of urban vegetation in climate change mitigation can happen through reducing cooling demand (by shading and evapotranspiration) and heating demand (by wind speed reduction), thereby reducing carbon emissions associated with urban energy use. For example, urban trees can provide cooling services for buildings, therefore reducing air conditioning demand (up to 30%) and related electricity needs which can indirectly contribute to GHG release (Nowak and Crane 2002).

Beyond mitigation benefits, urban ecosystems have more recently been recognized for their capacity to improve local well-being by contributing to air pollution reduction. Urban vegetation can trap pollutants on the leaves or indirectly improve air quality by regulating microclimates (e.g. by changing air temperature, modifying wind speed, changing the albedo, etc.) (Yang et al. 2005; Leung et al. 2011). Urban ecosystems (terrestrial and aquatic) can ameliorate the so-called Urban Heat Island effect (UHI) (Graves et al. 2001; Shin and Lee 2005) while in turn reducing UHI-related health hazards for urban residents resulting from heat stress and increased concentrations of temperature-dependent air pollutants (Solecki et al. 2005). In addition to mitigation and air quality benefits, urban ecosystems can also provide adaptation benefits, for example, flood control (Gill et al. 2007; Wilby and Perry 2006) and overall well-being and socio-economic benefits, from noise reduction to increased housing values (Groenewegen et al. 2006).

Productive landscapes such as urban and peri-urban agriculture can also contribute to attaining coupled sustainability and well-being benefits relevant for tackling climate change and biodiversity loss. Agricultural spaces in and around urban areas can facilitate access to fresh produce and traditional medicines, contribute to carbon management, and provide cultural services, community building, or innovative employment opportunities (Pearson et al. 2010; La Greca et al. 2011; Nicholls et al. 2020). Some approaches to urbanization highlight the importance of urban and peri-urban agriculture for sustainability and well-being, encouraging more compact city forms where the urban fringe is used for agricultural production, thus decreasing the need for industrialized production, extensive packaging, and lengthy distribution (Viljoen 2005). Besides being a source of ecosystem services, urban agricultural spaces, from home gardens to urban farms or agricultural parks, can also contribute to fostering (agro)biodiversity (Linares 1996; Galluzi et al. 2010; Quigley 2011; Moreno-Peñaranda 2012). Likewise, sustainable aquaculture and proper coastal fisheries management can also contribute positively to urban biodiversity while providing local foods, creating employment, and fostering technological innovation while reducing ecological footprints (Costa-Pierce et al. 2005; Yanagi 2005).

Furthermore, urban ecosystems can play an important role in protecting biodiversity within city boundaries and beyond. Rich biodiversity can be found within cities across continents and latitudes; some cities are located within areas of remarkably high biodiversity (so-called biodiversity hotspots) and many cities' protected areas lay within their borders and/or connect fragmented ecosystems, which increases ecological functionality and thus maximizes ecosystem services (SCBD 2012). Synergies are expected from aligning local biodiversity initiatives with national/global actions (Pisupati 2007). For example, the implementation of the Convention on Biological Diversity (CDB) at the city level presents opportunities for achieving local well-being and global sustainability outcomes (Puppim de Oliveira et al. 2011). Decision X/22 (adopted in 2010 by the Parties) recognized for the first time the importance of local actions for biodiversity as a support for countries' National Biodiversity Strategy and Action Plans (NBSAPs), which are the primary instruments used by national governments for implementing the CBD. Decision X/22 established the creation of the so-called Local Biodiversity Strategies and Action Plans (LBSAPs)—the local equivalent of NBSAPs—as a guiding strategy, complemented by specific actions to be adopted by cities in order to achieve optimal and realistic governance and management of local biodiversity and ecosystem services. Depending on the needs, capacities, and contexts of cities, biodiversity actions can vary, yet their effectiveness depends on integrating biodiversity issues across local development process, in a multi-scale, multi-sectoral, and multi-stakeholder approach (Avlonitis et al. 2013). Although local biodiversity actions tend to embrace processes within city boundaries, measures to reduce ecological footprints are becoming increasingly common in the local environmental agenda.

Yet designing urban policy instruments that improve local well-being *and* global sustainability in such comprehensive, multi-scale, synergic way can be an arduous task for cities, as different urban sectors and stakeholders have divergent interests and lack an integrative, holistic vision regarding target areas for action and scales of the policy intervention. The goal of this study is to identify existing strategies in cities which can potentially be conducive to generating extended sustainability and well-being co-benefits beyond climate, through the inclusion of ecosystem services and biodiversity components. The study combines the ecosystems services framework of the Millennium Ecosystem Assessment (MA 2005)—which sees (urban) ecosystems as providers of a variety of services that contribute to biodiversity, local well-being, and sustainability—with the urban development with co-benefits approach (Puppim de Oliveira 2013), which focuses on how to align climate, environmental, and development objectives in cities. By exploring the drivers, targets, actors, and policy instruments of existing urban strategies, together with the main barriers preventing their success and opportunities for further improvement, valuable lessons can be drawn for urban policy-making conducive to tackling both global and local dimensions of environmental problems.

11.2 Methodology

Brazil is a predominantely urban country, with over 84% of its population living in urban areas (Farias et al. 2017). This study focuses on the city of São Paulo, the largest metropolis in South America, and a pioneering city in Brazil regarding policy efforts to reduce local GHG emissions and, more generally, the implementation of innovative environmental policy initiatives at the local level in connection to broader sub-national, national, and international processes. Despite urbanization pressures, São Paulo still hosts within its municipal boundaries remnants of Atlantic Forest with significant biodiversity features, although some of the species are currently endangered or critically endangered. The city's average Ecological Footprint is 4.38 global hectares per capita[1] (WWF 2012), which is 49% larger than the Brazilian average, yet with a similar pattern of footprint composition among the various categories that compose the indicator, with a strong demand for grazing and agriculture land, as well as forests. Although São Paulo city (and its metropolitan area) is a powerful economic and financial hub in the country (also regionally and globally) inequality, poverty, and exclusion still affect a significant part of the local population. The city had around 12 million inhabitants in 2017 according to the Brazilian Institute of Geography and Statistics-IBGE, with over 20% of its residents living in 1241 slums or *favelas*. Challenges related to pollution, (irregular) urbanization, or food security are relevant for the local environmental agenda.

Through field visits and interviews with local policy-makers and relevant stakeholders in the environmental sector conducted in August 2011, a few selected initiatives implemented at the city level were identified according to their potential for coupling climate, ecosystem services, and biodiversity co-benefits. Interviewees included officers from different departments within the municipal government, from managers to personnel working in the field with local communities, together with institutions collaborating with national/international partners, as well as practitioners and community members. The selected initiatives include long and newly stablished ones at the time, with different target populations as well as diverse time and geographical scales, and varied budgetary and sectoral scopes. For each

[1]Ecological footprint, measured in global hectares per capita, refers to the equivalent total world area that would be needed if the entire population of the planet had the same consumption patterns as that particular city. In the case of São Paulo city, almost two and a half planets would be needed.

initiative, the study presents an overview, including main outcomes as well as main implementation challenges, followed by a brief analysis of their potential to contributing to ecosystem services and biodiversity co-benefits, which are summarized in a table comparing the selected initiatives with each other. The concluding section draws lessons from the city of São Paulo, which can be relevant to other (mega)cities confronting similar challenges.

11.3 Selected Initiatives Related to Climate, Ecosystems Services, and Biodiversity in São Paulo City

Four main local initiatives relevant for climate, ecosystem services, and biodiversity co-benefits were identified through the interviews. These were: (i) the City greenhouse gas (GHG) emissions inventory; (ii) the São Paulo City Green Belt Biosphere Reserve (GBBR); (iii) the Municipal Areas of Environmental Protection (*Areas de Protecão Ambiental Municipal*-APAs); and (iv) the Clean Agriculture (*Agricultura Limpa*) program.

11.3.1 Greenhouse Gas (GHG) Emissions Inventory: Land Use Change and Mitigation

São Paulo has been a pioneering city in Brazil regarding policy efforts to reduce local GHG emissions. In 2005 the city completed an inventory of local GHG emissions (São Paulo City 2005). Prior to this date, only two other GHG emissions inventories existed in Brazil that followed IPCC guidelines: one for the City of Rio de Janeiro (1998), and the national one (1994). Values of the 2005 inventory have been used as baseline for establishing reduction targets in the city (e.g. São Paulo City's Climate Change Policy approved in 2009 by the City Council, established GHG emissions reduction targets of 30% by 2012 from 2005 levels). The 2005 São Paulo's inventory measured GHG emissions by urban sectors—transport, energy, waste management, and land use change. The main types of land use in the city can be seen in Fig. 11.1.

According to the GHG emissions inventory, land use change in the city during the period 1997–2001 showed decreases in agricultural land use (−45 ha or 2.22% reduction), shrub areas (−1268 ha or 14% reduction), tree plantations (−151 ha or 3.83% reduction), and natural forests (−257 ha or 20% reduction). Land use types that showed increases were mainly urbanization (+1660 ha or 57% increase) and mining (+30 ha or 0.4% increase). Most of these land use conversions took place in the southern part of the city, where some of the most environmentally valuable areas are located, including two environmental protection areas (see Fig. 11.2). During the same time period, CO_2 emissions resulting from variation of carbon stocks related to

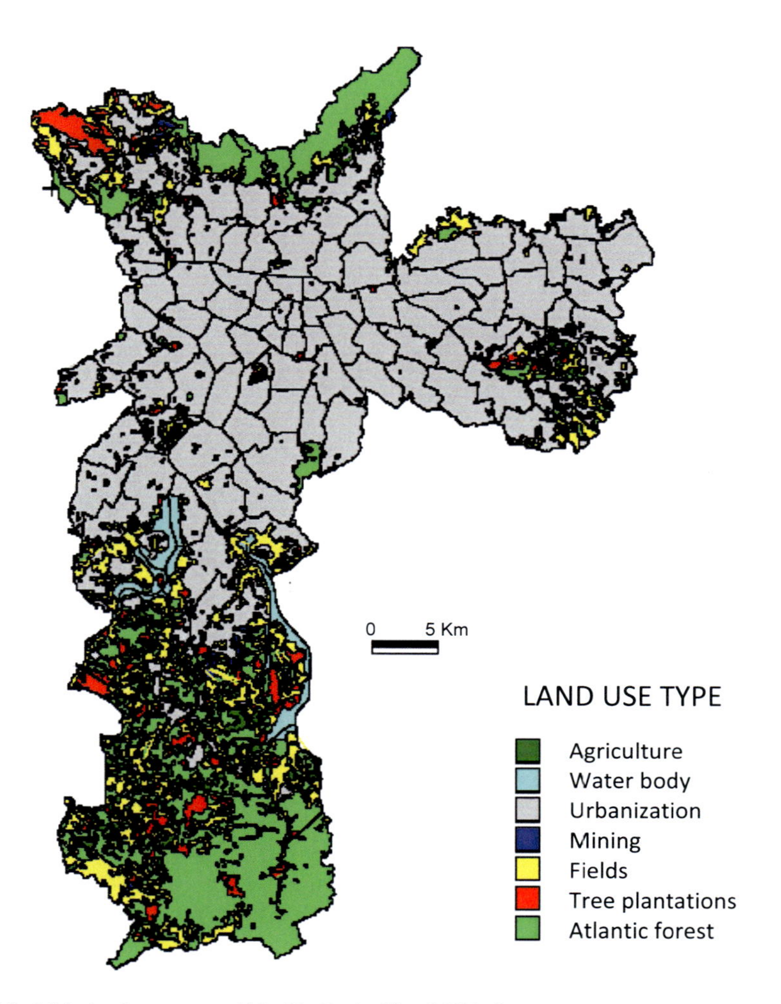

Fig. 11.1 Main land use types within São Paulo City (2005). Source: Department of Greenery and the Environment (SVMA)

changes of plant biomass cover were 6.7 kt CO_2 in plantations, 13.1 kt CO_2 in shrub areas, and 29.0 kt CO_2 in natural forests (totaling 48.7 kt CO_2 emitted). Regarding soils, only CO_2 emissions related to acidity correction with calcium carbonates were measured by the inventory. In 2003, 2.58 Gt of CO_2 were estimated as emissions resulting from the use of this practice in agricultural areas within São Paulo city. Overall, CO_2 emissions related to land use change for 2003 were estimated to be 51.3 Gt of CO_2 eq, with forests accounting for 57%, shrubs for 25%, tree plantations for 13%, and agricultural soils for 5% of these emissions.

Beyond land conversion, the role of agricultural activities in GHG emissions was also estimated in the inventory, including both energy consumption by agricultural

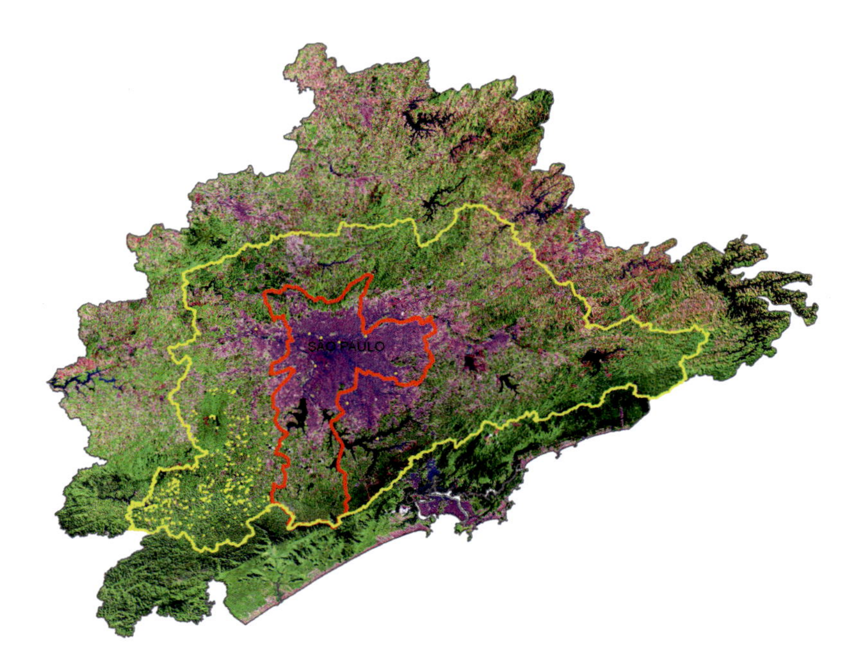

Fig. 11.2 Satellite map of the Green Belt Biosphere Reserve (GBBR) showing main land use types (2009): purple shades indicate urbanization; green shades vegetation, and dark shades water courses/bodies. São Paulo city is located at the center of the GBBR (red border), surrounded by its metropolitan area or Grande São Paulo (yellow border). Source: GBBR

activities and emissions related to enteric fermentation and waste management of farm animals. Emissions of GHG resulting from energy consumption were negligible (0.03%) when compared to that of transport (78.5%), residential (9.7%), industrial (7.2%), or commercial (2.6%) sectors. Regarding farm animals, total emissions were estimated at 0.783 Gg CO_2 eq (0.037 Gg of CH_4), with cows accounting for 77.2% followed by horses (16.9%) and chickens (3.6%).

After the completion of the inventory, the City engaged in an ambitious plan of local GHG emissions reduction. The deployment of methane capture mechanisms in the two main local waste disposal facilities led to a 20% reduction in total GHG emissions. Moreover, under the auspices of the clean development mechanism (CDM), the city sold over 800,000 local carbon credits resulting from the methane capture improvements in the Futures Market for 34 million Brazilian *Reais* (approximately 34 million USD) which, according to the interviewees, have been partly used for a variety of local environmental protection projects. Following the first Inventory, a second one took place in 2009, when a new Climate Change Municipal Policy was also enacted. The results of the second Inventory showed an increase in total GHG emissions (lead by the energy and waste sectors) above target levels, compelling the City to devise further mitigation actions (Sotto et al. 2019).

Opportunities and Barriers of Urban GHG Emissions Inventories as a Co-benefits Instrument
Local GHG emissions inventories that include land use change have the potential to provide relevant baseline data to better understand the role of urban ecosystems in the generation of climate co-benefits (e.g. carbon management by soils and vegetation). Although the São Paulo GHG Inventory was a pioneer effort to assess the role of local terrestrial ecosystems in local GHG emissions, it presents serious limitations regarding its capacity to highlight connections to broader ecosystem services provided by urban green areas beyond mitigation (e.g. pollution control, water filtration) and even more so to make connections to the biodiversity value of these areas. Moreover, when compared to other sectors such as transport, energy, or waste, the potential contribution of urban ecosystems to mitigation is markedly lower. This could lead to downplaying the need for designing mitigation instruments that include urban ecosystems as opposed to interventions in main contributing sectors (e.g. energy, waste).

Another limitation of local GHG emissions inventories is that they measure direct emissions related to land use change, but they do not quantify the emissions *saved* as a result of the provisioning of climate services by urban ecosystems, such as mitigation of the heat island effect resulting in lower cooling energy needs, or effects on wind patterns resulting in lower heating needs, to name a few. Moreover, only terrestrial ecosystems were considered in São Paulo's Inventory. Water bodies, including the two big reservoirs *Billings* and *Guarapiranga* located in the southern area of the city, as well as the main urban watercourses, were not assessed in terms of their climate services and biodiversity or their GHG emissions. This is especially relevant for the river system of São Paulo, particularly the Tietê river that crosses the city carrying significant loads of pollutants including organic matter, thus with the potential of significantly contributing to local GHG emissions.

Conversely, GHG emissions mitigation actions that were taken in the local waste management sector supported by inventory's findings (e.g. CDM projects) generated a significant inflow of financial resources to the City, which in turn reverted to local environmental actions (e.g. the Municipal Protection Areas mentioned in Sect. 3.3). Thus the Inventory was able to, indirectly, bring about biodiversity and ecosystems services' co-benefits for the City.

11.3.2 The São Paulo City Green Belt Biosphere Reserve (GBBR)

São Paulo city still hosts within its municipal boundaries significant biodiversity features. According to local inventories (Malagoli et al. 2008) the city hosts at least 3246 catalogued species, including over 2000 species of plants, 410 birds, and 92 mammals, some of them endangered and critically endangered. The city of São Paulo is located within the *Mata Atlântica* (Atlantic Forest) biome, a subtropical type

Table 11.1 Overview of the ecological and socio-economic relevance of the GBBR including the São Paulo city area (Rodrigues 2018)

Area	1,611,710 hectares
Urban land use	220,279 hectares
Designation	UNESCO: designation within the Biosphere Reserve of the Atlantic Forest (1994); individual designation (2017)
Management	Forestry Institute (*Instituto Florestal*) and São Paulo State Environment Secretary
Population	25 million (12% of Brazil's population)
Municipalities	78
Forested area	614,288 hectares
Plantation area (non-native tree species)	118,889 hectares
Main ecosystems	Dense ombrophile semideciduous Atlantic Forest (*Floresta Atlântica Ombrófila Densa and Semidecidual*), savanna (*cerrado*), contact savanna-Atlantic Forest, natural grasslands (*campos naturais*) high altitude forests, (*florestas de altitude*), sandbanks (*restingas*), mangroves
Watersheds	2 complete watersheds (*Alto Tietê* and *Baixada Santista*); 2 partial watersheds (*Sorocaba*, *Médio Tietê,* and *Piracicaba*)
Drinking water	8 systems supplying drinking water to over 20 million urban residents
Protected area	220,422 ha under "Conservation" designation + 50,000 ha under "Sustainable Use" designation
GDP	20% of Brazil's GDP is generated within GBBR area

of forest in South America spreading mostly in proximity to Brazil's Atlantic coast, originally extending from southern to northern latitudes of the country, yet severely deforested nowadays. Even though in the central urban area of São Paulo practically all the original Atlantic Forest vegetation has been replaced by urbanized or secondary formations, biodiversity still thrives in the southern (and some northern) parts of the City.

São Paulo city is located at the center of the so-called Green Belt Biosphere Reserve (GBBR) (*Reserva da Biopsfera do Cinturão Verde*), an area of high environmental value included in the UNESCO Man and Biosphere (MAB) program that extends over the surroundings of São Paulo metropolitan area as a true green belt (Fig. 11.2 and Table 11.1). The MAB program started in the early 1970s, when the international sustainable development agenda was still non-existent. Biosphere reserves aim at combining conservation and developmental goals, and many of them include urban areas within their boundaries—the so-called Urban Biosphere Reserves. According to GBBR's officials, the Reserve is "the result of a local mobilization process, with strong demands by the citizens to protect the local environment, especially from the threats of urbanization". Yet Biosphere Reserves are designated by national governments with the aim of experimenting in situ the integrated management of terrestrial and aquatic ecosystems and their biodiversity at a regional scale through the application of the MAB program concepts. Biosphere

Reserves are managed by the respective countries and all together constitute a Global Network of Biosphere Reserves. According to the rules governing the Reserves (UNESCO 2003), their functions include: (a) contributing to the conservation of landscapes, ecosystems, species, and genetic diversity; (b) fostering human and economic development in a sustainable way socially, culturally, and ecologically, and (c) supporting demonstrative projects, capacity building, and educational activities in environmental issues, together with research and monitoring of local, regional, national, and global dimensions of sustainable development.

In 2001, the Urban Group of the MAB program was established in order to create an appropriate framework for the creation of Biosphere Reserves in urban areas, to develop integrative urban and peri-urban sustainable development projects (UNESCO 2006). Urban Biosphere Reserves (UBR) have been initiated/established in cities of countries across the globe, including Cape Town, New York, Rome, Dar-es-Salaam, and Seoul, among others. Since its creation in 2001, the Urban component of the MAB program fosters the creation of new UBRs, while it supports existing UBRs. In Brazil, in addition to São Paulo, two other cities are involved in UBRs connected to the Atlantic forest Biome: Recife, in the northern state of Pernambuco, and Florianopolis, in the southern state of Santa Catarina. Urban Biosphere Reserves are defined as those Reserves "containing or adjacent to important urban areas, where ecological, socio-economic and cultural matters are greatly influenced by urban factors." UBRs are meant to "mitigate urbanization pressures and improve urban and regional sustainability."

Tackling Environmental and Socio-Economic Goals: Youth Eco-education
GBBR areas with the highest ecological value are for the most part located in the peri-urban space, which in the city of São Paulo (and its metropolitan area) often corresponds to areas with the highest level of social exclusion (Izique 2003). Acording to GBBR officials, the Reserve has tackled this challenge mainly through the so-called Youth Environment and Social Integration Program (*Programa de Jovens, Meio Ambiente e Integracão Social*—PJMAIS). The program, inspired by the FAO, was launched in 1996 with the support of UNESCO, combining environmental sustainability goals with social development objectives. PJMAIS offers eco-professional training for socially vulnerable youth between 15 and 21 years old residing within GBBR peri-urban areas. PJMAIS Teaching Centers (*Núcleos de Educacão Eco-profissional*) spread across GBBR's peri-urban area and are established through partnerships between regional (state) and local governments, community groups, and local stakeholders, while the executive branch of the GBBR manages technical, pedagogical, and marketing issues. Other partners not directly involved in the management but providing support at different levels include UNESCO, the Ministry of the Environment of Brazil, the Forestry Foundation (*Fundacão Florestal*) of São Paulo State, the United Nations Foundation, and the World Bank, among others. At the local level, training centers are managed by local governments in partnership with the private sector.

Reports from the GBBR show that, during the first 10 years of PJMAIS (1996–2006), over 1300 youth attended the program, obtaining standard secondary education degrees alongside eco-professional training in areas such as agriculture and forestry, sustainable tourism, recycling and art, and artisanal food processing. The program seeks to fully integrate the youth into society, using a teaching philosophy based on "the values of solidarity, self-esteem, citizenship, and appreciation for nature and fostering the development of entrepreneurial capacities for sustainable business development".

Sub-global Assessment (SA) of the Millennium Ecosystem Assessment (MA)
As part of the Millennium Ecosystem Assessment (MA) process, the so-called Sub-global Assessments (SGA) were established so that the same methodology used in the global evaluation could be used to evaluate particular regions (Millennium Ecosystems Assessment (MA) 2005). A total of 35 areas were selected worldwide as SGA sites from expressions of interest emerging from local authorities, and according to the relevance of regional data for contributing to the global assessment process. The GBBR was established as a SGA site mostly because of its geopolitical and socio-environmental significance, as an area heavily influenced by a large metropolis (Pires et al. 2010). The SGA process is a complex multi-institutional effort with a long execution timeline. The main findings of the SGA of the GBBR (Rodrigues 2018) are summarized on Table 11.2. Ecosystem services especially relevant for the co-benefits approach are those related to climate regulation, carbon sequestration, and pollution control. Vegetation and water bodies of the GBBR have a clear effect in mitigating the heat island effect originated in the densely urbanized areas (Fig. 11.3). Likewise, soils and vegetation play a key role in managing regional carbon pools. Effects on air quality are mediated through the capacity of above ground vegetation to retain pollutants.

Barriers and Opportunities of the GBBR as a Co-benefits Instrument
The effectiveness of the GBBR was assessed in 2005 according to the Indicators of Strategic Implementation established by the so-called UNESCO's *Seville Strategy* (UNESCO 2003). The main weaknesses identified at the GBBR were related to: (1) insufficient participation of the private sector; (2) lack of appropriate communication strategy; (3) funding constraints; (4) need for updated, stronger scientific and technical data; (5) lack of common vision among stakeholders and managers; and (6) effective stakeholder participation. Despite these challenges, interviewees point out that there are opportunities for further strengthening the effectiveness of the GBBR, including potential for inter-institutional and multi-stakeholder integration, and alternative management structures for advancing regional governance, provide an adequate forum for enhanced participation, transparent decision-making and conflict resolution.

11.3.3 Municipal Areas of Environmental Protection (APAs)

São Paulo has two Municipal Areas of Environmental Protection (*Areas de Protecão Ambiental Municipal*-APAs) (Fig. 11.4). The APA *Capivari-Monos* was established in 2001 (municipal Law 13,136 of 9 June 2001) and occupies a total area of 25,100

Table 11.2 Main ecosystem services provided by the GBBR including São Paulo city area (Rodrigues 2018)

Ecosystem service	Description and relevance for the GBBR
Supporting services	
Maintaining ecological processes and biodiversity	The Atlantic Forest is one of the most biodiverse biomes of the planet. The protection of its biodiversity relates to both ethical and human well-being concerns. The forests of the GBBR also act as effective ecological corridors. Soil formation, nutrient cycling, carbon absorption, pollination, resilience ecosystem services have been identified, among others
Provisioning services	
Water conservation (surface and underground)	Water bodies located within the GBBR provide drinking water for over 20 million people. If these resources are compromised, the impacts for the local population would be enormous. The preservation of forested areas is also directly linked to the quality of the water provided by the GBBR, which in turn has strong economic implications
Genetic resources (including pharmacological)	The Atlantic Forest biome where the GBBR is located hosts significant biodiversity resources which could provide valuable bio-materials for human well-being, including substances for pharmacological use and other uses of high economic importance
Food security	The GBBR hosts significant food production by smallholders, and it is becoming a reference in organic production at the national level. Moreover, agricultural activities in the urban and peri-urban area serve as a buffer and contribute to control urban expansion
Forest resources (timber and non-timber products)	Plantations located within the GBBR provide timber resources which are relevant for the local economy of São Paulo State. Natural forests also provide some timber products, although their role is more directly linked to biodiversity preservation, local well-being, and community involvement
Regulating services	
Climate regulation	GBBR ecosystems have a direct effect on regional climate regulation and mitigate the heat island effect originating from urbanized areas. They also affect local rain patterns, which could in turn be linked to flooding patterns in the urbanized areas (see Fig. 11.3)
CO_2 sequestration and pollution control	Secondary forests and tree plantations have an impact on carbon management at the local/regional level, and act as filters of the air pollution generated in urbanized areas. The total carbon stock in these types of forests reaches 23 million tons in CO_2 equivalents (C-CO_2).
Soil conservation and flood control	Prevents soil erosion, stabilizes areas sensitive to slides, and provides a permeable surface that minimizes flooding and hence protects urban dwellers and reduces damages to the urban infrastructure
Cultural services	
Leisure, recreation, esthetics	The city of São Paulo has very limited green space. Green peri-urban areas like the GBBR provide vital access to nature

(continued)

Table 11.2 (continued)

Ecosystem service	Description and relevance for the GBBR
	for urban dwellers (that otherwise will have to travel long distance to access nature) and thus have a positive effect on their physical and psychological well-being. Moreover, the area contributes to the esthetic value of the region, as it contains forests, coastal areas, mangroves, beaches, rural settlements
Historical and cultural value	The GBBR contains areas of significant historical value in Brazilian history, both for indigenous populations and after European colonization
Sustainable tourism	Many areas within the GBBR have a great potential to consolidate sustainable tourism options

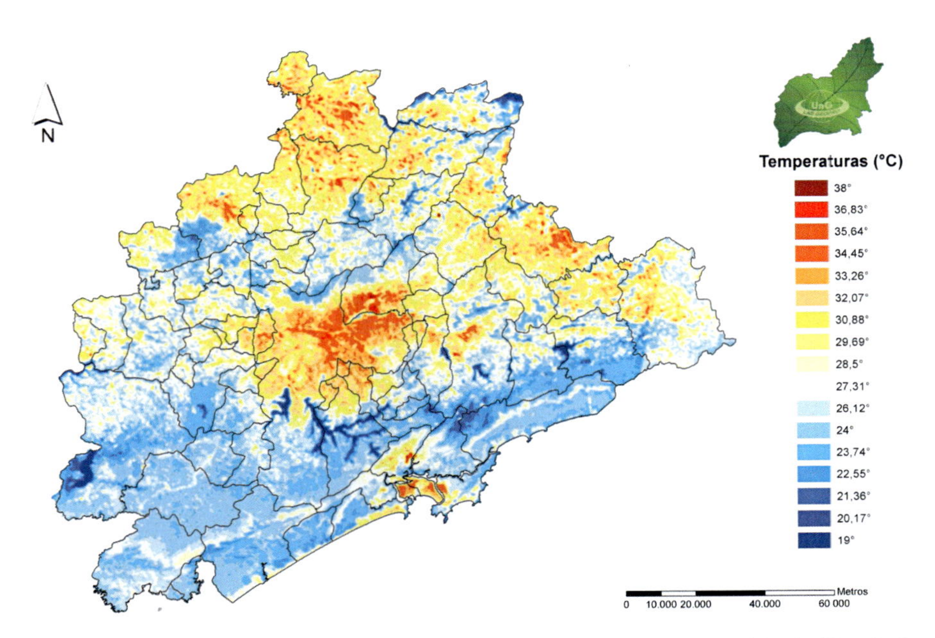

Fig. 11.3 Average temperatures within the GBBR showing the urban heat island effect (red, orange, and yellow shades) associated with urban land use (São Paulo city is located at the center of the image) Source: *Instituto Florestal, 2006*

hectares. The most recent APA, *Bororé-Colônia*, was created in 2006 (municipal Law 14,162 of 24 May 2006) and extends over 9000 hectares. Both APAs are listed as conservation category V by IUCN, which refers to areas that combine conservation with sustainable use of natural resources. The two APAs together occupy over 16% of the municipal area of the City and contain its two major drinking water reservoirs, providing 30% of the water consumed in São Paulo. The APAs are home to 70,000 urban residents, including original indigenous populations, long-established small family farmers (of European and Japanese descent), and rapidly

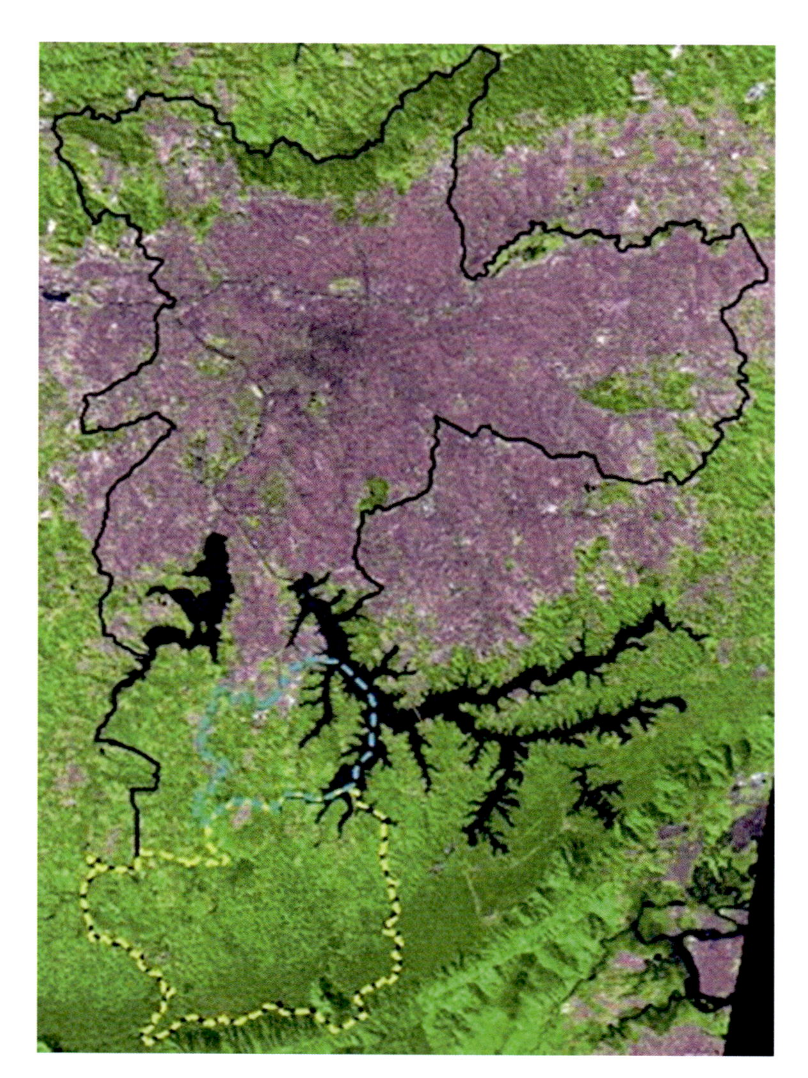

Fig. 11.4 Municipal Areas of Environmental Protection (APAs) in São Paulo city: APA *Capivari-Monos* (yellow outline) and APA *Bororé-Colônia* (blue outline). Source: Department of Greenery and Environment (Secretaria do Verde e do Meio Ambiente-SVMA) of São Paulo City, 2011

growing informal urbanization (*favelas*). According to 2003 data, 1,824,430 inhabitants of São Paulo city resided in 1241 *favelas*, accounting for 17.49% of the local population. According to APAs officials, the experience with the management of these areas shows that participation of local stakeholders in decision-making brings innovative solutions to tackle pressing conservation challenges.

Municipal laws such as the City Strategic Plan (*Plano Diretor Estratégico*) of 2014 contain provisions for environmental zoning aiming at ensuring adequate use of city areas key for local water production. The Strategic Plan reinforces the role of the APA approach, according to which only "sustainable use of natural resources" is allowed. City policies targeting these areas prioritize both the conservation of nature and the improvement of the livelihoods of local communities. The main components of the management plans governing socio-economic activities in the APAs include environmental education, tourism, handcrafts and cultural production, heritage protection, scientific research, restoration of degraded areas, and access and transportation.

APA *Capivari-Monos*
The APA *Capivari-Monos* comprises 251 km^2 (around one-sixth of total city area) in southern São Paulo. Crisscrossed by rivers and waterfalls, the area contains secondary formations and remains of primary forest of the Atlantic Forest biome, and hosts emblematic species such as the woolly spider monkey (*Brachyteles arachnoides*), the cougar (*Puma concolor capricorniensis*), and the South-American tapir (*Tapirus terrestris*). Historical and cultural values include two indigenous settlements of the Mbyá-Guarani community. The first inhabitants of the region that today comprises the municipality of São Paulo were the Mbyá-Guarani, a semi-nomadic indigenous group present in some regions of Brazil, Argentina, Paraguay and Uruguay, including the area where São Paulo City is located. Nowadays, around 282 indigenous families comprising over 1250 people live in two legally demarcated territories located inside the municipal boundaries of the city. The communities still preserve several of their traditional activities employing resources available locally, including handcrafts for everyday use in the house and the fields, together with spiritual rituals and medicinal practices.

The APA *Capivari-Monos* was the first of its kind to be established in 2001. A geo-environmental zoning mechanism created in 2004 regulates the use of land and natural resources in the area, including sub-areas of different degrees of protection under specific regulations. The APA has an encompassing Management Plan elaborated by the São Paulo City Department of Greenery and Environment (SVMA) which addresses both social and environmental issues, and it is implemented by a participatory council involving a variety of local/regional institutions and communities. The creation of the APA responded also to the decentralization of water resources management, prompting municipal authorities are to play an important role.

A series of activities and programs related to eco-tourism, handcrafts, heritage protection, reforestation, and ecological restoration involving local residents are being promoted in the APA *Capivari-Monos*. Reforestation activities are well established as a source of income for local residents and include numerous tree nurseries of native species as well as technical support for implementing projects in the field. Likewise, eco-tourism activities are being supported, which are especially relevant for indigenous groups. For example, the *Juruá Jaru Nhanderekoa* project, which focuses on enhancing ecologically and culturally sound tourism, promotes organized visits to the community and provides capacity for local indigenous tourist guides and a variety of locally made typical handcrafts. According to

APA officials and program participants, the project has increased local awareness about Mbyá-Guarani culture and its links to the local natural resources among (mostly urban) visitors, while improving local livelihoods and enhancing exchanges with other indigenous and non-indigenous communities in the area.

APA *Bororé-Colônia*

The APA *Bororé-Colônia* comprises two unique neighborhoods in the city. The *Bororé* area is located on the shores of the main water reservoir in São Paulo, and has remained relatively isolated from the rest of the city since no mainland access exists; instead, a raft transports people and vehicles between the two sides. The sense of disconnection from the main city is strong, as the area is known as the *Bororé* Island although it is in fact a peninsula. Bird watching and recreational water activities and sports have traditionally been practiced in *Bororé*. The other neighborhood within the APA, *Colônia Paulista*, contains numerous historical landmarks of European settlement in Brazil dating back from 1829, including some well-known religious sites such as the first protestant cemetery in the country. Agricultural activities—mostly small family farming—are still present in *Colônia Paulista*, conferring the area a distinctive rural-like landscape.

The *Bororé-Colônia* region is characterized by its high ecological value, yet under the threat of increasing informal urbanization. The APA was established in 2006 with the objective of protecting the water resources, biodiversity, and historical heritage of this area of the city which provides around 30% of the water consumed in the metropolitan region. In addition to environmental protection, the establishment of the APA aimed at improving the well-being of its residents. Some of the activities being promoted in the APA include organic agriculture, aiming at eliminating (or at least reducing) the use of agrochemicals by local smallholders, thus decreasing the risk of water pollution, while providing local farmers with alternative sources of income (see Sect. 3.4 about the City's Clean Agriculture program). Likewise, eco-tourism and recreational activities which are compatible with environmental preservation and that can provide local livelihoods are also being fostered in the APA *Bororé-Colônia*, including hiking, fishing, biking, and agri-tourism in family farms.

Challenges for the APAs

Despite being a key instrument for the management of local ecosystems, São Paulo's APAs are experiencing important challenges. According to São Paulo City's Department for Greenery and Environment, several activities explicitly unauthorized under the APA designation keep taking place in different locations within the APAs, having negative effects on local ecosystems. These activities include irregular urbanization (linked to the creation of slums) followed by disposal of wastewaters and solid wastes, and deforestation (whether through fires or direct tree cutting). The pressure of irregular urbanization is arguably one of the most significant challenges currently facing the APAs, as it connects directly to the failure of proper housing programs aiming at settling a growing low-income population largely unable to find regular housing alternatives. One of the major threats resulting from the current expansion of the informal urban fabric is the pollution of the water reservoirs (mainly through improper wastewaters disposal) which in turn puts São Paulo city and metropolitan area water provisioning at serious risk. The situation of indigenous populations within the APAs is also a cause of concern. The small size of the demarcated areas has been identified as one of the major challenges for the thriving of the communities, whose lifestyle is deeply connected to extensive areas and the biodiversity of the Atlantic Forest.

11.3.4 Co-benefits Linked to Urban Agriculture: The Clean Agriculture Program

Almost 15% of the area of São Paulo city (222 out of 1523 km^2) falls under the category of arable land. Yet as of 2003, only around 2% of the city's area (around 3600 ha) was actually under active cultivation (SVMA-IPT 2004). According to local inventories, there were 313 farming households within the city limits, with an average farm size of 12 ha. Most of these agricultural areas are located in the southern region of the city, next to the water reservoirs and within the protected spaces of the APAs and the GBBR.

The Clean Agriculture (*Agricultura Limpa*) program is a project launched by the Agriculture and Procuring Department of the City (*Secretaria de Agricultura e Abastecimento*) focusing on promoting "environmentally friendly" agriculture in the southern region of the city, where areas of high environmental value are located. According to the Agriculture and Procuring Department the southern region's ecosystem services—including climate regulation, water provisioning, and flood control, among others, are severely threatened by the expansion of irregular urbanization and agrochemical pollution. The Clean Agriculture program was established in January of 2010 as a response to these challenges, seeking to reinforce the traditionally rural character of this peri-urban area of São Paulo city, preventing urban expansion while transforming certain environmentally damaging practices by farmers into eco-friendly alternatives. The program targets local producers and offers technical and material support for conversion to chemical-free agriculture (without pesticides, but allowing chemical fertilizers) with the broader aim of preventing contamination of adjacent water bodies. Despite agricultural activities being widespread in the two environmental protection areas (APAs)—especially in *Bororé-Colônia*, an area of early European and later Japanese agrarian colonization—agricultural activities only started to be catalogued by the local government with the onset of the Clean Agriculture program in 2011.

The main driver of the Program was the effective protection of the water sources of the city by preventing agrochemical pollution caused by widespread agricultural activities in the area where water reservoirs are located. According to the Agriculture and Procuring Department, São Paulo city, with a population of around 12 million, has a hydrologic deficit (difference between water obtained from local sources and water consumed by the city) of almost 50%. Contributing to the conservation and restoration of the two main water reservoirs (*Billings* and *Guarapiranga*) together with the watershed of the major local river system (*Capivari-Monos*) is the main focus of the Clean Agriculture program.

Although the main policy objective of the Clean Agriculture program is to promote cleaner agricultural practices among local producers, supporting farmers to decrease farm abandonment and thus, in turn, controlling irregular urbanization in the southern region is a desired outcome of the Program. The southern region of São Paulo city has historically been an agricultural area, since the arrival of the first non-native settlers to the territory, being important for local and regional food provisioning. According to the Agriculture and Procuring Department, it was in the 1970s that local agriculture began to decline, partly due to the lack of appropriate extension programs and governmental support for local producers, together with the collapse of a major local agricultural cooperative operating in the region. As a result, and combined with dramatic urban population growth in São Paulo city and its metropolitan region, many farmers engaged in (often illegal) *loteamentos*—selling their land in parcels, which lead to the spread of urbanization, often irregularly, over former agricultural areas. The abandonment of agricultural activities has in turn lead to further irregular urbanization in the region, as the long-term settlers were no longer present to avoid invasions and re-selling of parcels. Some of these irregularly urbanized areas are in close proximity to local water reservoirs, further compromising proper environmental protection of local water resources from increasing pollution. The Agriculture and Procurement Department estimates that, while current urbanization rate in São Paulo city is around 1%, urbanization in the southern region is 8.5% per year. The Clean Agriculture program aims at reversing these urbanization trends by supporting local farmers to remain productive and by generating related employment opportunities for local residents (e.g. food processing).

The choice of pesticide free-agriculture (defined imprecisely as "organic" or "agroecological" agriculture in the Program) as the productive model for the local farmers is not only based on reducing pollution, but incorporates considerations related to obtaining local benefits in the form of increased well-being for the community, including food security through improved access to fresh and healthy produce. According to the Agriculture and Procurement Department, these social benefits are meant to contribute to overturning current socio-economic trends in the southern region of the city where agriculture and irregular urbanization coexist, which shows the lowest levels of Human Development Index in the municipality. Well-being benefits of the Program for the overall urban population of São Paulo city, beyond local water protection, are also desired Program outcomes, including "access to local, fresh, healthy produce," emphasizing the role of urban dwellers in the sustainability of their own city by means of purchasing locally grown, chemical-free products, not easily available through conventional market channels.

The legal framework under which the Clean Agriculture program operates contains both state and local regulations. At the state level, specific laws for the two areas containing the water reservoirs (N° 13,579, of 13 July 2009 and N° 12,233, of 16 January 2006) establish the need to "promote sustainable rural development in the region through the consolidation of local agroecological activities, thus minimizing agriculture-related pollution and controlling irregular urban growth". At the local level, a municipal decree (*Decreto* 45,853, of 27 April 2005) appoints the Procuring Department as one of the entities cooperating in the promotion of local agricultural production in the region, including vegetables, tree crops, and animal husbandry.

Main Achievements of the Clean Agriculture Program
The Clean Agriculture program has brought up not only the first systematic cataloguing of agricultural activities in the city but the establishment of a set of Agroenvironmental Best Practices Protocol (*Protocolo de Boas Práticas Agroambientais*) for clean agricultural production in the urban and peri-urban area, together with the creation of a branding scheme based on a certificate of origin for local products. Instruments for achieving policy targets include a series of interventions ranging from technical assistance and organizational support (for example, the creation of cooperatives and associations) to access to credit. With these instruments, the Program "incentivizes sustainable production conducive to environmental and socio-economic benefits, including job creation, income generation, and improved access to local, fresh produce."

A total of 403 active farmers (316 of which located in the Southern region) were identified by the first urban agriculture census completed by the city in 2011, with 39 farmers adopting the Agroenvironmental Best Practices Protocol that same year. The total aggregated area of the farms was 5187.06 hectares; of which 152.34 ha were devoted to perennial crops (shrubs, trees) and 1335.87 ha to seasonal crops (Table 11.3).

The Clean Agriculture program has promoted improvements in local environmental management. Dozens of organic conversion plans (*Planos de Conversão Agroecológica*) to transform farms using agrochemicals into "organic" farms have been approved over the years by an Executive Commission integrated by municipal and state representatives. This has been possible largely through improving extension services for organic conversion, which at the time of this field research averaged over sixty monthly consultations for the seven field agronomists working in the Program, which has expanded since then. The Program has also set up two eco-agricultural centers (*Casas da Agricultura Ecológica*) providing support for local farmers, each covering one of the two APAs, while an existing local agricultural cooperative (*Associação dos Agricultores Orgânicos de São Mateus*) and newly created ones are supporting local farmers to access credit through the national program by the Brazil Bank (*Banco do Brasil*). Access to local institutional and private markets (farmers' markets in the city center, restaurants, guest houses, public schools, etc.) are also being promoted by the Program. Widespread media coverage, together with the visibility of farmers' markets in downtown areas of the city have contributed to the favorable opinion about the Program among city-dwellers. An innovative urban agriculture project funded by the State Environmental Fund (*Fundo Estadual do Meio Ambiente*-FEMA) curerntly aims at converting all agricultural activities within city limits to chemical-free production; if successful, São Paulo will become the first city in the country to achieve such landmark.

Table 11.3 Main agricultural products cultivated in São Paulo city (2011). Source: Agriculture and Procurement Department (*Secretaria de Agricultura e Abastecimento*)

Products	Hectares
Flowers (cut)	255.30
Lettuce	172.29
Chuchu	166.32
Cabbage	134.48
Broccoli	116.71
Cauliflower	100.50
Beets	78.90
Shrubs	53.90
Carrot	24.60
Cebolinha	23.60
Banana	19.05
Kale	18.50
Herbs	17.88
Flowers	14.80
Spinach	10.60
Eggplant	10.00
Green peas	7.20
Kaki	2.50
Mushrooms	1.20
Cambuci	0.02
Chicory	0.50
Total	2144.6

Challenges of the Clean Agriculture Program

Despite receiving notoriety through the media, the Program faces important challenges. Perhaps the most crucial one relates to its capacity to foster not just "clean" but trully ecologically sustainable, agroecological practices across the farms. The elimination of all agrochemicals (fertilizers and pesticides) is unlikely to succeed unless a true effort is made to ensure the ecological integrity of the agroecosystem, including nutrient cycling (e.g. composting), polycultures, natural pest management, and agrobiodiversity. Moreover, aging of farmers, together with migration of local residents (particularly the youth) to more central areas of the city in search of job opportunities also jeopardizes the long-term success of the initiative.

11.4 Comparing the Co-benefits Potential of the Initiatives

The four environmental initiatives considered in the study—the GHG emissions inventory, the GBBR, the APAs, and the Clean Agriculture program—can potentially lead to generating (directly or indirectly) sustainability and well-being co-benefits by including ecosystem services and/or biodiversity. Table 11.4 presents a comparative analysis of the four initiatives regarding main actors involved, main

Table 11.4 Selected initiatives in São Paulo city and potential for aligning climate, ecosystem services, and biodiversity co-benefits

	Type of initiative			
	GHG emissions inventory	GBBR	Municipal Environmental Protection areas (APAs)	Clean Agriculture program
Stakeholders	Local (and national)	Local, national, international	Local	Local
Main driver	Urban sectors GHG emissions reduction	Biodiversity protection (Atlantic Forest)	Local ecosystems protection	Reduction of agrochemical pollution
Co-drivers	Air pollution reduction	Awareness rising Social inclusion	Urbanization control Sustainable, inclusive development	Urbanization control Sustainable, inclusive development
Target groups/ sectors	Urban sectors (public and private)	Locals and visitors Local youth Education sector	Locals and visitors Local residents	Peri-urban farmers Local residents
Main instruments	Data collection and assessment for the establishment of baselines	Ecosystem and biodiversity management Monitoring Outreach Environmental education	Monitoring Inspection Environmental education Incentives for local entrepreneurship	Incentives for organic farming conversion Access to credit, technical support, and local markets Inspection
Overall co-benefit potential	Indirect: limited GHG mitigation potential of land use change, yet gains from mitigation projects in other sectors can be invested in environmental protection	Medium: biodiversity rich Atlantic Forest ecologically connected beyond city limits, yet limited operational capacity at the local level	High: monitoring and inspection in areas with high biodiversity value managed at the city level in a context of urbanization pressure with a focus on inclusive development	High: mitigation of agrochemical pollution; control of informal urbanization; sustainable local livelihoods for inclusive development; creation of local production-consumption networks to expand benefits
Opportunities	From "land use" to "urban ecosystems" (e.g. include aquatic) Include urban ecosystems' climate services (e.g. heat island control)	Streamline the co-benefits dimension to strengthen the relevance of GBBR (e.g. control of urban heat island effect)	Expand sustainable local, inclusive livelihoods for indigenous and other vulnerable groups Link to other urban sectors and stakeholders	Scaling up local production-consumption networks (e.g. community gardens) Emphasize agroecological functionality and agrodiversity
Barriers	Markedly lower contribution of land use to GHG mitigation can downplay broader ecosystem services and biodiversity co-benefits	Lack of data (e.g. carbon stocks and flows, air pollution filtration) Limited operational capacity Dependence from supra-local actors	Lack of human and material resources Overload with "day-to-day" issues constrains systemic planning and effectiveness of programs	Lack of a truly agroecological approach to production (e.g. chemical fertilizers allowed) Aging of farmers

drivers and co-drivers, target groups, main policy instruments, as well as main opportunities for generating co-benefits and main barriers preventing success. Overall, the data show that the co-benefits linked to each initiative respond to the specific needs, capacities, and context of the city. Yet, the ability to integrate sustainability and well-being concerns across local development processes depends on adopting a multi-sectoral, multi-stakeholder approach, a finding consistent with urban biodiversity policy frameworks (Avlonitis et al. 2013).

Although the four initiatives involve (mainly) local stakeholders, the role of national actors is quite prominent in the GBBR, and somewhat relevant for the GHG emissions inventory; international actors are key in the GBBR. While linking to supra-local stakeholders does not necessarily condition the co-benefit potential of the initiatives, the GBBR case illustrates how complex, lengthy policy processes involving salient national and international actors can pose barriers to the operational capacity of the initiative at the local level. The main policy drivers differ across the four initiatives studied. While the Inventory has a clear GHG mitigation focus, climate outcomes (e.g. heat island control, carbon stocks) are secondary for the GBBR—mainly focused on biodiversity protection—and not addressed in the APAs, which are driven by local ecosystem protection and inclusive development. Climate is not specifically addressed by the Clean Agriculture program, which focuses on urbanization control and mitigation of agrochemical pollution. However, the creation of local production–consumption networks where local peri-urban farmers' produce is consumed mainly by city dwellers can have a positive impact on reducing GHG emissions associated with food miles. Pollution reduction drivers (air, water) are relevant across initiatives, while urbanization control objectives are particularly important for the Clean Agriculture program and the APAs. The broad range of co-benefits identified across the initiatives regarding timeframes (short vs log term), geographical scales (local vs regional vs global) and purpose (primary, secondary or unintended) mirrors the current debate in the urban climate action literature (Floater et al. 2016).

The target groups of the four initiatives range from prominent urban sectors (e.g. waste, transport, energy) in the case of the Inventory, to local residents, with an emphasis on vulnerable groups (e.g. local youth, peri-urban farmers, indigenous communities) in the other three cases. This is particularly relevant when considering the developmental context of São Paulo city, where the sustainability and well-being implications of co-benefits must be linked to promoting the inclusion of vulnerable groups. For example, air pollution reduction benefits resulting from GHG mitigation strategies targeting salient urban sectors (e.g. São Paulo city's CDM waste management project) may not have a clear social inclusion component. In contrast, co-benefits related to promoting the sustainable use of ecosystems and biodiversity in local livelihoods for farmers and indigenous communities, or local youth education and employment innitiatives promoted by the APAs or the Clean Agriculture program offer opportunities for inclusive development. Enviornmntal policy interventions must pay special attention to avoid worsening existing socio-spatial inequalities and overall negative impacts on vulnerable urban groups (Anguelovski et al. 2016). Main instruments for achieving desired policy outcomes also differ greatly across the four cases. While the Inventory focuses on data collection and assessment for the establishment of baseline GHG emissions across urban sectors to inform mitigation policies, the GBBR the APAs, and the Clean Agriculture program carry out environmental monitoring and inspection as well as outreach and environmental education activities. The APAs and the Clean Agriculture program also provide incentives for local entrepreneurship.

Taking into account the needs, capacities, and context of São Paulo city, the overall co-benefit potential of the APAs and the Clean Agriculture program are considered "High" when compared to that of the GBBR ("Medium") and the Inventory ("Indirect") in terms of their capacity to integrate sustainability and well-being issues across local development processes. Both the APAs and the Clean Agriculture program target social inclusion through the promotion of local livelihoods based on the sustainable use of local ecosystems and biodiversity. They also carry out inspection and monitoring, which allows for the early detection of problems and the establishment of action plans to address them. The multi-sectoral, multiscale approach of the APAs allows them to operate in conjunction with multiple entities across the local government and other urban sectors and stakeholders. This is particularly important in a metropolis like São Paulo city, where the magnitude and urgency of developmental challenges may result in downplaying biodiversity and ecosystem services priorities while, as the findings of this study suggest, they are fundamental for creating sustainable, inclusive local livelihoods. The Clean Agriculture program appears to have a more sectoral approach, focusing on the peri-urban farming communities, yet the linkage to the urbanization control dimension in connection to the APAs, together with the potential for scaling up sustainable local production–consumption networks that generate co-benefits (e.g. reduction of food miles) confers the Program a multi-scale, multi-stakeholder dimension.

The co-benefit potential of the GBBR is considered "Medium" mainly because, despite monitoring and protecting the rich biodiversity of the Atlantic Forest in connection to regional and international processes, it has a limited operational capacity to integrate sustainability and well-being across local development processes relevant for the São Paulo's context such as urbanization and social inclusion. The case of the Inventory is particularly interesting, since its co-benefit potential is "Indirect." On the one hand, the findings of the Inventory may have highlighted the limited GHG mitigation potential of "land use" as an urban sector compared to the waste, transportation, or energy sectors. Yet, economic gains from mitigation projects in waste management (e.g. through the Clean Development Mechanism) can revert into funding "land use" initiatives that in turn contribute to ecosystem services and biodiversity (e.g. the APAs).

As for the *opportunities* and the *barriers* faced by the different initiatives, which may in turn affect their co-benefit potential, the situation differs significantly across them. The technical capabilities created during the elaboration of the Inventory could be used to expand the study from a "land use" approach to an "urban ecosystems" one (including aquatic systems) to assess the full set of "climate services" provided by those urban ecosystems. However, the fact that the overall contribution of land use to GHG emissions is markedly lower than other urban sectors such as waste, transport, or energy could in fact downplay the relevance of land use as a climate co-benefits sector, and thus question the suitability of mainstreaming it. It is necessary to grow awareness among local governments and stakeholders involved in urban climate action about the contributions of land use initiatives (e.g. urban agriculture) to addressing local developmental (e.g housing, jobs), well-being (e.g. public health) *and* sustainability (e.g. adaptation, mitigation) needs (Dubbeling et al. 2019). For the GBBR, the main opportunities lay on streamlining the co-benefits dimension of biodiversity protection linked to the Atlantic Forest (e.g. supporting, provisioning, regulating, and cultural services, including but not limited to climate). Yet lack of accurate, reliable baseline data, limited operational capacity, and dependence from supra-local actors outside the local development process limit the potential of the GBBR.

For the APAs, opportunities center around the consolidation and expansion of sustainable local, inclusive livelihoods, particularly among indigenous communities and other vulnerable groups, allowing residents to benefit from the local ecosystems and the biodiversity they contain (e.g. tree nurseries, heritage restoration, agroecological farming, local tourism, etc.). Moreover, linking these experiences to other urban sectors and stakeholders could help mainstreaming the co-benefits approach in those sectors. For example, introducing decentralized, local composting of organic waste could generate income opportunities for APAs residents, while bringing an innovative approach to the City's waste management sector. The main challenges for the APAs are related to lack of human and material resources for carrying out their operations, which results in an overload with "day-to-day" issues that constrain systemic planning and the effectiveness of their programs. Last, the Clean Agriculture program has opportunities for scaling up local production–consumption networks by linking to other kinds of urban agriculture experiences in the city (e.g. allotment gardens, green roofs), together with emphasizing agroecological functionality and agrodiversity for extended co-benefits beyond reducing agrochemical pollution. The main challenges for the Program are connected to a lack of a systemic approach to sustainable agricultural production (e.g. organic chemical fertilizers allowed, non-native varieties used, etc.) which, together with low productivity and aging of farmers, jeopardizes the long-term viability of the Program. Agroecological strategies can improve the productivity of urban farms through organic soil management, biological pest control and optimal crop rotations (Altieri and Nicholls 2018), thus potentially securing further developmental, well-being and sustainability benefits. However, it is worth noting that prioritizing one benefit (e.g. biodiversity protection) will not necessarily yield further co-benefits, and could even result in negative outcomes (e.g. displacement of the urban poor). Ultimately, the success of urban co-benefits strategies related to ecosystem services and biodiversity will depend on their capacity to avoid a one-size-fits-all approach and, instead, seek a nuanced understanding of local developmental, sustainability and well-being processes that leads to synergies across sectors and initiatives (Colléony and Shwartz 2019).

11.5 Conclusion

Evidence from the four initiatives from São Paulo city examined in the study suggest that effective management of urban ecosystems to generate sustainability and well-being co-benefits requires a multi-scale, multi-stakeholder, and multi-sectoral approach. Initiatives that are able to achieve a balance between geographical scope, effective stakeholder participation, and involvement of multiple urban sectors show greater potential for creating co-benefits synergies. The specific needs, capacities, and context of the city must be taken into account when designing and implementing those initiatives, to ensure that they integrate sustainability and well-being issues across local development processes. In São Paulo city, where social inclusion is a paramount developmental need, mainstreaming a co-benefits approach in local environmental policy processes related to ecosystems and biodiversity can lead to attaining coupled sustainability and well-being benefits that, in turn, reduce exclusion and marginalization. Moreover, the study shows that co-benefits can emerge in connection to initiatives not specifically targeting them but as a result of efforts to tackling other urban challenges. For example, the urgency to reduce informal urbanization in areas of high environmental value can result in programs supporting sustainable, inclusive local livelihoods—for example, through chemical-free peri-urban agriculture.

Environmental challenges for today's cities have both global and local dimensions. While the former are directly connected to the sustainability of cities—that is, how to reduce urban ecological footprints—the latter are linked to the capacity of urban areas to provide adequate ecosystem services for the well-being of their residents. Proper management of urban ecosystems and the biodiversity they contain can yield a variety of coupled sustainability well-being benefits for cities—including climate change mitigation, heat and flood control, improved air quality, local food production–consumption networks, recreation, and spiritual fulfillment. The generation of sustainability and well-being co-benefits through urban ecosystem management is nevertheless a complex process, sometimes mediated through behavioral and cultural factors. This tremendous challenge requires innovative approaches to urban environmental policy that align different goals at different scales to improve engagement and effectiveness. Sectoralism and a lack of a common vision regarding how local ecosystems should be managed jeopardize the development of integrative policy responses to successfully tackle complex environmental challenges such as climate change and biodiversity loss.

Acknowledgements The author wishes to thank the United Nations University-Institute for the Advanced Study of Sustainability (UNU-IAS) for supporting this reserach, as well as the Department of Greenery and Environment (Secretaria do Verde e do Meio Ambiente-SVMA) of São Paulo city, the GBBR, and all the other interviewees who participated in the project, for the valuable information provided to elaborate this study.

References

Altieri M, Nicholls CI (2018) Urban agroecology: designing biodiverse, productive and resilient city farms. Agro Sur 46(2):49–60

Anguelovski I, Shi L, Chu L et al (2016) Equity impacts of urban land use planning for climate adaptation: critical perspectives from the Global North and South. J Plan Educ Res 36(3):333–348

Avlonitis G et al (2013) Local biodiversity strategy and action plan guidelines. https://cbc.iclei.org/tools/. Accessed 11 December 2020

Baró F, Chaparro L, Gómez-Baggethun JL et al (2014) Contribution of ecosystem services to air quality and climate change mitigation policies: the case of urban forests in Barcelona, Spain. Ambio 43(4):466–479

Cifuentes L, Borja-Aburto VH, Gouveia N et al (2001) Assessing the health benefits of urban air pollution reductions associated with climate change mitigation (2000-2020): Santiago, São Paulo, México City, and New York City. Environ Health Perspect 109(3):419–425

Colléony A, Shwartz A (2019) Beyond assuming co-benefits in nature-based solutions: a human-centered approach to optimize social and ecological outcomes for advancing sustainable urban planning. Sustainability 11(18):4924

Costa-Pierce BA, Desbonnet A, Edwards P, Baker D (2005) Urban aquaculture. CABI Publishing, Wallingford

Darby W, Kinney PL (2007) Health co-benefits of climate mitigation in urban areas. Curr Opin Environ Sustain 2:172–177

Dubbeling M, van Veenhuizen R, Halliday J (2019) Urban agriculture as a climate change and disaster risk reduction strategy. Field Actions Sci Rep 20:32–39

Elmqvist T, Folke C, Nyström M et al (2003) Response diversity, ecosystem change, and resilience. Front Ecol Environ 1(9):488–494

Farias AR et al (2017) Identificação, mapeamento e quantificação das áreas urbanas do Brasil. Embrapa Gestão Territorial. https://www.infoteca.cnptia.embrapa.br/infoteca/bitstream/doc/1069928/1/20170522COT4.pdf. Accessed 11 December 2020

Floater G, Heeckt C, Ulterino M et al (2016) Co-benefits of urban climate action: a framework for cities. London School of Economics and Political Science. https://www.c40.org/researches/c40-lse-cobenefits. Accessed 11 December 2020

Folke C, Jansson A, Larrson J, Costanza R (1997) Ecosystem appropriation by cities. Ambio 26:167–172

Galluzi E, Eyzaguirre P, Negri V (2010) Home gardens: neglected hotspots of agrobiodiversity and cultural diversity. Biodivers Conserv 19:2635–2654

Gill SE, Handley JF, Ennos AR, Pauleit S (2007) Adapting cities for climate change: the role of the green infrastructure. Built Environ 33(1):115–133

Graves HM, Watkins R, Westbury P, Littlefair PJ (2001) Cooling buildings in London. CRC Press, London

Groenewegen P, van den Berg A, De Vries S, Verheij R (2006) Vitamin G: effects of green space on health, well-being, and social safety. BMC Public Health 6(1):149

Hoornweg D, Sugar L, Trejos Gomez CL (2011) Cities and greenhouse gas emissions: moving forward. Environ Urban 23:207–227

IPCC (2007) Intergovernmental panel on climate change fourth assessment report. Cambridge University Press, Cambridge

Izique CO (2003) Mapa da ExcluSão. Pesquisa Fapesp. 83. São Paulo, Brazil

Jo HK (2002) Impacts of urban greenspace on offsetting carbon emissions for middle Korea. J Environ Manag 64:115–126

La Greca P, La Rosa D, Martinico F, Privitera R (2011) Agricultural and Green Infrastructures: the role of non-urbanised areas for eco-sustainable planning in a metropolitan region. Environ Pollut 159(8–9):2193–2202

Leung DYC, Tsui JKY, Chen F et al (2011) Effects of urban vegetation on urban air quality. Landsc Res 36(2):178–188
Linares O (1996) Cultivating biological and cultural diversity: urban farming in Casamance, Senegal. Africa 66(1):104–112
Malagoli LR, Blauth BF, Whately M (2008) Além do concreto: contribuiçõespara a proteção da biodiversidade paulistana. Instituto Socioambiental-ISA, São Paulo
Mayer H (2009) Air pollution in cities. Atmos Environ 33(24–25):4029–4037
McCormik K, Anderberg S, Coenen L et al (2013) Advancing sustainable urban transformation. J Clean Prod 50:1–11
Mcdonald RI, Kareiva P, Forman RTT (2008) The implications of current and future urbanization for global protected areas and biodiversity conservation. Biol Conserv 141(6):1695–1703
McHale MR, McPherson EG, Burke IC (2007) The potential of urban tree plantings to be cost effective in carbon credit markets. Urban Forest Urban Green 6:49–60
McKinney M (2002) Urbanization, biodiversity and conservation. BioScience 52:883–890
McPherson EG (1998) Atmospheric carbon dioxide reduction by Sacramento's urban forest. J Arboric 24(4):215–223
Millennium Ecosystems Assessment (MA) (2005) Ecosystems and human wellbeing: current state and trends assessment. Island Press, Washington DC
Moreno-Peñaranda R (2012) Biodiversity and culture, two key ingredients for a truly green urban economy: learning from agriculture and forestry policies in Kanazawa city, Japan. In: Simpson R, Zimmermann M (eds) The economy of green cities: a world compendium on the green urban economy. Springer, Dordrecht
Muller N, Werner P (2010) The case for implementing the convention on biological diversity in towns and cities. In: Muller N, Werner P, Kelcey JG (eds) Urban biodiversity and design. Wiley-Blackwell, Oxford
Nicholls E, Ely A, Birkin L et al (2020) The contribution of small-scale food production in urban areas to the sustainable development goals: a review and case study. Sustain Sci 15 (6):1585–1599
Nowak DJ, Crane DE (2002) Carbon storage and sequestration by urban trees in the USA. Environ Pollut 116(3):381–389
Pearson C, Pilgrim S, Pretty J (eds) (2010) Urban agriculture: diverse activities and benefits for city society. Earthscan, London
Pires BCC, Rodrigues EA, Victor RABM et al (2010) Evaluación Ecosistémica del Cinturón Verde de São Paulo, Brasil: una propuesta de gestión territorial en una reserva de la biosfera en ambientes urbanos. In: Rosas PA, Clüsener-Godt M (eds) Reservas de la Biosfera: Su contribución a la provisión de servicios de los ecosistemas. Experiencias exitosas en Iberoamerica. UNESCO, Paris, pp 31–48
Pisupati B (2007) Effective implementation of NBSAPs: using a decentralized approach. guidelines for developing sub-national biodiversity action plans. United Nations University Institute of Advanced Studies, Yokohama
Pouyat RV, Yesilonis ID, Nowak D (2006) Carbon storage by urban soils in the United States. J Environ Qual 35(4):1566–1575
Puppim de Oliveira JA (2013) Learning how to align climate, environmental and development objectives in cities: lessons from the implementation of climate co-benefits initiatives in urban Asia. J Clean Prod 58:7–14
Puppim de Oliveira JA, Balaban O, Doll CHN et al (2011) Cities and biodiversity: perspectives and governance challenges for implementing the Convention on Biological Diversity (CBD) at the city level. Biol Conserv 144:1302–1313
Quigley MF (2011) Potemkin gardens: biodiversity in small designed landscapes. In: Niemelä J et al (eds) Urban ecology: patterns, processes and applications. Oxford University Press, New York, pp 85–92

Raymond CM, Frantzeskaki N, Kabisch N et al (2017) A framework for assessing and implementing the co-benefits of nature-based solutions in urban areas. Environ Sci Pol 2017 (77):15–24
Rees W, Wackernagel M (2008) Urban ecological footprints: why cities cannot be sustainable- and why they are a key to sustainability. In: Shulenberger E et al (eds) Urban ecology: an international perspective on the interaction between humans and nature. Springer, New York
Rodrigues EA (ed) (2018) Ecosystem services and human well-being in the in the São Paulo City Green Belt Biosphere Reserve. Instituto Florestal. https://www.infraestruturameioambiente.sp.gov.br/institutoflorestal/2018/12/ecosystem-services-and-human-well-being-in-the-sao-paulo-city-green-belt-biosphere-reserve/. Accessed 11 December 2020
São Paulo City (2005) Inventário de Emissões de Gases de Efeito Estufa do Município de São Paulo. Secretaria do Verde e do Meio Ambiente-SVMA. São Paulo City, Brazil
SCBD (2012) Cities and biodiversity outlook. Secretariat of the Convention on Biological Diversity, Montreal, p 64
Shima N, Balaban O, Moreno-Peñaranda R (2011) Challenges and opportunities towards achieving environmental co-benefits through land use planning and management: experiences from Asian metropolitan cities. Proceedings of the 11th International Congress of the Asian Planning Schools Association, The University of Tokyo, Tokyo, 19–21 September 2011
Shin D, Lee K (2005) Use of remote sensing and geographical information system to estimate green space temperature change as a result of urban expansion. Landsc Ecol Eng 1:169–176
Solecki WD, Rosenzweig C, Parshall L et al (2005) Mitigation of the heat island effect in urban New Jersey. Glob Environ Change B 6:30–49
Sotto D, Philippi A, Yigitcanlar T et al (2019) Aligning urban policy with climate action in the Global South: are Brazilian cities considering climate emergency in local planning practice? Energies 12(18):3418
SVMA-IPT (2004) Instituto de Pesquisas Tecnológicas - GEO Cidade de São Paulo—Versão Preliminar, Secretaria Municipal do Verde e do Meio Ambiente, São Paulo City
TEEB (2010) The economics of ecosystems and biodiversity: mainstreaming the economics of nature: a synthesis of the approach, conclusions and recommendations of TEEB
UNDESA (2015) World urbanization prospects: the 2014 revision. population division, Department of Social and Economic Affairs. United Nations, New York
UNESCO (2003) Urban biosphere reserves in the context of the statutory framework and the Seville strategy for the world network of biosphere reserves
UNESCO (2006) Urban biosphere reserves—a report of the MAB Urban Group
Viljoen A (2005) Continuous productive urban landscapes (CPULs): designing urban agriculture for sustainable cities. Architectural Press, Oxford
Wilby RL, Perry GLW (2006) Climate change, biodiversity and the urban environment: a critical review based on London, UK. Prog Phys Geogr 30(1):73–98
Wilkinson P, Smith KR, Davies M et al (2009) Public health benefits of strategies to reduce greenhouse-gas emissions: food and agriculture. Lancet 374:1917–1929
Woodcock J, Edwards P, Tonne C et al (2009) Public health benefits of strategies to reduce greenhouse-gas emissions: urban land transport. Lancet 374:1939–1943
WWF (2012) The ecological footprints of São Paulo—state and capital and the footprint families. WWF-Brasil, Brasilia
Yanagi T (2005) Sato-Umi: new concept for the coastal sea management. Rep Res Inst Appl Mech 129:109–111
Yang J, McBride J, Zhou J, Sun Z (2005) The urban forest in Beijing and its role in air pollution reduction. Urban For Urban Green 3:65–78

Chapter 12
Evaluation of Habitat Functions of Fragmented Urban Forests for Wildlife: The Case of Kitakyushu City

Tohru Manabe, Minoru Baba, Kazuaki Naito, and Keitaro Ito

Abstract Urban forests with various shapes and areas exist in the mid-northern part of Kitakyushu City, Japan. These forests, which are dominated by evergreen canopy species, became fragmented largely in the 1960s. The differences in species composition between sample plots, as well as within the plots, were large for recently recruited small individuals in comparison with large ones. Thus, the effects of fragmentation on recruitment of woody species have recently become stronger. A molecular ecological approach also indicated that the effects of fragmentation on the genetic structure of one of the dominant woody species *Neolitsea sericea* had become severe. The effects of fragmentation on forest habitat functions were significant for large mammals such as wild boars as well as small mammals such as mice. For mid-sized mammals, however, the effects varied between species. Thus, to evaluate the habitat function of urban forests for wildlife, it is important to analyze the effects of fragmentation on various taxa using hierarchically related levels of biological organization.

Keywords Animal ecology · Fragmentation · Habitat function · Landscape ecology · Molecular ecology · Plant ecology · Urban forest

---

T. Manabe (✉) · M. Baba
Kitakyushu Museum of Natural History and Human History, Kitakyushu, Japan
e-mail: manabe@kmnh.jp

K. Naito
Graduate School of Regional Resource Management, University of Hyogo/Hyogo Park of the Oriental White Stork, Toyooka, Japan

K. Ito
Laboratory of Environmental Design, Faculty of Civil Engineering, Kyushu Institute of Technology, Kitakyushu-city, Fukuoka, Japan

K. Ito (ed.), *Urban Biodiversity and Ecological Design for Sustainable Cities*,
https://doi.org/10.1007/978-4-431-56856-8_12

12.1 Introduction

In Japan, secondary vegetation has profound relations with our daily lives, and has been maintained under traditional knowledge in suburban and rural districts. The vegetation has persisted under human impacts, and created its own particular ecosystems (Kamada et al. 1991). The rapid transformation of the main energy source from biomass to fossil fuels in the mid-1960s brought about changes in our lifestyle and in the socioeconomic values of secondary vegetation. These changes led to a large decrease in the area of secondary vegetation and a rapid change in its landscape structure from expansive areas to small fragmented elements (Iida and Nakashizuka 1995; Fukamachi et al. 1996). These changes have altered the community structure of abandoned secondary vegetation by the process of ecological succession (Kamada and Nakagoshi 1990, 1996; Hong et al. 1995; Manabe et al. 2003). Furthermore, increased occurrence of non-native species, as well as native species with strong shade tolerance, and changes of microclimatic conditions through edge effects has been reported in small fragmented forests (Murcia 1995; Yong and Mitchell 1994; Ishii et al. 2004; Tojima et al. 2004).

Urban forests provide various ecosystem services such as support for the recreational and aesthetic needs of people (Matsuoka and Kaplan 2008) and maintenance of water and air quality. Furthermore, urban forests provide various ecological functions for wildlife living in urban areas, acting as habitat, conduit, filter, source, sink, and so on (Forman 1995). Thus, the importance of urban forests should be considered from both basic and applied scientific viewpoints (McDonnell and Pickett 1990; Ishii et al. 2010). For example, a decrease in species diversity in small, fragmented forests has been reported for suburban and urban forests in Japan (Iida and Nakashizuka 1995; Murakami et al. 2004). In those studies, the habitat functions of urban forests were evaluated by using species diversity (the number of species). However, habitat functions should be evaluated from various hierarchical perspectives, as with the scientific term "biodiversity," which is used in relation to each genes, species, and ecosystems (or landscape) corresponding to the three fundamental and hierarchically related levels of biological organization.

In this chapter, we describe the spatiotemporal changes in urban forests in mid-northern Kitakyushu City (Fig. 12.1). In the next section, we describe, at both species (population) and gene level, the effects of fragmentation on the habitat function of the woody species of those forests. We also describe the effects on habitat function for mammals living in the urban area.

12.2 Study Site

Kitakyushu City has an area of 489.6 km^2 with a population of about 968,000 in 2013. The city includes many forests with an area estimated at 21,088 ha in 1984 and 19,994 ha in 2001 from digital land cover maps made using remote sensing

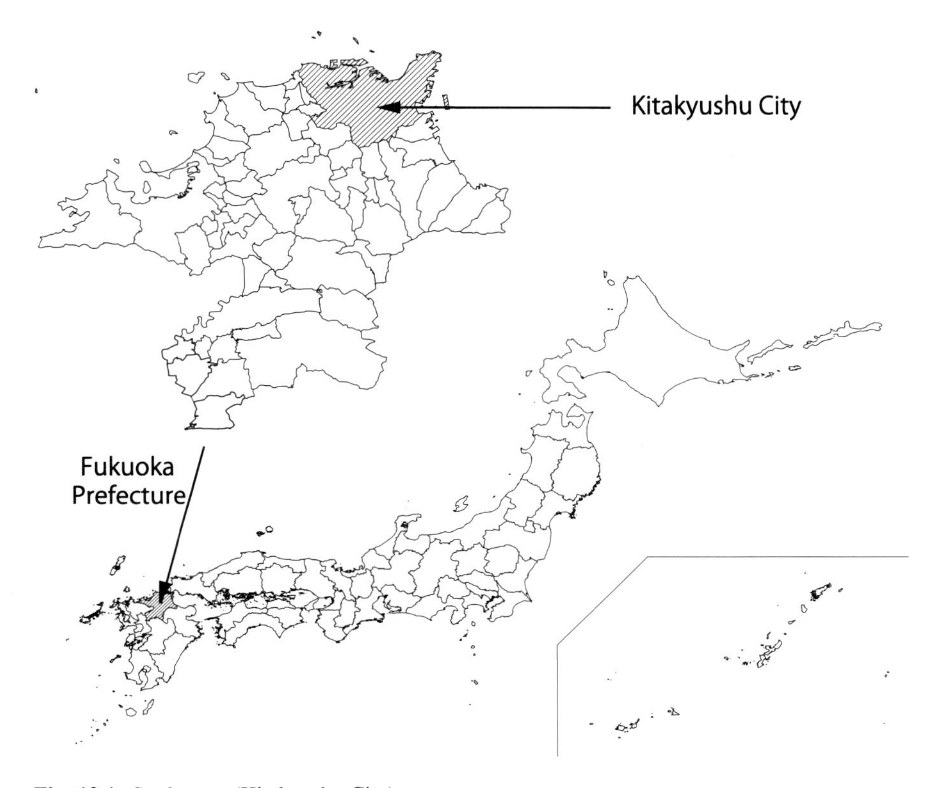

Fig. 12.1 Study area (Kitakyushu City)

(Landsat) data (Manabe et al. 2007). The potential natural vegetation is evergreen broad-leaved forest dominated by canopy species such as *Castanopsis cuspidata*, *Quercus salicina*, *Distylium racemosu*m, and *Persea thunbergii*, and evergreen subcanopy species such as *Cinnamomum japonicum*, *Camellia japonica*, and *Cleyera japoni*ca. Today, however, secondary broad-leaved forest is the dominant forest ecosystem, and natural vegetation can only be seen in some shrine/temple forests. In this study, *Castanopsis cuspidata* includes *Castanopsis cuspidate* var. *sieboldii*, because the species could not be consistently distinguished on account of the presence of hybrid and intermediate forms.

The target region of the study lies in the mid-northern part of the city (Fig. 12.2) where there are many secondary broad-leaved forests of various areas, shapes, and connectivities. The following four areas were chosen as study sites.

Yamada Green Park was used as an ammunition depot by the Japanese Army from the beginning of the 1940s until the end of World War II; it was then managed by the US Army and the Japan Self-Defense Force until 1995 when it became a park. The area has been free from severe human disturbance for more than half a century. Most of the area, about 345 ha, has therefore been covered by broad-leaved forests, and it has the largest and most continuous forest cover among the study sites

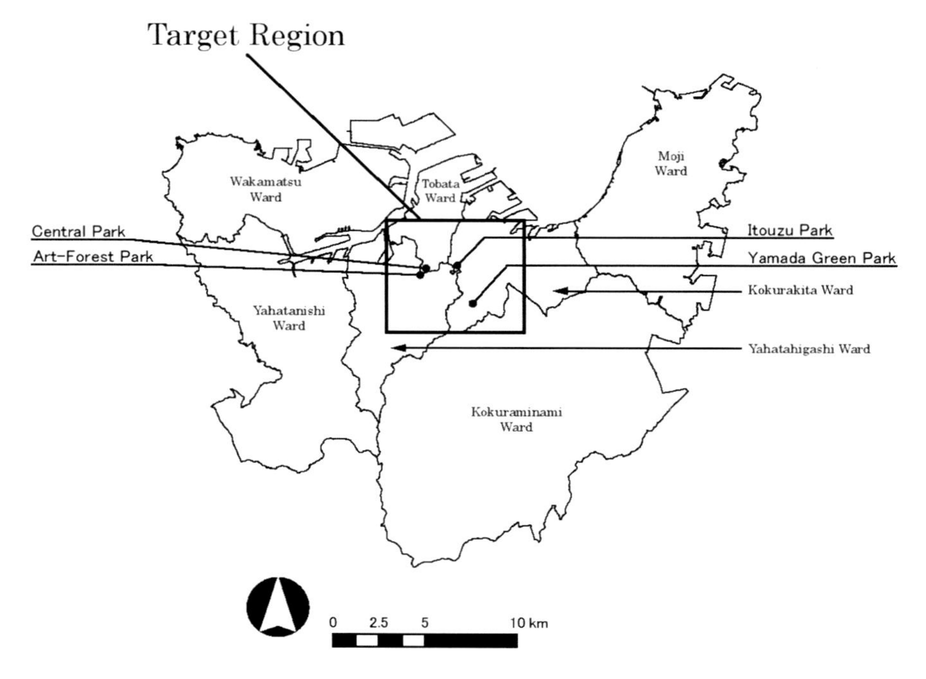

Fig. 12.2 Target region and study sites

Table 12.1 Characteristics of study sites and study plots

Study Sites	Yamada Green Park				Central Park		Art-forest Park	Itozu Park
Area (ha)	ca. 345				81.7		11.3	10.6
Study plot	Y1	Y2	Y3	Y4	C1	C2	AP	IP
Above sea level (m)	130	120	80	50	90	110	70	30
Slope direction	E	SSE	ENE	ESE	SE	SE	SE	S
Mean slope degree (^{O})	18.8	23.1	20.0	21.1	18.7	14.3	10.0	16.1
Vegetation in 1961[a]	TBF	TBF	TBF	TBF	SBF	SBF	SFG	TBF

[a]TBF, tall broad-leaved forests; SBF, small broad-leaved forests; SFG, small broad-leaved forests mixed with grasslands

(Table 12.1). Central Park (81.7 ha), which has established in 1940, is surrounded by small mountains. Art-forest Park (11.3 ha) surrounds the Municipal Art Museum established in 1974. Itozu Park, which has the smallest area (10.6 ha), was reestablished in 2002 as a zoo making use of secondary forests following the replacement of the recreation park with a zoo in 1998.

12.3 Changing Patterns of Urban Forests

The decrease in green space in urban areas was already a serious problem before the 1960s, the time of the fuel revolution. In this section, the changing patterns of green space in the target region between 1922 and 2000 are described.

The 1922, 1936, and 1960 digital vegetation maps of the target region were made from topographic maps published by the Geological Survey Institute of Japan. The 1961, 1974, and 2000 maps were made by using aerial photos. In those maps, green spaces were classified into cultivated lands and woodlands. Cultivated lands included paddy fields and other elements used for agriculture, and woodlands included broad-leaved and needle-leaved forests, bamboo forests, shrub lands, grasslands, and orchards. The areas of woodlands and cultivated lands in each year were calculated from those maps using Geographic Information System (ArcView 8.3). See Mitsuda et al. (2003) and Suzuki et al. (2004) for detailed descriptions of the study designs.

The area of cultivated lands decreased significantly between 1922 and 1936; however, between 1936 and 1948 there was no further decrease due to the need to grow food during World War II. After 1948, the cultivated area decreased progressively up to 2000, and the area in 2000 was no more than 20% of that of 1961. This decrease in cultivated lands after 1961 can be largely attributed to structural changes in industry and the socioeconomic changes in Japanese society.

Woodlands decreased slightly in the region between 1992 and 1960 (Table 12.2). This decrease can be attributed to two causes. First, woodlands had important socioeconomic value over this period. Second, some parts of the region had been designated Scenic Zones, which have scenic value and importance for conserving the quality of urban environment. These zones are conserved under the City Planning Law and other relevant laws (Manabe et al. 2007) enacted by the municipalities in 1936. Woodlands decreased substantially between 1961 and 1974, but increased slightly between 1974 and 2000. The large decrease of woodlands was mainly due to the fuel revolution in the 1960s. At that time, the value of woodlands greatly decreased, and a broad area of woodlands was cut all over Japan. In contrast, the slight increase in woodland area after 1974 resulted from planting in City Parks and the ecological succession of grasslands and shrub lands to forests following the fuel revolution. The same tendency was also reported in suburban areas (Manabe et al. 2003) and rural areas (Kamada and Nakagoshi 1990, 1996; Hong et al. 1995).

Table 12.2 Area (ha) of green space in the target region in each year. Areas were estimated by topographic maps (A) and aerial photographs (B)

Category	Year		
A	1922	1936	1960
Woodlands	2217.3	2125.8	2071.1
Cultivated lands	830.6	525.5	389.1
B	1961	1974	2000
Woodlands	1614.1	1265.7	1432.6
Cultivated lands	275.7	155.2	54.0

Hereafter, we focus on the ecological characteristics of urban forests in the case study area.

12.4 Community Structure of Urban Forests

Forests have various ecological functions such as habitat, conduit, filter, source, and sink (Forman 1995). Even though the focus here is on secondary forests in urban areas, they still play an important role for wildlife. The ecological functions of the forests depend not only on the structure of the forest community itself, but also on the landscape structure around the forests. In this section, the characteristics of the community structure of the urban forests at the study sites within the target region are described.

The study was conducted in 2004. The number of study plots, each 50 m × 20 m, was four for Yamada Green Park (Y1–Y4), two for Central Park (C1 and C2), and one each for Art-forest Park (AP) and Itozu Park (IP, Table 12.1). All stems of trees and shrubs taller than 2 m (mature stems) were identified by species and their diameter at breast height (DBH) were recorded. They were classified into two groups based on their vertical position: canopy stems (stems in the canopy layer) and understory stems (stems under the canopy layer). In this study, the canopy layer was taken to include foliage taller than 10 m.

Each of the six study plots (Y1, Y2, C1, C2, AP, and IP), each plot was divided into 10 meshes unites measuring 10 m × 10 m. One 5 m × 5 m quadrat and one 2 m × 2 m subquadrat were set in each mesh. All woody species, not including vines and forest floor species, of saplings (0.3 m–2.0 m) in each quadrat and seedlings (<0.3 m) in each subquadrat were identified and the number of individuals of each species counted.

Suzuki et al. (2004) reported that broad-leaved forests covered most of the Central Park area, although in 1961 the area consisted of various kinds of patches such as clear-cut sites, grasslands, shrub lands, and broad-leaved tall forests. In contrast, at that time, tall broad-leaved trees dominated Itozu Park, which was already isolated in 1961.

A total of 43 woody mature species grew in the plots: 32 were evergreen broad-leaved species and 11 were deciduous broad-leaved species. Evergreen broad-leaved species exceeded 75% of the cumulative basal area (BA) and 82% of stem number in all plots. Some community parameters such as the number of species and total BA of a plot were similar among all plots, except for C2. However, the density of the stems varied among plots (Table 12.3).

Two evergreen canopy species, *Persea thunbergii* and *Ilex integra*, two evergreen understory species, *Eurya japonica* and *Ligustrum japonicum*, and one deciduous understory species, *Ficus erecta*, occurred in all plots. Four evergreen subcanopy species including *Neolitsea sericea* occurred in seven plots, and two canopy species, *Castanopsis cuspidata* (evergreen) and *Rhus succedanea* (deciduous) and one

Table 12.3 Community parameters of each study plot

	Study plots							
	Y1	Y2	Y3	Y4	C1	C2	AP	IP
Number of species								
Evergreen trees	17	15	16	16	15	12	17	17
Deciduous trees	5	6	4	5	4	2	3	2
Total	22	21	20	21	19	14	20	19
Total basal area ($m^2\ ha^{-1}$)								
Evergreen trees	49.13	48.93	49.41	46.37	35.55	38.47	46.57	53.38
Deciduous trees	6.59	10.29	0.45	14.67	10.90	1.18	4.04	0.09
Total	55.72	59.22	49.86	61.04	46.45	39.65	50.61	53.47
Stem density (ha^{-1})								
Evergreen trees	3440	2840	2380	2660	2750	2230	2040	2100
Deciduous trees	200	260	510	410	460	150	310	40
Total	3640	3100	2890	3070	3210	2380	2350	2140

understory species, *Aucuba japonica* (evergreen), occurred in six plots. See Manabe et al. (2007) for a detailed description of the species composition in the plots.

Castanopsis cuspidata had the largest cumulative BA in four of the plots. Evergreen species of the family Lauraceae had the largest cumulative BA in the other plots. Among them, *Persea thunbergii* had the largest BA in three plots and *Neolitsea sericea* in one plot. The plots were classified into Castanopsis type (Y2, Y3, C1, and IP) and Lauraceae type (Y1, Y4, C2, and AP) based on the different dominants in the canopy layer.

Among the 42 species of saplings in the quadrats, 33 were evergreen broad-leaved species. Eight species including *Cinnamomum japonicum*, *Castanopsis cuspidate,* and *Neolitsea sericea* occurred in more than four plots (Table 12.4). Mature stems of these species also frequently occurred in the study plots. The number of species per plot varied from seven to 26. Among the 31 species of seedlings in the subquadrats, only nine were evergreen broad-leaved species. Eight species occurred in more than four plots. Among them were typical pioneer deciduous species such as *Zanthoxylum ailanthoides* and *Mallotus japonicus*, which occurred more frequently than saplings or matures trees of these species. The number of species per plot varied largely from three to 25.

The difference in species diversity (the number of species per plot) among plots became larger as the stage of life history became younger. Furthermore, the difference in species composition became larger as the stage of life history became younger. These findings suggest that differences in microenvironmental conditions between study sites, as affecting recruitment and/or establishment, recently became larger.

There were five evergreen species that occurred frequently through all life stages: *Castanopsis cuspidata*, *Persea thunbergii*, *Neolitsea sericea*, *Cinnamomum japonicum,* and *Eurya japonica*. These species have therefore been able to maintain their population at all the study sites. On the other hand, the seedling density of some

Table 12.4 The density of saplings (250 m^{-2}) and seedlings (40 m^{-2}) of dominant tree species in each study plot

Species	Study plots						Frequency[a]
	Y1	Y2	C1	C2	AP	IP	
Cinnamomum japonicum							
Saplings	57	1	210	2060	42	26	6
Seedlings	13	0	85	240	9	31	5
Castanopsis cuspidata							
Saplings	2	8	225	10	6	19	6
Seedlings	2	60	470	0	6	607	5
Neolitsea sericea							
Saplings	73	23	55	10	7	1	6
Seedlings	42	18	175	1510	28	3	6
Persea thunbergii							
Saplings	14	6	5	0	2	3	5
Seedlings	31	28	40	10	37	19	6
Camellia japonica							
Saplings	0	0	10	10	6	3	4
Seedlings	0	0	0	0	2	4	2
Eurya japonica							
Saplings	18	3	35	0	2	43	5
Seedlings	9	2	5	0	0	6	4
Aucuba japonica							
Saplings	18	2	0	60	12	0	4
Seedlings	8	0	0	0	0	0	1
Ligustrum japonicum							
Saplings	27	7	20	0	0	31	4
Seedlings	0	0	0	0	1	41	2
Xanthoxylum ailanthoides							
Saplings	0	0	0	0	0	4	1
Seedlings	4	2	535	0	0	6	4
Symplocos lucida							
Saplings	4	0	0	0	1	7	3
Seedlings	3	1	30	0	0	14	4
Daphniphyllum teijsmannii							
Saplings	30	0	0	0	0	0	1
Seedlings	25	1	5	0	0	9	4

[a]The number of plot that the species occurred

pioneer species, which usually recruit at the disturbed sites such as canopy gaps in old-growth forests or clear-cut sites, were especially high at some plots, such as in Itozu Park, the smallest site. Small sites may have recently been affected by the strong edge effects.

These urban forests, dominated by evergreen broad-leaved species, showed some differences in mature species composition from old-growth evergreen broad-leaved forests. For example, some evergreen species such as *Neolitsea sericea* and *Cinnamomum camphora* were dominant in these urban forests, but are found less often in old-growth forests in western Japan (Tanouchi and Yamamoto 1995; Manabe et al. 2000). In contrast, some evergreen canopy species such as *Distylium racemosum* and *Quercus salicina* and some evergreen subcanopy species such as *Cleyera japonica* and *Camellia japonica* were less dominant in the urban forests, although these species are dominant in old-growth forests.

These community characteristics of the urban forests may be attributed to fragmentation and past human disturbance in the study area, although the kind and degree of the effects vary between forests.

12.5 Genetic Diversity of *Neolitsea sericea*

Fragmentation of habitats for wildlife is one of the most serious factors causing a decrease of biodiversity (Wilcox and Murpphy 1985). Habitat fragmentation is especially damaging for species such as large mammals and birds, which need large habitats and live at low population densities. Furthermore, gene flow of some plant species among fragmented habitats may be maintained through their seed dispersal, and/or pollen flow if their habitats are less fragmented. Indeed, the pollens of certain plant species can be dispersed hundreds of meters (Chase et al. 1996). However, genetic diversity and gene flow of plants are still unknown factors in urban areas where their habitats have been rapidly and severely fragmented. In this section, we describe the genetic diversity at each habitat and genetic relations among individuals of plant species, *Neolitsea sericea*, using amplified fragment length polymorphism (AFLP), in order to understand habitat function of fragmented urban forests at the gene level.

Neolitsea sericea (Family Lauraceae) is a dioecious subcanopy evergreen species and occurs widely in secondary forests in the target region (Fig. 12.3; cf. Sect. 12.4). The species flowers in autumn, and ripens in the following autumn. Pollen and seeds

Fig. 12.3 Study species (*Neolitsea sericea*)

Table 12.5 The number of samples, bands, and polymorphic bands in each study plot

Study sites	Study plots	Mature trees			Saplings		
		Samples	Bands	Polymorphic bands	Samples	Bands	Polymorphic bands
Yamada Green Park	Y3	3	27	5	9	39	29
	Y4	6	30	22	22	47	42
Central Park	C2	30	44	39	10	41	33
Art-forest Park	AP	6	28	10	2	24	7

are dispersed by small species of Diptera and birds, respectively. Many frugivorous birds can easily move among fragmented forests if these forests are not widely separated; thus, the genetic diversity of bird dispersal plants could be used as an index of the degree of forest fragmentation. In order to assess the degree of forest fragmentation to the bird dispersal plants, *Neolitsea sericea* was chosen for this study.

Leaves from target species of mature trees growing at the study plots were collected and also from as many saplings in the quadrats within the plots as possible. This sampling was conducted at two plots in Yamada Green Park (Y3 and Y4), one plot in Central Park (C2), and one plot in Art-forest Park (AP). In total, 88 samples, 45 from mature trees and 43 from saplings were collected.

DNA was extracted from the sampled leaves using the methods described by Stewart Jr and Via (1993). Restriction enzyme treatment and ligation of extracted DNA were performed using the kits of Applied Biosystems Inc. (ABI) and Invitrogen. Next, the primer pair CTG-ACC, which was chosen through preliminary amplification, was also amplified selectively using a thermal cycler (Gene Amp PCR System 9700). Then, band patterns were determined from the amplified samples using a DNA sequencer (ABI Prism 310).

To evaluate the genetic similarity among individuals, the Jaccard index was calculated, using the data on the existence or nonexistence of the bands at each genetic locus, for each individual. These genetic similarities were used for obtaining clusters of genetically similar individuals by the group average method. Le Progiciel R v4.0 was used for these analyses.

Sixty-one bands, with the range of 35–232 bp, were detected using AFLP. The average number of bands per sample was 24.1. The maximum and the minimum number of bands was 47 for saplings in Y4 and 24 for saplings in AP (Table 12.5). Among them, polymorphic bands ranged from five (mature trees in Y3) to 42 (saplings in Y4). More than 90% of bands were polymorphic for the mature trees in C2 and the saplings in Y4. Most of the mature trees and saplings were distinguishable by their band patterns.

There were no clusters that were formed by all individuals in a single plot, indicating that the genetic diversity in each plot did not decrease for the species (Fig. 12.4). However, there were some clusters that were formed by genetically similar individuals. This tendency was remarkable in Central Park as small clusters

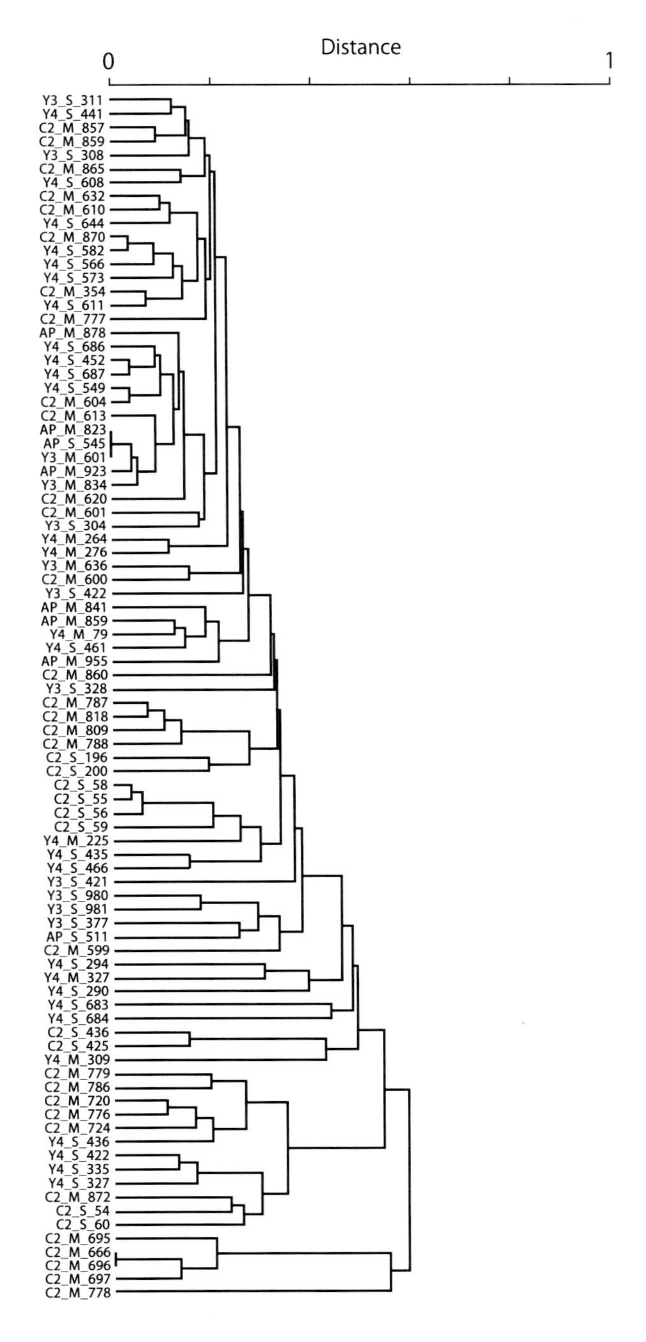

Fig. 12.4 Dendrogram for 88 samples based on cluster analysis using group average method. The code at the left side indicates each individual analyzed: plot (Y3, Y4, C2, AP)_growth stage (M: mature or S: saplings) _individual number. For example, Y3_S_311 indicates the sapling with its number of 311 in Y3

were formed by four or five genetically similar mature trees and saplings in C1. Other small clusters were also formed by the saplings in Y3 and Y4. Thus, small patches formed by saplings that might have derived from a single parent tree or genetically similar parent trees were recognized.

In the fragmented habitats, genetic diversity and gene flow are affected by the amount of seeds and pollen dispersed from other habitats. Furthermore, the amount of seeds and pollen dispersed from other habitats are affected not only by biological factors such as the mobilization capacity of seed dispersers and pollinators, but also by the landscape structure such as the area, shape, and connectivity of habitats and the surrounding matrix, and by the time since the habitats were fragmented.

The existence of small patches formed by genetically similar saplings suggests that the effects of habitat fragmentation on the genetic structure of *Neolitsea sericea* may have grown in recent years, from diverse condition when the mature had established in the study sites. Furthermore, the existence of these patches arising from genetically similar mature tree in Central Park may indicate that habitat fragmentation affected the genetic structure of species earlier in Central Park than in larger Yamada Green Park. Further studies, combined with investigation of the fauna, particularly the behavior of seed dispersers and pollinators, are needed to clarify the effects of fragmentation on the genetic structure of plant species.

12.6 Mammal Fauna in Urban Forests

The evaluation of fauna as well as flora in urban forests is important for a complete understanding of the habitat function of the forests. In addition, the interactions between plants and animals such as seed dispersal and grazing play important roles in the forest dynamics. Understanding fauna in urban forests is, therefore, a key aspect of evaluating the habitat functions and sustaining mechanisms of the forests.

Although 27 faunal species from 11 families are known in Kitakyushu City (Arai et al. 1992), the species living in specific urban areas are not known. In this section, we describe the mammal fauna of Yamada Green Park, Art-forest Park, Central Park, and Itozu Park, which was evaluated by camera trap method.

Photographs of wildlife were taken using cameras with an infrared motion sensor (Trail-Master, Goodson & Associates Inc.; Fieldnote II, Marif Co., Ltd), which are triggered by the body temperature of the wildlife (Table 12.6). The cameras were set at heights of 0.8–1.6 m, and were arranged to look downward or obliquely forward. Grains were used as inducement baits for wildlife at some photographic points in November and December 2004.

The number of photographs taken was 2687 over a total of 3339 days × cameras from November 2004 to February 2006. Among these, 1200 photos were deemed valid (Table 12.7, Fig. 12.5). The mammals photographed were Japanese badgers (*Meles anakuma*), Japanese martens (*Martes melampus*), weasels (*Mustela* spp.), raccoon dogs (*Nyctereutes procyonoides*), wild boars (*Sus scrofa*), Japanese hares (*Lepus brachyurus*), mice, domestic dogs (*Canis familiaris*), and domestic cats (*Felis catus*). Weasels (*Mustela* spp.) may include both the native Japanese weasel (*Mustela itatsi*) and non-native Siberian weasel (*Mustela sibirica*), because the

Table 12.6 Study periods in each study site

Study sites	Study periods	Inducement baits
Yamada Green Park	Nov. 19. 2004–Dec. 2, 2004	Partly yes
	Feb. 4, 2005–Feb. 23, 2005	No
	May 11, 2005–Jan. 31, 2006	No
Central Park	Jan. 29, 2005–Feb. 28, 2006	No
	May 12, 2005–Feb. 28, 2006	No
Art-forest Park	Dec. 9, 2004–Dec. 28, 2004	Partly yes
	Jan. 29, 2005–Feb. 12, 2005	No
	May 12, 2005–Feb. 28, 2006	No
Itozu Park	May 25, 2005–Feb. 28, 2006	No

Table 12.7 The number of photographs, valid photographs, valid days on which photographs were taken, and the photographic efficiency

Study sites	The number of photographs	The number of valid photographs	The number of valid days	The photographic efficiency
Yamada Green Park	1165	757	864	0.88
Central Park	451	125	780	0.16
Art-forest Park	476	184	1152	0.16
Itozu Park	595	134	543	0.25
Total	2687	1200	3339	0.36

species could not be consistently distinguished. Most of the mice were considered to be the large Japanese field mouse (*Apodemus speciosus*), although the other species might be included. In addition, some bird species, such as crows (*Corvus* spp.) and oriental turtledove (*Streptopelia orientalis*), some insects such as hornets and swallowtail butterflies, and humans taking a walk were also photographed.

Photographs from Yamada Green Park showed the greatest number of mammal species among the study sites (Table 12.8). The photographic efficiency, which was defined as the number of varied photographs divided by the number of varied days on which photographs were taken, was also the highest (0.88) among the study sites. Japanese badgers occurred most frequently in Yamada Green Park photographs followed by Japanese martens and wild boars. In Art-forest Park, five mammals—Japanese badgers, Japanese martens, weasels, raccoon dogs, and mice—were found, although the photographic efficiency was low (0.16). The most abundant mammal was again the badger. In Central Park, Japanese martens, weasels and mice were found; the photographic efficiency was also low (0.16). In Itozu Park, the photographic efficiency was slightly higher (0.25) with weasels being the most often photographed mammal and Japanese martens also evident. Only two species were found at all study sites: Japanese martens and weasels.

Fig. 12.5 Examples of mammals photographed by camera trap method. (**a**) Raccoon dog (*Nyctereutes procyonoides*) November 21, 2005 at Yamada Green Park. (**b**) Japanese badger (*Meles anakuma*) November 21, 2005 at Yamada Green Park. (**c**) Japanese hare (*Lepus brachyurus*) November 12, 2005 at Yamada Green Park. (**d**) Japanese marten (*Martes melampus*) June 16, 2005 at Itozu Park. Photographed date and time are printed at the lower right

Japanese martens can use arboreal space effectively as their habitat, and eat plentiful vegetables (Tatara and Doi 1994). Those findings indicate that secondary forests are important habitat for Japanese martens. Japanese martens were even found in Itozu Park, the smallest of the study sites, which faces onto a trunk road. The home range of the Tsushima marten (*Martes melampus tsuensis*), a subspecies of the Japanese marten, extends over 70 ha (Tatara 1993), suggesting that the Japanese martens found in Itozu Park may use the forests not only there but also in nearby Art-forest Park, and that the species can use fragmented forests as their habitats when that fragmentation is not severe.

Field mice were not found in Itozu Park. This was attributed to the relatively low mobilization capacity of mice (Hirota et al. 2004) and the fact that Itozu Park, where most of the area has been used as a zoo, has extensive artificial environments.

Japanese badgers and raccoon dogs were found in both Yamada Green Park and Art-forest Park. However, they were not found in Central Park, which is much larger than Art-forest Park. Japanese badgers were found more than raccoon dogs in the study sites, although generally badger populations in Japan have been decreasing except on Shikoku Island (Abe et al. 2005). Kaneko (2002) reported that the home range of Japanese badgers was 10.5–40.8 ha at Hinode-machi in Tokyo, and that area was thought to be small for the species. Only a few badger individuals can live

Table 12.8 The number of photographs of wildlife taken by trail cameras in each study site

Study sites	Japanese badgers	Japanese martens	Weasels	Raccoon dogs	Wild boars	Japanese hares	Field mice	Domestic dogs	Domestic cats	Humans	Birds	Insects	Unknown
Yamada Green Park	296	108	3	54	102	30	13	0	0	0	70	65	16
Central Park	0	20	12	0	0	0	2	0	15	0	34	41	1
Art-forest Park	23	17	5	15	0	0	14	1	7	5	66	29	2
Itozu Park	0	33	38	0	0	0	0	0	23	0	23	12	5

in Art-forest Park, and maintenance of the badger population may be difficult if the home range area reported by Kaneko (2002) is necessary. Further information such as metapopulation structures over a wider area is needed in considering the long-term population dynamics of the species in the study sites.

Wild boars and Japanese hares were not found in the study area except at Yamada Green Park. Wild boars, relatively large mammals, may need lager habitats such as Yamada Green Park for maintaining their population. On the other hand, the reason that Japanese hares, a small mammal, were not observed in the forests might be attributable to the type of vegetation, which is not suitable for hares to feed on except at Yamada Green Park, while the density of hare predators such as domestic cats, domestic dogs, Japanese martens, and weasels may be high in the other forests.

From this analysis, the habitat function for mammals is the highest in Yamada Green Park, because it is the largest of the sites studied and has extensive mountains behind the park. The other three sites are contiguous to each other, even though they are fragmented by artifacts such as roads. This landscape ecological characteristic suggests that the medium-sized mammals can move between the forests although mice cannot do so. On the other hand, these three sites are not suitable habitat for some mammal species, such as Japanese badgers, that depend on specific site characteristics.

12.7 Conclusion

Urban forests with various shapes and areas still exist in our target region of mid-northern Kitakyushu City. These forests, which are now dominated by evergreen canopy species, became fragmented and rapidly isolated in the 1960s. The species diversity and differences in composition of woody species among these forests were much larger for the young cohorts, consisting of smaller individuals, than the old cohorts. Thus, the effects of fragmentation on recruitment and/or establishment of woody species have increased recently. The molecular ecological approach applied in this study also indicated that the effects of fragmentation on the genetic structure of common woody species *Neolitsea sericea* have also become more severe in recent years. Furthermore, the habitat function of these forests for mammals differed among species. For example, the effects were severe not only for large mammals such as wild boars, which require large areas for living, but also for small mammals such as field mice, which have relatively small mobilization capacity. On the other hand, the effects varied between species for the medium-sized mammals.

Indirect Effects of Fragmentation on Woody Species Populations
The forests in the study sites were classified as Castanopsis type (Y2, Y3, C1, and IP) and Lauraceae type (Y1, Y4, C2, and AP). The number of adults (canopy trees and understory trees), saplings, and seedlings were counted for *Castanopsis cuspidata*, *Persea thunbergii*, and *Neolitsea sericea* in six plots (Y1, Y2, C1, C2, AP, and IP)

Table 12.9 The density (0.1 ha^{-1}) of adults, saplings, and seedlings of three dominant tree species at Castanopsis type plots and Lauraceae type plots

	Castanopsis type plots			Lauraceae type plots		
Species	IP	C1	Y2	AP	C2	Y1
Adults	80	39	28	27	5	0
Saplings	76	900	32	24	40	8
Seedlings	15,175	11,750	1500	150	0	50
Adults	8	5	38	9	15	52
Saplings	12	20	24	8	0	56
Seedlings	475	1000	700	925	250	775
Adults	0	36	5	6	36	36
Saplings	4	250	92	28	40	292
Seedlings	75	4375	450	700	37,750	1050

following the methods described above. Here, canopy trees and understory trees were defined as individuals in the canopy layer and those under the canopy, respectively.

For these three species, the density of seedlings was high in the plots in which the density of conspecific adults was also high (Table 12.9). This tendency was strong for *Castanopsis cuspidata* than for the other two species. This is partly attributable to the effects of the density of adults as seed sources. The success of seedling recruitment must be stronger for the trees bearing barochory seeds (*Castanopsis cuspidata*) than those bearing anemochory and zoochory (*Persea thunbergii* and *Neolitsea sericea*) seeds within small areas such as the plots studies.

Among Castanopsis type forests (Y2, C1, and IP), however, the density of *Castanopsis* seedlings was much higher in C1 and IP than in Y2. C1 and IP showed very small population densities of mice, which are the main consumers of *Castanopsis* seeds, indicating that the relatively low seedling density of *Castanopsis* was a contributing factor to the low density of mice, which was also affected by fragmentation of the habitats. Therefore, the population dynamics of *Castanopsis cuspidata* could be affected by forest fragmentation indirectly in the target region.

Thus, we conclude that it is important to analyze the effects of fragmentation on various taxa from hierarchically related levels of biological organization in order to evaluate the habitat function of urban forests for wildlife.

Acknowledgements The studies described in this chapter were partly supported by a Grant in Aid of Science Research (90359472) from the Japan Society for the Promotion of Science (head of project: T. Manabe) and projects for contributions to the region from the Ministry of Education, Culture, Sports, Science and Technology and the Kyushu Institute of Technology (head of project: K. Ito). We thank the members of the Yamada Green Park and Itozu Park for permitting these studies. We also thank D. Isono, T. Umeno, J. Kaku, H. Kashima, T. Suzuki and D. Hashimoto for their assistance.

References

Abe H, Ishii N, Itoo T, Kaneko Y, Maeda K, Miura S, Yoneda M (2005) Guide to the mammals of Japan. Tokai University Press, Tokyo (in Japanese)

Arai S, Izawa M, Ono Y (1992) Mammals of Yamada Park in Kitakyushu City. In: Ota M et al (eds) Nature of Yamada Park, Kitakyushu City, Japan. Kitakyushu Mus. Nat. Hist, Kitakyushu, pp 223–244

Chase MR, Moller C, Kesseli R, Bawa S (1996) Distant gene flow in tropical trees. Nature 383:398–399

Forman RTT (1995) Land mosaics –the ecology of landscape and region. Cambridge Univ Press, Cambridge

Fukamachi K, Iida S, Nakashizuka T (1996) Landscape patterns and plant species diversity of forest reserves in the Kanto Region, Japan. Vegetatio 124:107–114

Hirota T, Hirohata T, Mashima H, Satoh T, Obara Y (2004) Population structure of the large Japanese field mouse, *Apodemus speciosus* (Rodents: Muridae), in suburban landscape, based on mitochondria D-loop sequences. Mol Ecol 13:3275–3282

Hong SK, Nakagoshi N, Kamada M (1995) Human impacts on pine-dominated vegetation in rural landscape in Korea and western Japan. Vegetatio 116:161–172

Iida S, Nakashizuka T (1995) Forest fragmentation and its effects on species diversity in sub-urban coppice forests in Japan. For Ecol Manag 73:197–210

Ishii HT, Iwasaki A, Sato S (2004) Seasonal variation of edge effects on the vegetation, light environment and microclimate of primary, secondary and artificial forest fragments in south-eastern Hyogo prefecture. In: IUFRO international workshop on landscape ecology 2004. FFPRI, Tsukuba, pp 64–67

Ishii HT, Manabe T, Ito K, Fujita N, Imanishi A, Hashimoto D, Iwasaki A (2010) Integrating ecological and cultural values toward conservation and utilization of shrine/temple forests as urban green space in Japanese cities. Landsc Ecol Eng 6:307–315

Kamada M, Nakagoshi N (1990) Patterns and processes of secondary vegetation at a farm village in southwestern Japan. Jpn J Ecol 40:137–150 (in Japanese with English synopsis)

Kamada M, Nakagoshi N (1996) Landscape structure and the disturbance regime at three rural regions in Hiroshima prefecture, Japan. Landsc Ecol 11(1):15–25

Kamada M, Nakagoshi N, Nehira K (1991) Pine forest ecology and landscape management: a comparative study in Japan and Korea. In: Nakagoshi N, Golley FB (eds) Coniferous forest ecology from an international perspective. SPB Academic, The Hague, pp 43–62

Kaneko Y (2002) Inner structure of the badger (*Meles meles*) home range in Hinode-town. Jpn J Ecol 52:243–252 (in Japanese with English synopsis)

Manabe T, Nishimura N, Miura M, Yamamoto S (2000) Population structure and spatial patterns for tree in a temperate old-growth evergreen broad-leaved forest in Japan. Plant Ecol 151:181–197

Manabe T, Kashima H, Ito K (2003) Stand structure of a fragmented evergreen broad-leaved forest at a shrine and changes of landscape structure surrounding a suburban forest, in northern Kyushu. J Jpn Rev Tech 28:438–447

Manabe T, Ito K, Isono D, Umeno T (2007) The effects of the regulation system on the structure and dynamics of green space in an urban landscape: the case of Kitakyushu City. In: Hong SK, Nakagoshi N, Fu B, Morimoto Y (eds) Landscape ecological applications in man-influenced areas. Springer, Dordrecht, pp 291–309

Matsuoka RH, Kaplan R (2008) People needs in the urban landscape: analysis of landscape and urban planning contributions. Landsc Urban Plan 84:7–19

McDonnell MJ, Pickett STA (1990) Ecosystem structure and function along urban-rural gradients: an unexpected opportunity for ecology. Ecology 71:1232–1237

Mitsuda Y, Manabe T, Ito K, Kashima H, Suzuki T (2003) The methods of making the digital vegetation maps by using digital orthophotographs –in the case of the Yamada Green Park in

Kitakyushu City. Bull Kitakyushu Mus Nat Hist Hum Hist Ser A 1:57–65 (in Japanese with English abstract)
Murakami K, Maenaka H, Morimoto Y (2004) Factors influencing species diversity of ferns and fern allies in fragmented forest patches in the Kyoto City area. Landsc Urban Plan 70 (3–4):221–229
Murcia C (1995) Edge effects in fragmented forests: implications for conservation. Trends Ecol Evol 10:58–62
Stewart CN Jr, Via LE (1993) A rapid CTAB DNA isolation technique useful for RAPD fingerprinting and on the PCR applications. BioTechniques 14:748–750
Suzuki T, Manabe T, Ito K, Umeno T (2004) Analysis of landscape changes by using digital vegetation map in mid-northern region in Kitakyushu City. Bull Kitakyushu Mus Nat Hist Hum Hist Ser A 2:79–85 (in Japanese with English abstract)
Tanouchi H, Yamamoto S (1995) Structure and regeneration of canopy species in an old-growth evergreen broad-leaved forest in Aya district, southwestern Japan. Vegetatio 117:51–60
Tatara M (1993) Social system and habitat ecology of the Japanese marten *Martes malampus tsuensis* (Carnivore: Mustelidae) on the Islands of Tsushima. Ph D Dissertatiom, Kyushu University
Tatara M, Doi T (1994) Comparative analysis on food habits of Japanese marten, Siberian weasel and leopard cat in the Tsushima Islands, Japan. Ecol Res 9:99–107
Tojima H, Koike F, Sakai A, Fujiwara K (2004) Plagiosere succession in urban fragmented forests. Jpn J Ecol 54:133–141 (in Japanese with English abstract)
Wilcox BA, Murpphy D (1985) Conservation strategy: the effects of fragmentation on extinction. Am Nat 125:879–887
Yong A, Mitchell N (1994) Microclimate and vegetation edge effects in a fragmented podocarp-broadleaf forest in New Zealand. Biol Conserv 67:63–72

Chapter 13
The Effects and Functions of Spatial and Structural Characteristics of Shrine Forests as Urban Green Space

Naoko Fujita

Abstract In order to promote biodiversity and ecological networks in an urban area, it is important to understand both the area's spatial and structural characteristics and its cultural, historical, and social characteristics. The objective of this study is to analyze the effects and functions of spatial and structural characteristics of shrine as urban green space by focusing on the spatial composition and form of slopes in the center of Tokyo and evaluating their green space possession. An investigation was conducted on the preservation of green areas in Shinto shrines built on a steep slope within the JR Yamanote Line loop in the center of Tokyo's metropolitan area in Japan. An analysis reveals that Shinto shrines located in hilly topography were found to be particularly good for possession of green space in accordance with the angle of inclination of the slope or the location of the shrine forest space. Although the number of this type of Shinto shrine is small, the shrine forest space is formed around all sides of the building, which is the reason for high green space possession. Moreover, there was high possession of green spaces in cases where the building was located on the concave knick line, but this was because the shrine forest spaces were formed to the rear of the buildings and these shrine forests are recognized as confined spaces or considered sacred. Our study also found that linear green spaces, a private approach road connecting the shrine to a local street or public road, were able to have the potential of the function as corridors for connectivity to other green spaces in urban area.

Keywords Sacred forest · Urban green space · Shinto shrine · GIS · Tokyo · Japan

N. Fujita (✉)
Department of Environmental Design, Faculty of Art and Design, University of Tsukuba, Tsukuba, Japan
e-mail: fujita.naoko.gf@u.tsukuba.ac.jp

K. Ito (ed.), *Urban Biodiversity and Ecological Design for Sustainable Cities*,
https://doi.org/10.1007/978-4-431-56856-8_13

13.1 Introduction

In urban areas where high-density land use and high environmental load are concentrated, the breeding and living environment for wildlife is limited to green spaces (Grimm et al. 2000; Ministry of Environment 2002). Therefore, urban forests and other green spaces are the best means for judging the structure and diversity of an ecosystem and assessing its condition and function (Godefroid and Koedam 2003; Zhao et al. 2010). As such, the preservation of cultural or natural green spaces and the planned creation of public green spaces is a common theme in modern urban development, and significant resources have been devoted to the analysis of environmental potential and green space value (Antrop 2005; Balram and Dragićević 2005; Breuste 2004; Iwami et al. 1987). This research has shown that, historically, large-scale green spaces have been preserved, while small-scale green spaces have undergone transformations to accommodate urban land use. Consequently, green spaces tend to be fragmented and reduced by modern uniform spatial development (Chace and Walsh 2006; Fujita and Kumagai 2004; Organization for Landscape and Urban Green Technology Development 2006). As a result, the amount of natural space where diverse wildlife can breed has been greatly reduced in urban areas (De Chant et al. 2010; Organization for Landscape and Urban Green Technology Development 2006). Tokyo, the largest and most heavily urbanized city in Japan, is an excellent example of this effect. Moreover, the decline in green spaces in cities results in environmental deterioration, such as the impoverishment of flora and fauna, the heat island effect, heavy rain in urban areas, and the loss of traditional culture and history—the last one being the very foundation on which such spatial preservation was cultivated (Ishii et al. 2010; Koga 2002; Kyakuno 2005; Mikami 2005; Mikami et al. 2005; Okazaki and Kato 2005; Uchida et al. 2002).

Forman (1995) and Dramstad et al. (1996) suggested the principles of landscape ecology. To obtain an ecosystem-dependent design solution, biodiversity is an essential natural capital that must be reassessed (Dramstad et al. 1996; Morimoto 2011). According to Forman's (1995) "form-and-function" principles of landscape ecology, "compact forms are effective in conserving resources" and "convoluted forms enhance interactions with the surroundings." An important characteristic influencing interactions between the matrix and a patch is boundary shape (Forman and Godron 1986). In addition to these basic definitions, urban biodiversity is also shaped by social or cultural contexts (Nilon 2011). Riitters et al. (1995) focused on the relationship between the spatial and structural features of the landscape and the range of potential ecological functions, using multivariate factor analysis to identify patterns. Common axes and dimensions of structure were identified. Previous analyses have shown that increasing distances between patches significantly reduces landscape connectivity, and that matrix elements have a lesser effect on landscape connectivity than the effects of habitat elements (Goodwin and Fahrig 2002).

Among the researchers employing this methodology are Fujita and Kumagai (2004), who analyzed the changes in green spaces at the center of Tokyo from the perspective of uneven distribution of topography and green spaces. Their research indicates that the reduction of green spaces owing to spatial development accompanying urbanization has led to an uneven distribution of topography and green spaces. In addition, they evaluated the survival characteristics of green spaces on slopes. Slopes and green spaces have been studied since the 1980s (Kim et al. 1989; Tabata et al. 1984; Tabata and Watanabe 1986), and interest in such studies in Japan has increased through a series of publications called Edo-Tokyo studies (Jinnai 1985; Maki 1980). In recent years, such studies have focused on the changes in the scenery of slopes and steps (Matsumoto and Tonuma 2004; Randall 2004).

The range of definitions and the recognition of urban biodiversity is also shaped by sociocultural contexts (Nilon 2011), with cultural processes expressed through design and planning decisions that shape the pattern and distribution of urban biodiversity (Millard 2010). Alban and Berwick (2004) have described this influence on Shintoism and Buddhism, religions common in Japan and both of which attribute a spiritual and divine nature to trees. Japanese temples traditionally comprise small gardens or huge parks that are designed and maintained in keeping with religious tenets, with special attention given to trees. Likewise, some religious institutions own huge forest estates, and although the way in which these areas are managed has changed significantly with the economic development of the country, it nonetheless continues to be emblematic of a distinctive cultural vision of nature.

Many shrine forests have survived urban development only because they are regarded as sacred places (Hashimoto et al. 2006) or are protected by law after being designated as natural monuments or parts of scenic zones (Suzuki et al. 2004; Fujita et al. 2005; Umeno et al. 2006; Manabe et al. 2007). Imanishi et al. (2005) attempted to conserve all plant species found in the Kyoto shrine forests for researching the herbaceous plant species richness and species distribution patterns. They found that the largest forest contained over 50% of the tree species and small patches contained the Red-Listed species.

Another factor affecting the urban green space is its nature-oriented design. Morimoto (2011) suggested that the biodiversity is not only a resource of culture but also a product of it. One of the characteristic forms of biodiversity nurtured by culture in Kyoto is Japanese gardens, which play an important role in providing an urban wildlife habitat for ferns and mosses. Because shrines are religious facilities, there is little concern about them being destroyed or demolished as easily as other types of land uses or facilities. Shrine vegetation provides recreational opportunities, wildlife shelter, and floristic biodiversity. Consequently, utilizing shrine forests strategically and intensively as part of landscape design in urban greening plans may be a desirable means of improving biodiversity and ecosystem services. In this sense, we can say that urban shrines can help promote urban biodiversity. Species richness in shrine forests in southern Japan is decreasing with a decline in forest area, and many rare species, such as ferns, moss, and orchids, are getting extinct (Ishida et al. 2005).

Shintoism, an indigenous Japanese religion characterized by the worship of ancestors and a belief in nature spirits, holds that a god resides in everything; temples and shrines to the unified divinity or to individual deities are the religion's typical places for worship. Every community, no matter how small it may be, usually has one or several shrines; there are thought to be over 80,000 such shrines in Japan. *Shasoh*, which refers to the wooded area of a Shinto shrine or the woodland surrounding it (Ueda 2001), is a term that refers to a forest or woodland that is biologically healthy—a concept that embraces the Japanese approach to the natural world (Fujita et al. 2005), specifically as a space that can be defined from diverse perspectives (Fujita et al. 2007). Almost all shrines are located in forested areas (Fig. 13.1). Even small shrines built in extremely urbanized areas, such as central Tokyo, tend to have small woods on their grounds (Fig. 13.2). However, at present, there is little interest in shrine forests for urban green space planning, and there has been no strategical incorporation of such spaces in green urban development because Shinto shrines are categorized as religious institutions and are therefore exempt from city planning decisions and general urban analysis. This situation makes it difficult to understand Shinto shrines as meaningful components of urban biodiversity and ecosystem services.

It is considered that analysis of urban Shinto shrines in their role as urban green spaces—taking urban planning aspects as well as ecological aspects into consideration—is a field that offers considerable scope for further research. Urban Shinto shrines have the potential to operate as urban green spaces, with quantifiable ecological benefits (Fujita and Kumagai 2007), and it is therefore necessary to reevaluate these spaces as such. This study aims to analyze the effects and functions of spatial and structural characteristics of shrine as urban green space by focusing on the spatial composition and form of slopes in the center of Tokyo.

13.2 Data and Methods

13.2.1 Study Area

An area within the JR Yamanote Line was selected as the subject of this study (Fig. 13.3). The Yamanote Line is 34.5-km long. It passes through the main areas of Tokyo's center and encompasses an area of 6400 ha. The area within the Yamanote Line is characterized by its undulating topography. The Yamanote Uplands, located within this area, include the Ueno, Hongo, Toyoshima, and Yodobashi uplands, and the valley areas defined by four rivers, and the area extends to the end of Musashino Daichi (the Musashino uplands). The natural environment of the area under consideration is mainly classified as urbanized and partly residential with extensive green space (Ministry of the Environment Biodiversity Center of Japan 2008).

A survey of special plants and vegetation conducted in this study identified woodlands around shrines and temples, in public gardens, in universities, and in parks and similar areas. A survey of flora and fauna distribution focused on insects, large trees in woodlands, trees at shrines and temples, trees in universities, and trees in parks. Urban areas are often characterized by low biodiversity, the introduction of nonnative species, and simplified species composition (Grimm et al. 2000; McKinney 2002; Lundholm and Marlin 2006). It was found that these research areas show

Fig. 13.1 Examples of Shrines surrounded by forest at Ōji shrine in Kita-ku, Tokyo prefecture, Japan (above), and at Hatonomori Hachiman shrine in Shibuya-ku, Tokyo prefecture, Japan (below). Shinto is an indigenous primordial Japanese religion characterized by practices such as worship of ancestors and belief in nature spirits. Photo by N. Fujita

characteristics of typical urban ecosystems with a reduced number of different species and reduced population of each species. In other words, compared with the natural ecosystem, species at the top of the ecosystem are either reduced or not breeding, and species that are poor adapters or nonadapters to human development

Fig. 13.2 An example of a small shrine built in an extremely urbanized area at the Kiyokawa Inari shrine and Umemori Inari shrine in Taito-ku, Tokyo prefecture, Japan. Even small shrines in extremely urbanized areas have trees on their grounds. Photo by N. Fujita

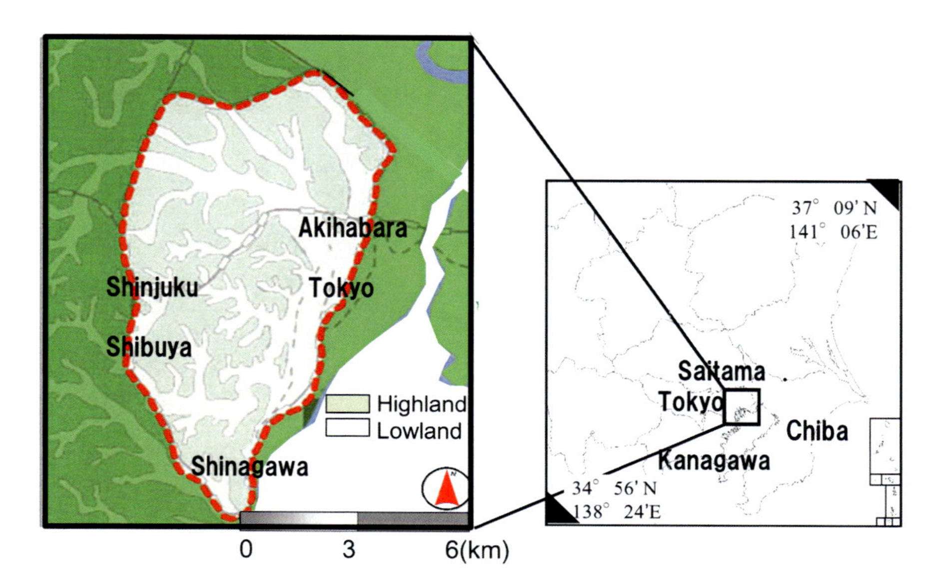

Fig. 13.3 Area researched. The research area is within the JR Yamanote line loop, which is indicated by the dotted line

have been eliminated. As a result, naturalized plants and naturalized animals are commonly found in such areas.

13.2.2 Extraction by GIS of Shinto Shrines Located on Slopes

In Tokyo, shrines are randomly distributed across the landscape, but tend to be accompanied by certain geographical features such as small hills, spring water, streams, and rivers, which are objects of nature worship (Fujita and Kumagai 2007; Ishii et al. 2010). Geographic factors such as elevation, slope, and distance have a significant impact on the changes in forest landscape, and the integrative geographic environmental factors index is valuable in describing landscape dynamics (Liang and Ding 2010). In this study, we used two criteria to define Shinto shrines located on slopes: (1) they had to be located on slopes as defined by microtopography[1] and (2) they had to be affiliated with the local Shinto shrine agency.

First, with regard to criterion (1), the following procedure was followed. Topographical index data were prepared according to "Numerical Map on a 5 m Mesh (Elevations)" published by the Geospatial Information Authority of Japan in 2003. The digital data, obtained using an aerial laser scanner, provide details down to the microtopographical level. This enables the extraction and comparison of microtopographical differences in the case of flat areas where there are few undulations, as compared to mountainous terrain or narrow areas. Therefore, these data were effective in extracting the microtopography of the shrines that were the subject of this research. Of these maps, "Tokyo Metropolitan Wards," published on December 1, 1993, was chosen for this study.

The topographical indices used were elevation, slope, Laplacian, and the openness of the site; they were used because they are effective for extracting microtopography in comparatively narrow areas such as those within the Yamanote Line. According to Fujita and Kumagai (2006), the Laplacian topographical index can express differences in topography when the differences cannot be clearly determined using the slope topographical index, and the openness index helps us understand topography over a wide area. To determine the relationship between green spaces around shrine forests and topography, we need to understand the trend in relation to the openness of the sites. In addition, using the slope and the Laplacian helps determine connectivity in detail.

[1]The definition of microtopography used in this study is taken from the New Geographical Dictionary (Fujioka 1979): "Small-scale topography (undulations, expanses) that can sufficiently enter the range of human observation, but are small enough that the topography cannot be sufficiently expressed on normal topographical maps. Topography with relative heights 10^{-2}–10^{2} m, widths and lengths 10^{3}–10^{3} m."

1. Topographical index data for elevation: Elevation is the distance from the sea level. The original DEM data from the "Numerical Map on a 5 m Mesh (Elevations)" was used to prepare the elevation topographical index data. The data were imported into GIS and, following georeferencing, processing was performed by the automatic positioning mosaic process such that it was possible to deal with a single image.
2. Topographical index data for slope: The slope is an index of the angle with respect to the horizontal and is expressed as 0–90°. For preparing the slope topographical index data, DEM data were imported into GIS, georeferencing was carried out, and on the basis of a raster image produced by a mosaic process, a shaded diagram of the slope and slope direction was produced using a raster process.
3. Topographical index data for Laplacian: Laplacian is an index that emphasizes sites with a large incline, with convex upwards expressed as "−," concave upwards as "+," and flat as "0." To prepare the Laplacian topographical index data, DEM data were imported into GIS and georeferencing was carried out similar to the preparation of the slope topographical index data. Then, on the basis of a raster image produced by a mosaic process, the data were prepared from a spatial filter close to 4 (close) by a filter analysis of the raster process.
4. Topographical index data for openness: openness is an index of the degree of openness to the ground, and is expressed as 0–360°. To prepare the openness topographical index data, DEM data were imported into GIS and georeferencing was carried out similar to the preparation of the slope topographical index data. On the basis of a raster image produced by a mosaic process, analysis was carried out using Spatial Manipulation Language (SML), a programming language in which we can write scripts for manipulating geographical spatial data objects. Next, with regard to criterion (2) mentioned above, we identified Shinto shrines affiliated to the Local Shinto Shrine Agency. First, information on all of the Shinto shrines located within the Yamanote Line was obtained. As of 2010, there was no digital data for individual Shinto shrines available. The sources of digital data containing Shinto shrine information, although not for individual Shinto shrines, include "Detailed Numerical Information (Land Use 10 m Mesh) in 1994" and "Numerical Maps 25000 (Place Name, Public Facilities) in 2001," both published by the Geospatial Information Authority of Japan. However, these data are not categorized as analysis data, and thus, we could not use these data to extract Shinto shrine point data. Therefore, we used the Tokyo Shinto Shrine Directory (1986), which is the most comprehensive document for Shinto shrines in Tokyo. First, the information from this directory was input into Excel, the input data were temporarily converted into comma-separated values, and address matching was carried out. This process was performed using the Center for Spatial Information Science address matching service provided by the Tokyo University Center for Spatial Information Science. On the basis of this data, point data were imported and georeferencing of the imported data was carried out.

Finally, the results of (1) and (2) were overlaid, and Shinto shrines that coincided with both (i.e., Shinto shrines both located on slopes as defined by microtopography and affiliated with the Local Shinto Shrine Agency) were extracted.

Of the Shinto shrines extracted, those that had a green space were selected for study. In this regard, "having a green space" was determined as follows: the Shinto shrine point data on the existing GIS and aerial photographic images was overlaid, and Shinto shrines for which green coverage provided by tall trees could be recognized were selected. Aerial photographs with a resolution of 50-cm and 24-bit full-color digital images taken from April to June 2003 were used in this analysis. Tall tree green coverage included single trees and rows of trees in addition to woodlands. In addition, as the judgment was made from aerial photographs, the state of the floor of the woodland, or in other words, the intensity of human use, was not evaluated. The following in situ surveys were carried out for the Shinto shrines that were selected for this survey.

13.2.3 Survey of the Present Situation of Shinto Shrines Located on a Slope and Having Green Spaces

Although shrines are owned by local parishes allied with Japan's Association of Shinto Shrines and are managed and maintained by Shinto priests, the temptation to employ the spaces for financial or other gains is tremendous. Traditionally, many shrine forests have been maintained with minimum vegetation management because it was believed that the natural state of the forest would best be preserved by minimizing human intervention (Hashimoto et al. 2007; Ishii et al. 2010). The degree of management therefore depends largely on the consciousness, knowledge, or preference of the relevant priest or parish (Hashimoto et al. 2007). Accordingly, field surveys and interview surveys were conducted with regard to the status and management of green spaces (Table 13.1). The field surveys collected data on the status of the green spaces and constituent elements other than green spaces within the boundaries. First, regarding the status of the green spaces, visual surveys were conducted of the forest physiognomy, vegetation, the presence or absence of seedlings, and the presence or absence of notable species. Regarding the constituent elements within the boundaries, visual surveys were conducted for matters related to the constituent elements of the Shinto shrine, vacant spaces, and the presence or absence of facilities other than the constituent elements. We plotted these elements on our maps. The vacant spaces of the constituent elements within the boundaries and the approach roads to the shrine are elements that are difficult to interpret from aerial photographs captured by geographical information systems (GIS); therefore, in order to supplement the spatial information, actual measurements were taken during the in situ survey, which assisted in the calculation of areas.

Table 13.1 Contents of in situ survey

Category	Sub-category	Sub-sub-category	Detailed category	Notes
Survey item	State of green space	Woods	–	[a]
		Plants	–	[b]
		Planted from seed or not	–	[c]
		Notable species present or not	–	[d]
	Constituent elements within boundary	Shrine related	Main building, worship building, Inari, other buildings	–
			Torii, hand washing place	–
		Vacant space	Vacant places, approach roads	–
		Other facilities	Public buildings, firefighting stores, kindergartens, gateball fields, parks, memorials	–
Hearing survey	Items regarding specific relationships	National, city, or ward designations	Memorials, historical remains, protected areas, etc.	–
	Items regarding green spaces	Establishment	–	–
		State of preservation/change	–	–
		State of management	–	–
		Awareness	–	[e]

[a]Woods are named using representative plants
[b]All plants in the green space were recorded
[c]Carried out in order to determine the possibility of natural renewal
[d]Growth of chusan palm, Japanese aucuba, and fatsia, as characteristic urbanized trees, were indicated. The area that was the subject of this research is a typical urban area, so the presence or absence of chusan palm and Japanese aucuba as notable species was recorded
[e]Historical value, ecological value, spiritual satisfaction, etc.

The data obtained in these in situ surveys were processed by converting the positional distributions into spatial information, which was then plotted on digital diagrams within the boundaries; we then performed the analysis.

13.2.4 Analysis of Green Space Possession

In this study, we analyzed the possession of green spaces containing trees in woodland areas, single trees, or rows of trees within the boundary of a Shinto shrine. To analyze green space possession, each location was first georeferenced on an aerial photographic image (1 file, 2 × 1.5 km) and then imported. We then carried out a mosaic process. Next, detailed illustrations (e.g., sketches, layout drawings, and building drawings) of the Shinto shrines were obtained. However, since the sources

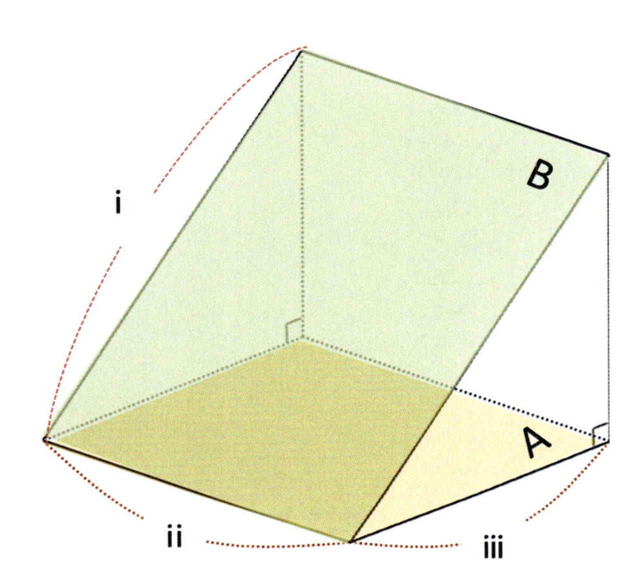

Fig. 13.4 The concept of area calculation in this research. This study focused on Shinto shrines located on slopes, and as such, analysis was not conducted using the plane area ($A = ii \times iii$) with conventional topographical maps. Rather, the study performed an analysis wherein the land area from the actual shape was calculated ($B = i \times ii$)

of these drawings were different, the method of preparation was not uniform and there were issues with scales or elevations not being accurate. A combination of the aerial photographic images, land use drawings, and 25,000-scale topographical maps that were already corrected was used as the reference layer for georeferencing. Next, these photographic images, drawings, and maps were overlaid on the georeferenced Shinto shrine drawings, and polygons representing the shapes of the sites of the Shinto shrines and those representing the shapes of the shrine forests within the sites were prepared. The constituent elements obtained in the in situ survey were then plotted on the illustrations. The relative position of green spaces was determined on the basis of the distribution and area of the main shrine buildings and the approach roads, and by calculating the area of the shrine forest as a percentage of the area of the Shinto shrine. Further, since this study focused on Shinto shrines located on slopes, the analysis did not employ the plane area (Fig. 13.4a) with conventional topographical maps; rather, the land area was calculated from the actual shape (Fig. 13.4b). In addition, the topographical points of inflection were divided into a convex knick line (also known as a break line, wherein the slope increases in obliquity to form a convex outer surface) and a concave knick line (where the slope decreases in obliquity to form a concave outer surface) (Fig. 13.5; Tamura 1990). These were determined by the change in the Laplacian value in this study. Using this process, information regarding the Shinto shrines located on slopes was reproduced in a three-dimensional form that more closely depicted the space within the boundaries of the Shinto shrines.

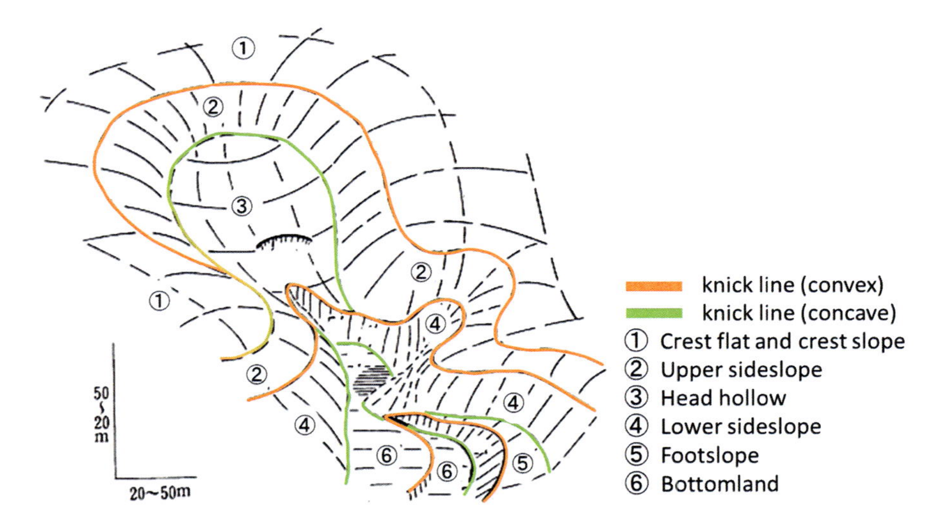

Fig. 13.5 Explanatory diagram for knick lines (convex and concave) based on a frame format of a microtopography of hilly terrain area by Tamura (1990)

13.3 Results

13.3.1 Shinto Shrines Located on Slopes in the Center of Tokyo

Table 13.2 shows the total number of Shinto shrines within the research area (125; 15.5% of the total number within the Metropolitan Tokyo area), the number of those Shinto shrines in green spaces (44; 35.2% of the 125 shrines), and the number of those Shinto shrines located on slopes (27; 21.6% of the 44 shrines in green spaces). We found that 68% of the shrines are built on slopes in Tokyo (Fujita and Kumagai 2007). In other words, they are more likely to be built where land is developed for urban purposes. This tendency was not observed among temples and parks. These results show that shrines have been evaluated as spaces that may become rich green spaces. Since shrines are evenly distributed and located in all areas, they become the most familiar green space in any area, and, compared with other facilities, they have strong ties with geographical features. For example, the Tokyo metropolitan area is

Table 13.2 Summary of Shinto shrines that were the subject of the research

No. Shinto shrines in the research area	
No. Shinto shrines within Yamanote line loop	125 (100%)
No. of these Shinto shrines having green spaces	44 (35.2%)[a]
No. of these Shinto shrines located on slope	27 (21.6%)[a]

[a]Numbers within parentheses are percentages of the number of Shinto shrines within the Yamanote line

coated with asphalt and concrete; shrines and their forests, on the other hand, are ideal habitats for insects, grass, and moss. Moreover, the inorganic environment at the top and bottom of the slopes differs.

13.3.2 Green Space Possession of Shrine Forest Spaces

The results for the spatial constituent elements are shown in Fig. 13.6. Some Shinto shrines contain large or old trees that are considered to be sacred (44.4%), and linear green spaces were mainly the lines of trees on the sides of the approach roads, which are found in all Shinto shrines. With regard to the constructed facilities that have no relationship to the original Shinto shrine, 26.7% were found to be public buildings; 9.6%, firefighting stores; 3.7%, kindergartens; 9.6%, gateball fields; 27.4%, parks; and 17.8%, memorials. The forests around such facilities include many rare plant communities, which are remnants of regional endemic vegetation. The large heritage trees are important habitats for birds, insects, and small animals in urban areas, and are designated as natural monuments by local and federal governments.

Using the areas within the boundaries of Shinto shrines located on a slope and having a green space and the area of the green space derived this way, the green space spatial composition patterns were obtained in accordance with the location of the green space within the Shinto shrine (Fig. 13.7). The results show that the Shinto shrines can be broadly classified into two categories on the basis of position and slope: those in which the main building of the Shinto shrine is on the concave knick

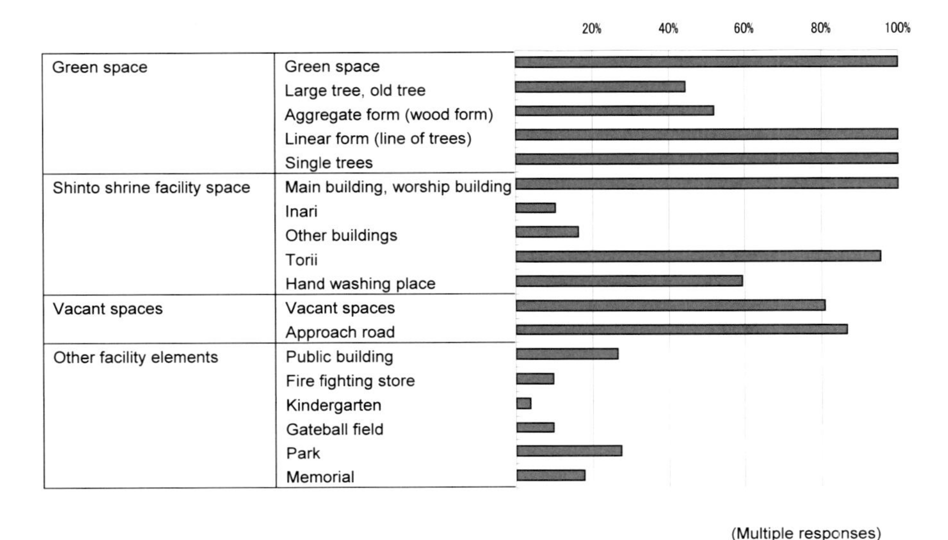

Fig. 13.6 Shinto shrine spatial constituent elements by rate of appearance at all Shinto shrines

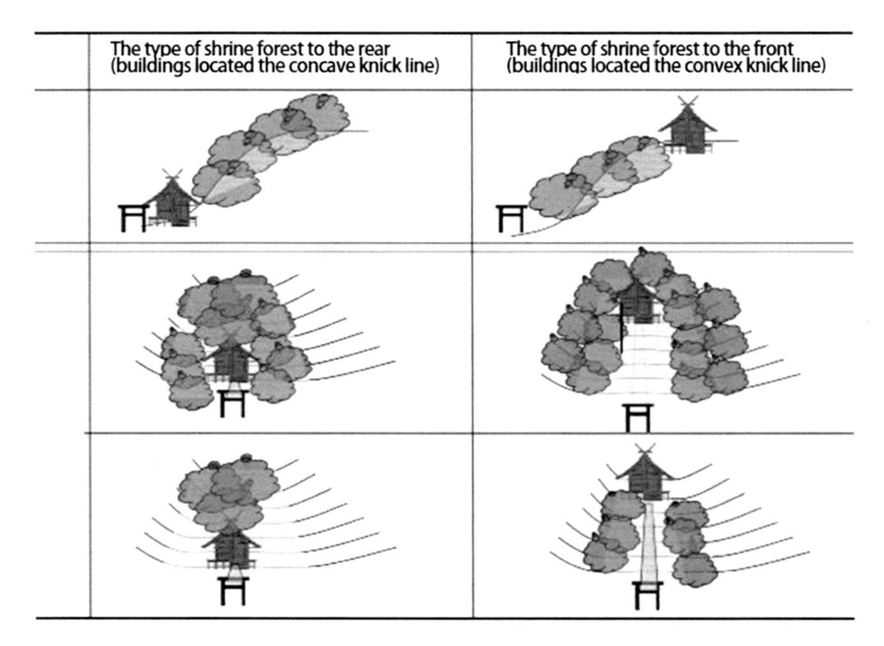

Fig. 13.7 Percentages of Shinto shrines according to the compositional pattern of green space within the boundary

line side with the green space to the rear (41.7%) and those in which the main building is on the convex knick line side with the green space (shrine forest) to the front (58.3%). In addition, 11.8% of the Shinto shrines have shrine forest cover to the rear and the sides, and 29.9% of the Shinto shrines have shrine forest cover only to the rear. Where buildings are located close to the convex knick line, 19.4% of shrines have forest cover to the rear and the sides within the boundary and 38.9% have forest cover only to the sides within the boundary.

The selected Shinto shrines revealed the following: First, Shinto shrines located on hilly topography have relatively high possession of green space (Fig. 13.8a) because the shrine forest space is formed on all sides of the main buildings (mean = 58.8 ± 16.7%). Likewise, buildings on the concave knick lines (Fig. 13.8b) have relatively high possession of green space. This is because the shrine forest space was formed to the rear of the building (mean = 31.5 ± 8.1%). Conversely, the buildings on the convex knick line side of the slope either had high or low green space possession (Fig. 13.8c). The reason for the difference may lie in the relationship between the approach road and the inclination of the slope for each type (mean = 28.6 ± 13.4%). In Shinto shrines where the building is on the convex knick line side, the *torii* (shrine gate symbolizing the sacredness of the ground inside, composed of two pillars connected by two beams at the top and usually placed next to the chamber) is found close to the concave knick line and the approach road extends from there to the building. With areas having high possession of green space, the approach road is provided on a steep slope and the shrine forest space is formed to

Fig. 13.8 Shinto shrines located on hilly topography (left) are classified as a group with relatively high possession of green space. Shrines located on the concave knick lines (center), tend to have relatively high rates of green space. Cases in which the building is on the convex knick line side of the slope show either high or low green space possession (right)

the outside of the approach road. On the other hand, shrines with low possession of green space tend to be located on gently inclined slopes with rows of trees provided on the approach roads only, with other buildings or another site situated outside.

13.4 Discussion and Conclusions

Since Shinto shrines are categorized as religious institutions and their sites are managed by private individuals (priests), their green spaces are commonly considered as private gardens. This makes it difficult to understand shrines as meaningful components of urban biodiversity and ecosystem services. Most shrine forests are facing anthropogenic disturbance in varying degrees (Tabata et al. 2004), and many shrine forests do not have a good enough vegetation management plan. However, as noted earlier, the benefits of recognizing them as such are manifold. We need to systematically promote a general understanding of the importance of managing shrines as green spaces for urban biodiversity. If shrine priests become aware of the value that forests offer and take responsibility for the maintenance and management of shrine vegetation, they will identify unwanted species as threats and try to eliminate them. This approach is simple and reasonable.

There are many types of "potential" ecological spaces in urban areas, for example, public/private parks and gardens, street plantations, green walls, green roofs, rain gardens, and ecological parking. These green corridors, which can be considered as core patches and sources of biodiversity, serve as passageways for small animals, birds, and other creatures. According to Ignatieva et al. (2011), it is important to understand such characteristics in order to raise awareness of the importance of ecological networks, because urban ecological networks are the only corridors for wildlife activity in urban areas. They also highlighted that the planning and design of an ecological network formed at the beginning of the twenty-first century is seen as multidisciplinary, involving all kinds of "potential" ecological spaces within the city. Our study found linear green spaces (lines of trees) in all Shinto shrines (Fig. 13.9). These linear green spaces were on the sides of the *sandow*, a private approach road connecting the shrine to a local street or public road. From the landscape ecological perspective, the *sandow* can contribute to the space by serving as a corridor for connectivity and wildlife movement to facilitate urban biodiversity. Considering this, Fujita and Kumagai (2006) analyzed the relationship between shrine forests and topography from the perspective of the connectivity of green spaces, and they proposed evaluating the outdoor spaces of Shinto shrines as urban green spaces. This concept of connectivity was also proposed by Koshizawa (1998). In fact, these studies are part of a trend toward understanding urban green spaces using the concepts of spatial layout and networks. Shinto shrines and their approach roads are gradually being considered as ecological corridors, stepping-stones toward functioning ecological networks. Recently, Ignatieva et al. (2008) have proposed new models of urban ecological networks that respect, conserve, and enhance natural processes. Such models are expected to improve biodiversity, aesthetics, and cultural identity, and serve as an important framework for creating sustainable cities.

Our findings on spatial constituent elements have several historical and cultural implications. A relatively large number of the Shinto shrines were found to contain large or old trees that were considered sacred (44.4%), and the distribution of vegetation from seedlings tended to correlate with topography rather than with green space area (Fig. 13.6). Further, we found that potentially native species such as castanopsis and oak were growing within the boundaries, and ginkgo biloba and other main garden species were seen more in the green spaces (Table 13.3). Considering these findings together with the characteristics of seedlings of trees (Fig. 13.10) in the green spaces within the boundaries, we can see that the incidence of growing Chusan palm (15 shrines) and Japanese Aucuba (12 shrines), which are characteristic of urban green spaces, is significant. However, from the interview surveys as well as documentary records (Tokyo Shinto Shrine Directory 1986), we found that many Shinto shrines suffered damage during the Second World War. In addition to buildings, some shrine forests were destroyed by fire. Therefore, although spatial continuity can be observed, it is clear that, in several cases, the existing trees were planted by the chief priests (the custodians of the Shinto shrines) or the shrine parishioners during the postwar restoration.

As stated, Shinto shrines located in hilly topography were found to be conducive to the development of green spaces, depending on the angle of inclination of the slope or the location of the shrine forest space. Although the number of such Shinto shrines is small, the shrine forest space is formed around all sides of the building, which is why such shrines have high possession of green space. We infer that Shinto shrines were established in places where it was believed that a spirit dwelled on that ground or where the establishment was influenced by the worship of Mount Fuji[2] a

[2]A religious belief that worships Mount Fuji as a mountain god. After the founding of Sengen Shinto shrine, from the Heian Period onwards, mountain ascetics and disciples spread among the

Fig. 13.9 Examples of linear green spaces on the sides of the *Sandow*. The approach roads shown here connect local streets or public roads and shrine buildings at Hikawa shrine in Shibuya-ku, Tokyo prefecture, Japan (above), and at Hikawa shrine in Shinagawa-ku, Tokyo prefecture, Japan (below). *Sandow* functions as a corridor for connectivity and wildlife movement for urban biodiversity. Photo by N. Fujita

Table 13.3 Correlation between each element in the green space within the boundary

	Green space area	Difference in elevation	Topography	Slope angle	Vegetation from seeds	Chusan palm	Japanese aucuba
Green space area	1	0.201	−0.118	−0.201	−0.282	0.032	0.289
Difference in elevation		1	0.39	−0.123	0.337	0.307	0.382
Topography			1	0.075	0.544	0.361	0.339
Slope angle				1	−0.119	−0.249	−0.146
Vegetation from seeds					1	0.875++	0.526
Chusan palm						1	0.643+
Japanese aucuba							1

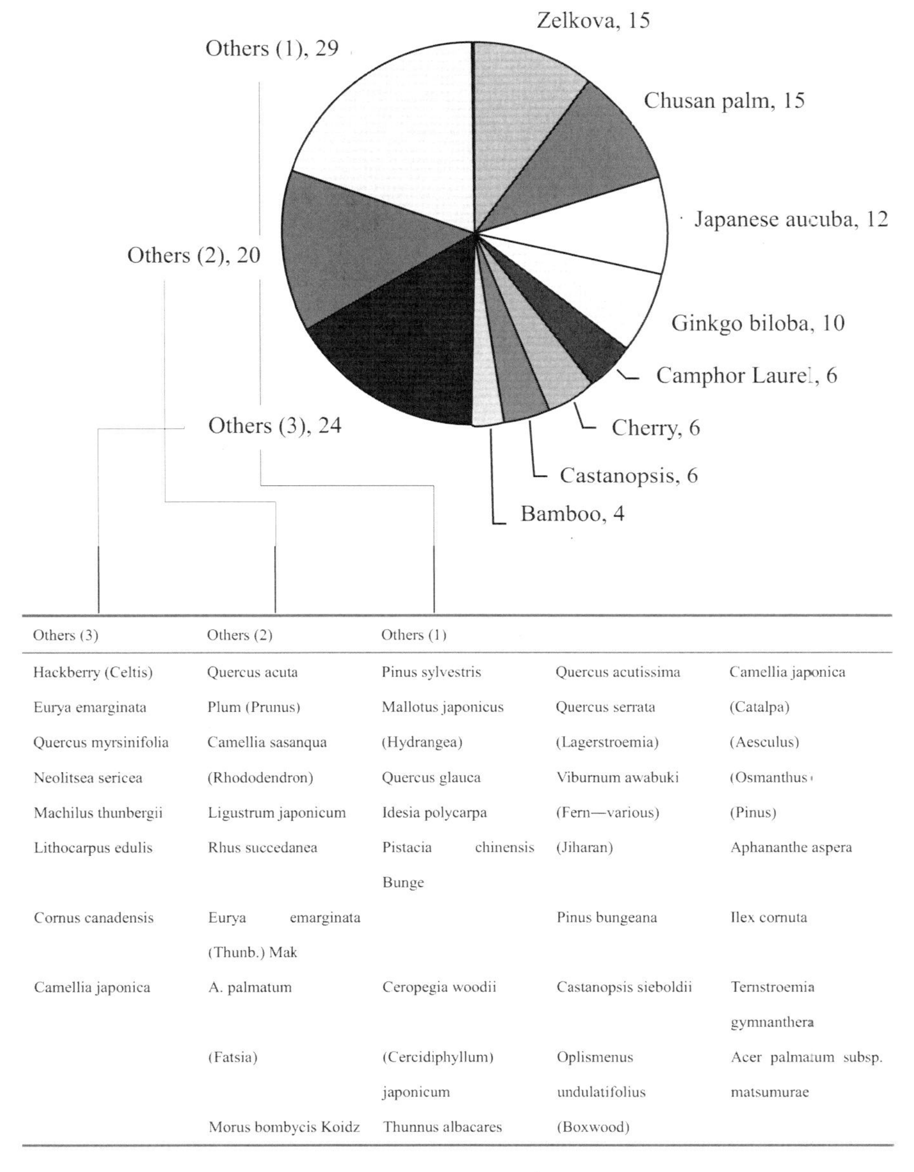

Others (3)	Others (2)	Others (1)		
Hackberry (Celtis)	Quercus acuta	Pinus sylvestris	Quercus acutissima	Camellia japonica
Eurya emarginata	Plum (Prunus)	Mallotus japonicus	Quercus serrata	(Catalpa)
Quercus myrsinifolia	Camellia sasanqua	(Hydrangea)	(Lagerstroemia)	(Aesculus)
Neolitsea sericea	(Rhododendron)	Quercus glauca	Viburnum awabuki	(Osmanthus)
Machilus thunbergii	Ligustrum japonicum	Idesia polycarpa	(Fern—various)	(Pinus)
Lithocarpus edulis	Rhus succedanea	Pistacia chinensis Bunge	(Jiharan)	Aphananthe aspera
Cornus canadensis	Eurya emarginata (Thunb.) Mak		Pinus bungeana	Ilex cornuta
Camellia japonica	A. palmatum	Ceropegia woodii	Castanopsis sieboldii	Ternstroemia gymnanthera
	(Fatsia)	(Cercidiphyllum) japonicum	Oplismenus undulatifolius	Acer palmatum subsp. matsumurae
	Morus bombycis Koidz	Thunnus albacares	(Boxwood)	

Fig. 13.10 Trees and plants in the green spaces within the boundaries. Botanical names in parentheses denote genus distinctions, which may include a variety of species

popular object of worship in the study area during the Edo period. Moreover, buildings located on the concave knick line there had high possession of green spaces, but this was because the shrine forest space was formed to the rear of the building and these shrine forests are recognized as confined spaces or were frequently regarded as sacred by the Shinto shrines. We therefore infer that these factors have played a role in the green space possession of such Shinto shrines.

Connecting large and small shrine patches would promote an understanding of the value of biodiversity and ecological networks. For instance, despite being located in the heart of highly urbanized Tokyo, the woodland forest at Meiji Jingu Shrine is expansive and serene. From the results of our surveys, several characteristics had pointed out on Meiji Jingu Shrine (Hamano 2005, 2010; Koshimizu et al. 2001). The 72-ha man-made forest is an essential element for maintaining the majesty of the shrine. Work on the Meiji Shrine forest began in 1920; before that, most of the land was farmland or meadow, with only about 1/5th being forested and mostly comprising a range of small trees. The basic objective of the plan was to create a "climax forest" that would be maintained through natural regeneration, that is, a forest that would have its own cycle. Plans were drawn in anticipation of changes in forest composition. The first to be planted were pioneer species such as red pine, followed by a mixture of various types of conifers, deciduous broadleaf trees, and evergreens. Finally, evergreen broadleaf species that grew naturally in the Tokyo area were introduced. This method was considered highly innovative for the day. The topography and soil were maintained as they were before planting, and landscape alterations were kept to a minimum. The cultivation plan called for the transition to be completed in 50–100 years. The idea was to have humans create the conditions during the initial period of vegetational transition, but to let nature take its course after that to complete the forest. From the beginning of planting to the present day, the forest has been managed by specialists at the shrine. A large number of trees were needed to create the shrine's forest, and most of them were collected through donations. Because the trees had to be appropriate for a shrine forest, elaborately sculpted trees from private gardens, foreign species, fruit trees, and trees with elaborate blossoms were rejected. Ultimately, 95,559 donated trees were accepted. An additional 8222 trees were provided by government organizations, 2840 trees were purchased, and 15,951 native trees were transplanted, bringing the total number to 122,572 trees, representing 365 species that were used in the project. With the exception of the prevalence of camphor trees, this process, which took place over the course of 80 years since creation, led to the formation of an evergreen broadleaf forest that is identical to a natural forest. The management method has always been to let nature take its course, with only minimal necessary intervention, such as removing dead trees and branches. While leaf-eating insects inhabit the forest,

people. During the Edo period, the Fujikou religion became organized and obtained many followers in the Kanto region. Throughout the Kanto region, Fuji mounds that resembled Mount Fuji were constructed as places of worship and as alternatives to the mountain. We can still find worshipers climbing Mount Fuji wearing white ceremonial robes and carrying a staff while chanting the "purification of the six roots of perception."

there is no need to kill them with insecticide, because they help maintain a healthy balance in the food chain that has developed there. It was decided to prohibit the raking of fallen leaves in the forest, prohibit human entry, move leaves that had fallen on walkways into the forest, and watch out for fires and flames. Conversely, small shrines in central Tokyo, such as those that appear in this study, are usually decorated with a few slender trees such as *hisakaki* (*Eurya japonica* Thunb.), and their grounds are often cultivated to resemble Japanese gardens. From a biological perspective, it is not appropriate to call them "forests" or "natural habitat." Nevertheless, when viewed from a different perspective, in these spaces, one can sense the strong beliefs that lead people to maintain the shrine by planting trees even on the smallest grounds or at considerable costs. Of the facilities found on the shrine grounds that had no relationship to the original Shinto shrines, 26.7% were public buildings, 9.6% firefighting-related, 3.7% kindergartens, 9.6% gateball fields, 27.4% parks, and 17.8% memorials (Fig. 13.6). These facilities were constructed on sites originally meant to be preserved as open/green spaces, so they are considered "intruders" into natural biodiversity and habitat.

Bierwagen (2005) suggested that landscape patterns that promote ecological connectivity facilitate the spread of undesirable organisms or processes. This view led to an investigation into how urban forms can be used to predict ecological connectivity and assist in prioritizing urban landscapes for conservation activities and risk management. Bierwagen found that qualitative categories are not adequate for describing ecological connectivity: multivariate descriptions are much better predictors, with the significant synthetic variables being urban area, number of urban patches, urban patch extent, level of aggregation, and perimeter-area fractal dimension. The dominance of area as a differentiating variable led to the development of a new urban connectivity index that used a combination of urban area and state population size.

In this study, to determine the possession of green space as a function of the location of the shrine forest space, we carried out a survey of Shinto shrines located on slopes within the Yamanote Line in the center of Tokyo, and determined the possession of green space by analyzing the distribution of constituent elements within the boundaries. However, it seems that it is too easy to build facilities that are not connected to the original function of the Shinto shrine on the shrine forest or on vacant green spaces around the shrine. This should be noted as an issue of concern. The results of the interview surveys also made it clear that transformations have occurred in places that were shrine forests or vacant spaces until now. In other words, the decline in the site areas and the increase in the number of buildings within the sites has led to a decline in shrine forest spaces. Therefore, if, for example, we focus on only the tree area, we would consider the green space within the boundaries of Shinto shrines to have been reduced, owing to the large amount of non-tree green space. The shrine forest as an independent green space would not be able to perform its function. However, although shrine forests are private lands, they differ from land available for commercial facilities or private housing, and they retain the character of Shinto shrines. Therefore, while the scale of the green space of each individual shrine forest is small, collectively, they are superior to other green spaces because of the continuity of ownership and location. We believe that shrine forests have a latent value, and using shrine forest cover more positively in future urban green space planning will contribute to urban green spaces.

Because shrines are religious facilities, it is not easy to destroy or damage them as easily as other types of land or facilities can. Consequently, the strategic and intensive use of shrine forests as part of landscape design in urban greening plans may be one desirable means of improving biodiversity and ecosystem services. In this sense, we can say that shrines in cities can promote urban biodiversity. This view is shared by Ishii et al. (2010), who suggested that shrine forests can be used as stepping-stones in the urban green space network, whereas spatially clustered temple forests can be integrated to form large green spaces.

The results of this study are significant for the promotion and introduction of Shinto shrine spaces for understanding the characteristics of their elements and functions for green spaces.

References

Alban N, Berwick CK (2004) Forests and religion in Japan: from a distinctive vision of trees to a particular type of forest management. Revue Forestiere Francaise 56(6):563–572

Antrop M (2005) Why landscapes of the past are important for the future. Landsc Urban Plan 70 (1-2):21–34

Balram S, Dragićević S (2005) Attitudes toward urban green spaces: integrating questionnaire survey and collaborative GIS techniques to improve attitude measurements. Landsc Urban Plan 71(2-4):147–162

Bierwagen BG (2005) Predicting ecological connectivity in urbanizing landscapes. Environ Plan B Plan Des 32(5):763–776

Breuste JH (2004) Decision making, planning and design for the conservation of indigenous vegetation within urban development. Landsc Urban Plan 68(4):439–452

Chace JF, Walsh JJ (2006) Urban effects on native avifauna: a review. Landsc Urban Plan 74 (1):46–69

De Chant T, Hernando GA, Velázquez SJ, Kelly M (2010) Urban influence on changes in linear forest edge structure. Landsc Urban Plan 96(1):12–18

Dramstad WE, Olson JD, Forman RTT (1996) Landscape ecology principles in landscape architecture and land-use planning. Harvard University Graduate School of Design. Island Press/ ASLA, Washington, DC

Forman RTT (1995) Land mosaics: the ecology of landscapes and regions, 2nd edn. Cambridge University Press, New York

Forman RTT, Godron M (1986) Landscape ecology. Wiley, New York

Fujioka K (1979) New geographical dictionary. Taimeidou, Tokyo

Fujita N, Kumagai Y (2004) Research on the changing landscape at the center of Tokyo and the uneven distribution of wooded areas. J Jpn Inst Landsc Arch 67(5):577–580

Fujita N, Kumagai Y (2006) Characteristics of an urban green space layout, with a focus on shrine forest space, by analyzing the topography and connectivity of green spaces using GIS. J City Plan Inst Jpn 41(3):373–378

Fujita N, Kumagai Y (2007) Research into the differences in the distribution of Shinto shrines, temples, and parks within cities using GIS. Landsc Ecol 12(1):9–21

Fujita N, Ono R, Kumagai Y (2005) Research into the meaning of "shrine forest" and change in its significance with regard to the preservation of historical sites, scenic spots, and natural monuments. J Jpn Inst Landsc Arch 68(5):417–420

Fujita N, Kumagai Y, Shimomura A (2007) Research into the concept of green spaces in relation to the outdoor spaces of Shinto shrines, by comparing the synonyms for "shrine forest". J Jpn Inst Landsc Arch 70(5):591–596
Godefroid S, Koedam N (2003) Identifying indicator plant species of habitat quality and invisibility as a guide for peri-urban forest management. Biodivers Conserv 12:1699–1713
Goodwin BJ, Fahrig L (2002) How does landscape structure influence landscape connectivity? Oikos 99:552–570
Grimm N, Grove JM, Pickett STA, Redman CL (2000) Integrated approaches to long-term studies of urban ecological systems. Bioscience 50(7):571–584
Hamano C (2005) The Meiji Jingu Shrine: forest compositions. Arch Conserv 26(3):35–39
Hamano C (2010) Meiji Jingu forest: present conditions and future outlook. Garden Deity 4:140–138
Hashimoto D, Ito K, Manabe T, Isono D, Umeno T (2006) Basic study on the distributional patterns and ecological characteristics of shrine/temple forests in Kitakyushu City. Kyushu. J For Res 59:56–59
Hashimoto D, Ito K, Iijima S (2007) The consciousness of Shinto priests for the management of shrine forests in urban areas. Landsc Ecol Manage 12:45–52
Ignatieva M, Stewart G, Meurk C (2008) Low impact urban design and development (LIUDD): matching urban design and urban ecology. Landsc Rev 12:60–73
Ignatieva M, Stewart G, Meurk C (2011) Planning and design of ecological networks in urban areas. Landsc Ecol Eng 7:17–25
Imanishi A, Imanishi J, Murakami K, Morimoto Y, Satomora A (2005) Herbaceous plant species richness and species distribution pattern at the precincts of shrines as non-forest greenery in Kyoto city. J Jpn Sci Reveget Tech 31(2):278–283
Ishida H, Hattori T, Takeda H (2005) Comparison of species composition and richness among primeval, natural and secondary lucidophyllous forests on Tsushima Island, Japan. Veg Sci 22:1–14
Ishii HT, Manabe T, Ito K, Fujita N, Imanishi A (2010) Integrating ecological and cultural values toward the conservation and utilization of shrine/temple forests as urban green space in Japanese cities. Landsc Ecol Eng 6(2):307–315
Iwami R, Kawakami H, Lu B (1987) Environmental evaluation of green spaces and measurement of their value based on the potential concept. J City Plan Inst Jpn 22:13–18
Jinnai H (1985) Spatial anthropology of Tokyo. Chikuma Shobo, Tokyo
Kim S, Tashiro Y, Tabata S (1989) The characteristics of the distribution of green covered areas according to the scale in high-density urban areas. J Jpn Inst Landsc Arch 53(1):1–9
Koga Y (2002) The reduction in woodland vegetation and changes in land use form and industrial structure: a study of rhododendron dilatatum in the Boso Peninsula. Pap Environ Inf Sci 16:363–368
Koshimizu H, Kondo M, Hamano C, Kobayashi T, Shibata S (2001) Forestry concepts and techniques by landscape architects: a case study of the forest around Meiji Jingu Shrine. J Jpn Inst Landsc Arch 65(2):143–150
Koshizawa A (1998) Urban planning for water and green, and its concepts. Shintoshi 52(12):28–37
Kyakuno T (2005) The effect of urban land coverage and land use on the day and nighttime heat island phenomenon during summer. J City Plan Inst Jpn 40:679–684
Liang G, Ding S (2010) Integrative analysis of geographic environmental factors on forest landscape dynamics: a case study of Luoning County, Yiluo River Basin. Shengtai Xuebao 30 (6):1472–1480
Lundholm JT, Marlin A (2006) Habitat origins and microhabitat preferences of urban plant species. Urban Ecosyst 9:139–159
Maki F (1980) Appearing and disappearing City. Kajima Institute Publishing SD, Tokyo
Manabe T, Ito K, Isono D, Umeno T (2007) The effects of the regulation system on the structure and dynamics of green space in an urban landscape: the case of Kitakyushu City. In: Hong SK, Nakagoshi N, Fu B, Morimoto Y (eds) Landscape ecological applications in man-influenced areas. Springer, Berlin, pp 291–309

Matsumoto Y, Tonuma K (2004) Transformation in the visual appearance of slopes in the center of Tokyo, changes in the land use in Edo, Tokyo, and changes in scenery. J Arch Plan 577:119–126
McKinney ML (2002) Urbanization, biodiversity, and conservation. Bioscience 52:883–890
Mikami T (2005) The urban heat island phenomenon and its formative factors: research in the case of Tokyo. J Geogr 114(3):496–506
Mikami T, Daiwa H, Ando H (2005) Research into localized heavy rain in summertime within Tokyo. Annual Report of the Tokyo Metropolitan Research Institute for Environmental Protection. pp 33–42
Millard A (2010) Cultural aspects of urban biodiversity. In: Müller N, Werner P, Kelcey JG (eds) Urban biodiversity and design. Wiley, Oxford, pp 56–80
Ministry of Environment (2002) New national strategy for biodiversity. Gyosei, Tokyo
Ministry of the Environment Biodiversity Center of Japan (2008) Biodiversity Information Systems, Natural Environment Conservation Basic Survey. http://www.biodic.go.jp/J-IBIS.html
Morimoto Y (2011) Biodiversity and ecosystem services in urban areas for smart adaptation to climate change: "Do you Kyoto"? Landsc Ecol Eng 7:9–16
Nilon CH (2011) Urban biodiversity and the importance of management and conservation. Landsc Ecol Eng 7:45–52
Okazaki J, Kato K (2005) The effect of the fragmentation of urban green spaces on birds in the breeding season. Proc Center Environ Inf Sci 19:353–358
Organization for Landscape and Urban Green Technology Development (2006) Guidebook for green urban regeneration. Gyosei, Tokyo
Randall A (2004) Linked landscapes: creating greenway corridors through conservation subdivision design strategies in the northeastern and central United States. Landsc Urban Plan 68 (2-3):241–269
Riitters KH, O'Neill RV, Hunsaker CT, Wickham JD, Yankee DH, Timmins SP, Jones KB, Jackson BL (1995) A factor analysis of landscape pattern and structure metrics. Landsc Ecol 10(1):23–39
Suzuki T, Manabe T, Ito K, Umeno T (2004) Analysis of landscape changes by using distal vegetation map in mid-northern region in Kitakyushu city. Bull Kitakyushu Nus Nat Hist Hum Hist Ser A 2:79–85
Tabata S, Watanabe K (1986) Research into the transformation of green coverage structure in urbanized green spaces. J Jpn Inst Landsc Arch 49(5):287–292
Tabata S, Igarashi M, Shirako Y (1984) Research into the urban structure of Edo and Tokyo from the viewpoint of green spaces. J Jpn Inst Landsc Arch 47(5):298–303
Tabata K, Hashimoto H, Morimoto Y, Maenaka H (2004) Growth and dynamics of dominant tree species in Tadasu-no-mori forest. J Jpn Inst Landsc Arch 67:499–503
Tamura T (1990) How to understand the elements of micro natural environment: microtopography. In: Matsui T, Takeuchi K, Tamura T (eds) Natural environment of hilly terrain area: their characteristic and conservation. Kokon Shoin, Tokyo, pp 47–54
Tokyo Shinto Shrine Directory (1986) Tokyo Shinto Shrine Agency. In: Tokyo
Uchida T, Yokouchi M, Okada T (2002) Research into urban planning for the environment: investigation of methods for promoting biodiversity in major cities. J City Plan Inst Jpn 37:787–792
Ueda A (2001) What is a shrine forest? In: Ueda A (ed) Revival of the tutelary woods: the beginning of the study of shrine forests. Shibunkaku, Kyoto, pp 3–33
Umeno T, Ito K, Manabe T (2006) Research on the influence of a city planning system for changing land covering using satellite images: a case study in Kitakyushu city. Kyushu J For Res 59:42–46
Zhao M, Escobedo F, Staudhammer C (2010) Spatial patterns of a subtropical, coastal urban forest: implications for land tenure, hurricanes, and invasives. Urban Forest Urban Greening 9 (3):205–214

Chapter 14
Collaborative Management of Satoyama for Revitalizing and Adding Value as Green Infrastructure

Hayato Hasegawa, Tomomi Sudo, Shwe Yee Lin, Keitaro Ito, and Mahito Kamada

Abstract The management of abandoned *Satoyama* which is the Japanese name for socio-ecological production landscape has become a difficult problem for biodiversity conservation in Japan. In order to conserve *Satoyama* and regional biodiversity, restoration of the relationships with local people is important. This study aims to discuss how we can sustain local people's interaction with *Satoyama* in the present time. The study site Mt. Omine, which can be described as a typical *Satoyama* landscape, is located in Fukutsu city, Fukuoka, Japan. We investigated changes in the relationships between local people and Mt. Omine and the ecological condition of the abandoned forest on Mt. Omine. We found that ecosystem services provided from Mt. Omine declined in association with the reduction in use of the forest. In the abandoned forest, Moso-bamboo (*Phyllostachys edulis*) has expanded and dense bamboo thickets have formed. However, new collaborative management has been started in order to implement city plans for the environment. The collaborative management project is trying to obtain ecosystem services through discussing how to utilize *Satoyama* and conducting workshops. The result of the workshops implies that the values of Mt. Omine for the present generation of local people are different from the past; however, people could obtain ecosystem services and conserve biodiversity through the management which is conducted in the same way as in the past. In addition, the process of restoring *Satoyama* could provide opportunities for local people to realize the values and functions of the surrounding natural environment. Therefore, the stakeholders of *Satoyama* restoration should explore its values in the present situation and learn from the past methods and wisdom in the relationship between people and *Satoyama*.

H. Hasegawa (✉) · T. Sudo · S. Y. Lin
Faculty of Civil Engineering, Kyushu Institute of Technology, Kitakyushu, Fukuoka, Japan

K. Ito
Laboratory of Environmental Design, Faculty of Civil Engineering, Kyushu Institute of Technology, Kitakyushu-city, Fukuoka, Japan

M. Kamada
Graduate School of Technology, Industrial and Social Sciences, Tokushima University, Tokushima, Japan

K. Ito (ed.), *Urban Biodiversity and Ecological Design for Sustainable Cities*,
https://doi.org/10.1007/978-4-431-56856-8_14

Keywords Collaborative forest management · Ecosystem Services · Green Infrastructure · *Satoyama*

14.1 Introduction

Satoyama is a Japanese rural and peri-urban landscape comprising a mosaic of ecosystem types, including secondary forests, agricultural lands, irrigation ponds, and grasslands, along with human settlements (Duraiappah and Nakamura 2012). As the result of using natural resources for daily life, *Satoyama* provides rich biodiversity and various ecosystem services (ES) to local people (Katoh et al. 2009). For instance, *Satoyama* provided charcoal, firewood, timber and forest materials (Provisioning services); climate, water quality and disaster control (Regulating services); primary production, nutrient cycling, soil formation and habitat structure (Supporting services); cultural heritage, sense of identity, tourism, walking and recreation (Cultural services) (Saito and Shibata 2012). Through this situation, the *Satoyama* could be regarded as one element of the green infrastructure, which is a practical application that creates or enhances multiple ecosystem services for land development and infrastructure building towards biodiversity conservation and sustainable development. However, the relationship between people and *Satoyama* has been lost due to economic growth and lifestyle changes (Morimoto 2011) and underuse of natural resources is one of the factors that has caused crucial biodiversity loss in Japan (Ministry of the Environment 2012).

In abandoned *Satoyama*, the periodical cutting on trees was stopped and created monotonous landscapes by the progression of vegetation succession (Kamada 2017). The expansion of bamboo forest is a great concern for biodiversity decline, landslide disaster, and landscape changes (Someya et al. 2010). Moso-bamboo (*Phyllostachys edulis*) was introduced from China to Japan in the 1700s and harvested for edible bamboo shoots (Shinohara et al. 2019). Bamboo culms were also used for building materials and other commodities, however, the imports of cheaper bamboo shoots were increased and the use of plastic products became generalized, many areas of bamboo forests were abandoned and became dark, because the tall and dense bamboo prevented sunlight from reaching the forest floor (Suzuki and Nakagoshi 2011).

In order to conserve *Satoyama* and regional biodiversity, ecosystem management should be promoted. In some areas of Japan, *Satoyama* restoration projects have been undertaken. In the study site, Fukutsu city, forest management in *Satoyama* as an implementation of the city's plan has been conducted. Since 2014, Fukutsu city government has worked with local people and the Laboratory of Environmental Design of Kyushu Institute of Technology and established the Second Environmental Master Plan and Regional Biodiversity Strategy in 2017. In these plans, collaborative forest management is required because of the expansion of bamboo forest. After these plans were established, local people and the Laboratory of

Environmental Design started collaborative forest management in Mt. Omine as a model area for *Satoyama* restoration in Fukutsu city.

This study aims to discuss how we can sustain local people's interaction with *Satoyama* in the present time. The overall objective of this research is to set out an evaluation of changes in ecosystem services and the ecological condition of the abandoned *Satoyama*. From implementation of the collaborative forest management, the processes of management and the ecosystem services which local people obtain from Mt. Omine were described in order to discuss present values for local people to become involved with *Satoyama*.

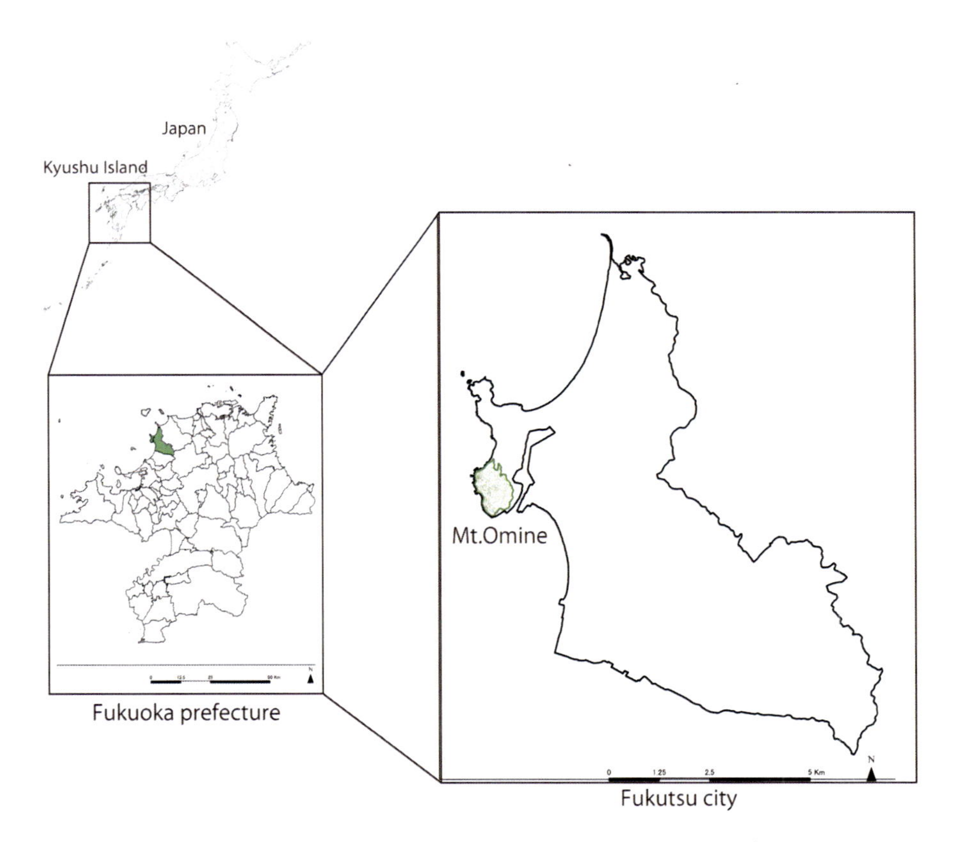

Fig. 14.1 The location of study site: Mt. Omine in Fukutsu city, Fukuoka Prefecture, Japan

Fig. 14.2 The windsock and decoration using bamboo for traditional event; Tsuyazaki Gion Yamakasa (photo by Keitaro Ito)

14.2 Study Site

The study site, Fukutsu city, is located in Fukuoka prefecture, Kyushu, Japan (Fig. 14.1). The total area of Fukutsu city is 52.3 km^2, and its 31.6% is agricultural land, 27.5% is forest, 12.0% is housing land, 8.4% is road, and 2.8% is open water. The population is 65,831 (Fukutsu city 2019), and it has been steadily increasing. This city is popular with its location in the suburbs of a large city and a lot of nature remaining. Fukutsu city was formed by the merger of two towns, Fukuma district and Tsuyazaki district in 2005. Tsuyazaki district has a lot of culture and nature surrounded by sea, mountains, tidal flats, and rice fields. Fishing and agriculture are also thriving in this area. From the Edo period to the Meiji period, the salt industry which is useful in lowland areas prospered and as a result, many merchant houses, restaurants, and Japanese Inns were built in this area (Fukutsu city 2017).

One of the most traditional and important events in Tsuyazaki district is "Tsuyazaki Gion Yamakasa." This event has been continued for about 300 years, and people carry a large float to pray for the repelling of epidemics and disasters. In this event people use Madake-bamboo (*Phyllostachys bambusoides*) for windsock

Fig. 14.3 Landscape of Mt. Omine. Its elevation is 114.5 m (photo by Hayato Hasegawa)

Fig. 14.4 The abandoned *Phyllostachys edulis* (Moso-bamboo) forest in Mt. Omine (photo by Hayato Hasegawa)

and decoration of the float (Fig. 14.2). Every year, they collect the highest bamboo in Mt. Omine.

Mt. Omine is located in Tsuyazaki district (Fig. 14.3). People live at the foot of Mt. Omine and they had used there as a *Satoyama*. However as mentioned above, the bamboo forest has been abandoned and expanded (Fig. 14.4). In Fukutsu city's plans, Mt. Omine was described as one of the important *Satoyama* landscapes in Fukutsu city and forest management was required.

14.3 Method

14.3.1 Evaluating the Changes of Ecosystem Services

To evaluate the changes in the relationship between local people and Mt. Omine from past to present, semi-structured interview surveys were conducted with five local people who have lived over 60 years in Tsuyazaki district. Interviewees were asked about the usage of Mt. Omine and value for them in the past and present. Two of the interviewees (A and B) own land in Mt. Omine and live at the foot of the mountain. Two other interviewees (C and D) who have moved from other cities about 60 years ago have a lot of knowledge about plants. The fifth interviewee (E) has lived in Tsuyazaki district and he is about two decades younger than the other interviewees. The collected data was transcribed to text data and structured according to the types of ecosystem services with reference to Tomita (2007).

14.3.2 Evaluating the Ecological Conditions of Bamboo Forest

The expansion of Moso-bamboo on Mt. Omine has serious ecological implications, highlighted as one of the issues in the Fukutsu city's plan. To evaluate how much Moso-bamboo forests were expanded on Mt. Omine, aerial photograph analysis was conducted. The changes of Moso-bamboo area were analyzed from aerial photographs of Mt. Omine in 1948, 1981, and 2010 provided by the Geographical Survey Institute in Japan.

In expanded area, individual numbers of Moso-bamboo were recorded within two grids 20 m × 20 m; Grid 1 and 2. In addition, species compositions of forest floor vegetation were recorded within 16 sub grids of 5 m × 5 m in Grid 1 and 2 for evaluating forest current conditions. Next to the Moso-bamboo forest, coppice forest of *Lithocarpus edulis* (Matebashii in Japanese) has been established (Fig. 14.5). Trees with many sprouts indicate that these plants used to be cut in the past. In order to obtain basic information to consider future usages of these plants, we recorded the

Fig. 14.5 The coppice of *Lithocarpus edulis* (Matebashii) in Mt. Omine (photo by Hayato Hasegawa)

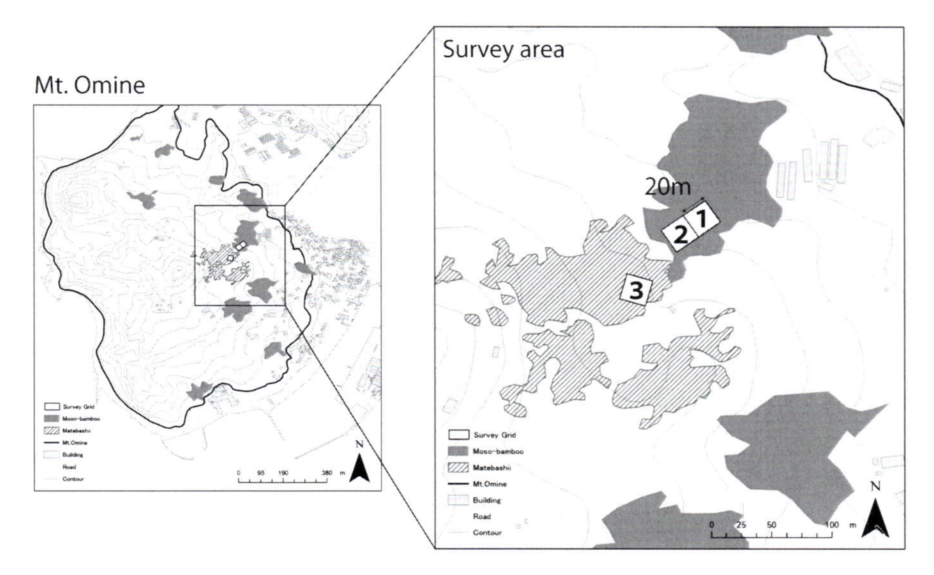

Fig. 14.6 The location of survey grids 1, 2 for *Phyllostachys edulis* (Moso-bamboo) and C for *Lithocarpus edulis* (Matebashii in Japanese) forest

number of trunks within a quadrat of 20 m × 20 m; Grid 3. Locations of grids for vegetation surveys are summarized in the map of Fig. 14.6.

14.4 Result

14.4.1 Ecosystem Services in Past and Present for Local People

According to the interviewees A and B, some trees from Mt. Omine had been used as firewood and timber for building their houses in the past. They also obtained money by selling pine trees as firewood. When they used natural resources from the forest, there was almost no bamboo forest. Accordingly, they had maintained Mt. Omine for obtaining provisioning services (firewood and timber). Although A and B recognize the Moso-bamboo expansion and the necessity for maintenance, it is difficult for them to do it alone.

From the interview of interviewees C and D, they indicated flora and landscape of Mt. Omine in the past. They remember a lot of researchers came for observation and Mt. Omine was called as "a treasure trove of plants." The species which the interviewees could remember were shown in Table 14.1, as well as their habitats. Some species need bright or dried environment in forest or forest edge for growth. According to C and D, many species have disappeared because of construction of the road and park. They appealed to the local government to preserve these plants in the past, however, it was unaccepted. We walked with C and D in the area and found that some species have remained there. However, their populations are much smaller than in the past, according to C and D. In the past, C and D obtained cultural services (enjoying flower landscape) and Mt. Omine supplied supporting services (native species habitat).

Table 14.1 The native species in this region revealed in the interview survey and its habitats

No	Speices	Habitats
1	*Lysimachia clethroides*	Bright forest, Bright glassland
2	*Gentiana zollingeri*	Bright forest, wet area
3	*Melampyrum roseum* var. *japonicum*	Forest edge
4	*Chloranthus fortunei*	Forest edge
5	*Cimicifuga simplex*	Glassland
6	*Cephalanthera falcata*	Interior of forest
7	*Chloranthus serratus*	Interior of forest, dark area
8	*Cymbidium virescens*	Interior of forest, dried area
9	*Gentiana thunbergii*	Interior of forest, dried area
10	*Aeginetia indica*	lives with *Miscanthus sinensis*
11	*Aster spathulifolius*	On the rocks at the coast
12	*Tripterospermum japonicum*	Shade of tree
13	*Cephalanthera erecta*	Shade of tree

Table 14.2 The plants that local people used in the past and the value they derived from them

Past value	Plant species
Firewood	*Chamaecyparis obtusa, Cryptomeria japonica, Lithocarpus edulis, Pinus.* Spp.
Timber	*Chamaecyparis obtusa, Cryptomeria japonica*
Enjoying flower landscape	*Aeginetia indica, Aster spathulifolius, Cephalanthera erecta, Cephalanthera falcata, Chloranthus fortunei, Chloranthus serratus, Cimicifuga simplex, Cymbidium virescens, Gentiana thunbergii, Gentiana zollingeri, Lysimachia clethroides, Melampyrum roseum* var. *japonicum, Tripterospermum japonicum,*
Collecting edible wild plants	*Dioscorea japonica, Diospyros kaki, Ficus erecta, Lactarius hatsudake, Phyllostachys edulis, Pyrus pyrifolia, Stauntonia hexaphylla, Vaccinium bracteatum, Pueraria thunbergiana Benth. Smilax china* L.
Materials for festival	*Phyllostachys bambusoides*
Materials for fishery and agriculture	*Phyllostachys edulis*
Plaything for children	*Eriobotrya japonica,*
Making basket	*Sinomenium* Spp.

When E was a child, about 50 years ago, children in that area played on Mt. Omine. In the past, there were no snacks and play toys like nowadays, Mt. Omine was a meaningful place where they played by collecting plants and also ate some edible plants through exploration play. The knowledge on plants was learned from older children. Children at that time obtained cultural services (playing, collecting edible plants, learning about plants). He has recognized children no longer play and eat edible plants in Mt. Omine in these days.

The plants coming up from interview surveys and their values for local people in the past are described in Table 14.2. The local people who were raised in surrounding of Mt. Omine used various natural resources for daily live; Mt. Omine supplied diverse ecosystem services to local people. Before the 1960s, *Satoyama* had been maintained through landowners' activities to obtain provisioning services. Various cultural services could be provided in the situation, from Mt. Omine to local people. Around 1960, a fuel revolution took place, and gas and electricity replaced the resources in daily lives. After the 1960s, people no longer got provisioning services from Mt. Omine, while children still played in the forest and they obtained cultural services. However, almost all ecosystem services declined in association with the reduction in use of the forest, lifestyle changes, and construction of infrastructures. Change of relationship between local people and Mt. Omine was described as ecosystem services changes (Table 14.3).

14.4.2 The Ecological Characteristics of the Study Site

By analysis of aerial photographs, the Moso-bamboo forest has expanded from 0.21 ha to 5.17 ha in 62 years (Fig. 14.7). Average trunk number of Moso-bamboo is 7925/ha. On the forest floor 33 species were found (Table 14.4), and almost all

Table 14.3 The ecosystem services changes for local people from interview survey

	Past ES →	Present ES
Provisioning Services	Firewood; Foods; Ceeeping plant; Timber; Plaything; Spring Water; Ship Material; Festival Material	Spring Water; Ship Material; Festival Material
Cultural Services	Playing (catching creatures, exploration); Collecting edible wild plants; Learning plants; Enjoying Flower Landscape	
Supporting Services	Native species habitat	

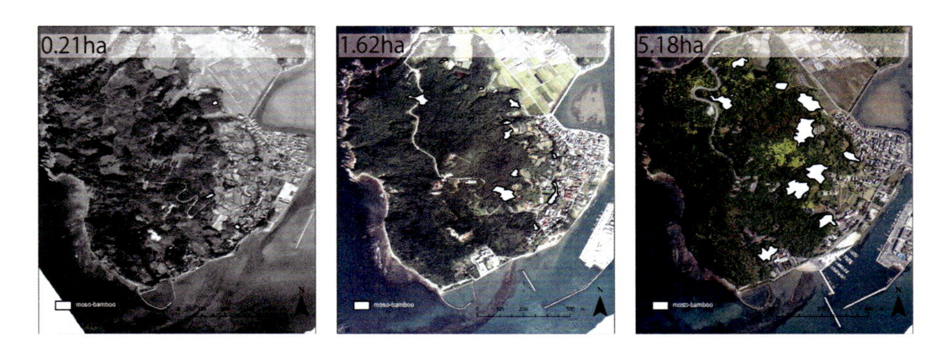

Fig. 14.7 The expansion of *Phyllostachys edulis* (Moso-bamboo) forest area in 1948, 1981, and 2010, respectively

plants were seedling or sapling. In coppice forest next to the Moso-bamboo forest, 27 individuals of *Lithocarpus edulis* were there, and average number of sprouts was seven and the maximum is 23 per one stump. In the interview survey, landowners told that local people used *Lithocarpus edulis* as firewood in the past.

Table 14.4 The recorded species and individual number in 16 grids

No	Sprecies	grid	1-1	1-2	1-3	1-4	1-5	1-6	1-7	1-8	2-1	2-2	2-3	2-4	2-5	2-6	2-7	2-8
		Indivual numbers	29	7	8	20	8	12	18	18	39	17	5	17	12	22	1	11
1	*Quercus acuta*							1				1				2		
2	*Maackia amurensis*								1		1							
3	*Ficus erecta*			2	2	3			1	2	1	1	1	1	1	1		
4	*Dendropanax trifidus*				1	4	1	3		1	6	2	1	3	1	2		1
5	*Symplocos kuroki*															2		
6	*Zelkova serrata*													1				
7	*Castanopsis cuspidata*									1								
8	*Quercus serrata Murray*									1								
9	*Quercus myrsinifolia*																	2
10	*Neolitesea sericea*		9	1	1	3	2	2	1	4	8	3				1	1	
11	*Castanopsis sieboldii*								1									
12	*Machilus thunbergii*		1				1											
13	*Ligustrum lucidum*					4					3							
14	*Eurya japonica*		2				1	1				2						
15	*Daphniphyllum teijsmannii*													1	4			
16	*Elaeocarpus sylvestris*		1		1									1				
17	*Lithocarpus edulis*								1				1			1		
18	*Aphananthe aspera*		2		1	2				1	2							
19	*Ilex integra*										2					3		
20	*Ardisia japonica*		3						1	1	6		1					
21	*Camellia japonica*		1			1				4				3				1
22	*Cinnamomum yabunikkei*		2	1			1	1	6			1		2	1	1		2
23	*Daphniphyllum macropodum*				1				5			2		1				
24	*Aucuba japonica*						1							1				
25	*Damnacanthus indicus*																	3
26	*Pittosporum tobira*								1									
27	*Ligustrum japonicum*		2			3				2	8	2		1	1	1		1
28	*Aucuba japonica*			1				1				1						
29	*Euonymus japonicus*										2							
30	*Callicarpa mollis*				1													1
31	*Oplismenus undulatifolius*		1															
32	*Arisaema ringens*			2									1		1			
33	*Liriope muscari*		5				1	3		1		2		2	3	8		

14.5 The Collaborative Restoration Project of Mt. Omine

The collaborative restoration project was started in 2017 as implementation of the Second Environmental Master Plan and Regional Biodiversity Strategy in Fukutsu city. The management members, a professor, and students of the Laboratory of Environmental Design and one local person have been working together since the process of formulating Fukutsu city's plan. Hence they recognized the necessity of forest management in Mt. Omine, and discussed management ways. After establishment of the plans, they started collaborative management for restoring the relationships between local people and Mt. Omine and conserving biodiversity.

At first, some management members asked the landowners to allow cutting trees and holding events in the forest. Then a workshop was conducted for local people including landowners, young people, and children to share the forest condition and ecological problems of Mt. Omine. Following that, several workshops for cutting trees, walking in the forest, and exchanging opinions on the future forest conditions were promoted. The participants were local people who have been interested in conserving the regional environment and create the forest where they can use. They enjoyed cutting bamboo and recreation in the forest, and children played on the withered bamboo (Fig. 14.8a, b). Some participants in the workshop offered to join to the management; they are living in the region and working on regional development with their specialties such as arts, education, and community building. Then the members started monthly meetings to discuss how to promote forest management in this region (Fig. 14.8c).

At the second stage, density control of Moso-bamboo and *Lithocarpus edulis* is required to make bright forest to restore forest floor vegetation. The management members discussed how to create a social system for reusing and circulating unused natural resources. And then they promoted workshops to produce some forest products such as bamboo charcoal and shiitake mushrooms (Fig. 14.8d), as activities relating to tree cutting. The management members continue to discuss how to obtain economic benefits from these forest products.

Purposes of participants are to enjoy cutting bamboo, communicating with other participants and contributing to conservation of Mt. Omine. It can be said that the abandoned forest functions as a place to interact people with nature. Participants in the workshop obtained cultural services (ecological education, recreation) and provisioning services (bamboo charcoal and shiitake mushroom). And a resultant of the activities, the forest structure has been restored and floor vegetation has become recovering.

14.6 Discussion

Enhancing multifunctionality of green infrastructure by linking ecological and social processes will maintain or improve human well-being (Wang and Banzhaf 2018). Green infrastructure can be a tool for building social capital, because its initiatives

Fig. 14.8 (**a**) Conservation together sitting in a circle, (**b**) Children who are playing on the withered *Phyllostachys edulis*, (**c**) Monthly meeting with management members, (**d**) Cultivating Shiitake mushroom with *Lithocarpus edulis* (photo by Hayato Hasegawa)

bring together people with different perspectives and it helps to forge bonds where they did not previously exist (Benedict Mark and McMahon 2006). Social capital is defined as *resources embedded in a social structure which are accessed and/or mobilized in purposive actions* (Nan 1999). Social capital refers to social networks, the norms of reciprocity and trustworthiness among individuals (Putnam 2000). In order to obtain multiple ecosystem services from *Satoyama* and function as green infrastructure for human well-being, building social capital for nature restoration is important.

In the study site, the *Satoyama* restoration project was started as an implementation of the Environmental Master Plan and Regional Biodiversity Strategy. The researchers of the Laboratory of Environmental Design have shared and discussed about the problems and necessity of biodiversity conservation with Fukutsu city's officers and local people for years before the implementation of *Satoyama* restoration. These processes are intended to enhance social networks and trust among stakeholders and increase social capital for local people's involvement in *Satoyama* restoration in Fukutsu city. Benedict Mark and McMahon (2006) mentioned that for social capital to exist, people's involvement is a necessity, not only participation in community activities, but also in decision making. The plans and their formulation process with various participants could be opportunities for building communities that can become involved in biodiversity conservation.

Figure 14.9 shows the changes of interrelationship between social system and Mt. Omine. The ecosystem services that local people are trying to obtain from Mt. Omine have been changed from past to present. Kamada (2017) pointed out that the domain of *Satoyama* has changed from private to commons and people are trying to retrieve a way of life harmonious with nature by collaboration with a wide range of participants. The management members of Mt. Omine restoration project are neither landowners nor foresters, who plant and cut trees to produce timber. Nevertheless, they are working on regional development with their specialties and trying to restore the system of *Satoyama*. Fukamachi (2017) reported that the local young people who have moved into the area organized and started preservation of the local *Satoyama* with the aim of bringing about enjoyment, understanding, and a willingness in their immediate vicinity in Kyoto prefecture. In Mt. Omine, local people particularly recognized cultural services such as educational and recreational values through the discussion and the implementation of *Satoyama* restoration. For forest restoration and sustainable use, local people's awareness of the benefits and ecosystem services by means of participation processes and environmental education are a necessity (Zerbe et al. 2020). The process of forest management in Mt. Omine could enhance people's awareness of the benefits and ecosystem services from Mt. Omine. The values for the present generation of local people are different from the past; however, people could obtain ecosystem services and use them for their well-being again through the restoration of relationships between people and nature.

The *Satoyama* management also should be considered more for realization of biodiversity conservation. In this study, the past flora and the present condition of abandoned forest were described. Bamboo forest situated adjacent to low or transitional vegetation requires more immediate management than in mature secondary forest (Suzuki and Nakagoshi 2011). Suzuki et al. (2008) suggested a bamboo forest management by continuously cutting and density control will restore forest floor vegetation. Thus, Mt. Omine restoration project would promote biodiversity conservation by continuously cutting bamboo and density control; however, the project would be inadequate for biodiversity conservation. The next challenges for the project will be to select priority areas for density control based on the survey and to share the visions of the forest restoration with the management members.

14.7 Conclusion and Future Issues

This chapter has attempted to describe the changes of relationships between people and *Satoyama*, the ecological conditions of abandoned forest, and the process of collaborative forest management as implementation of city plans. The Japanese traditional landscape, *Satoyama* supplied multiple ecosystems for local people in each region (Katoh et al. 2009). In Fukutsu city, city plans for the environmental and

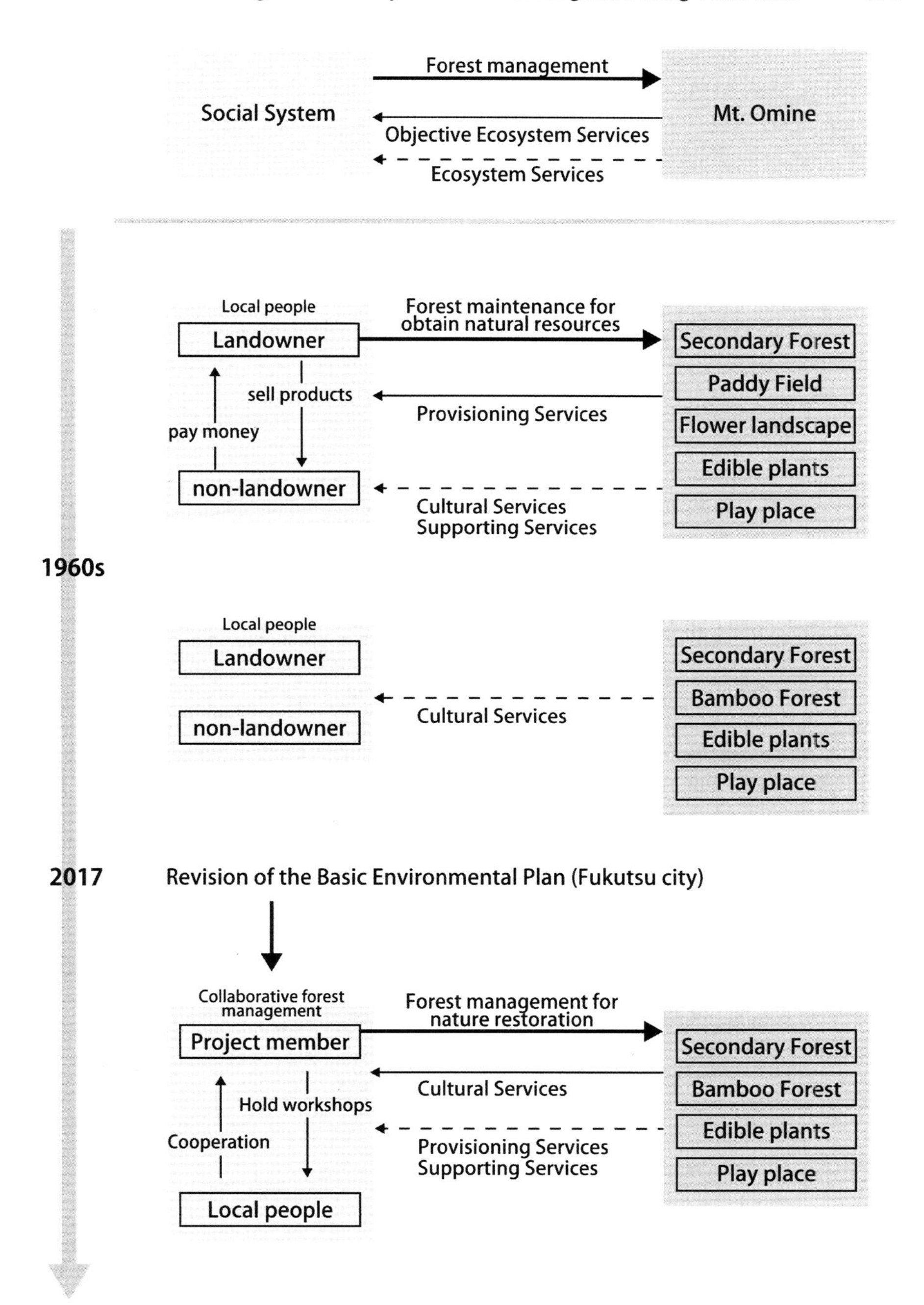

Fig. 14.9 The difference of interrelationship between social system and ecosystem services which people trying to obtain from Mt. Omine. The ecosystem services which is the purpose of management are described as objective services in this figure

the process of their formulation could help to build communities that can implement *Satoyama* restoration contributing to regional biodiversity conservation. In association with city environmental management and plans, continuing discussion how to restore and utilize *Satoyama* as green infrastructure with various participants has become more urgent in present days. Re-recognizing traditional landscapes as green infrastructure could be given more consideration in city environmental management for regional biodiversity conservation and re-connecting people to nature. Furthermore, how we can invite local people to practical nature restoration activities is also important. The evaluation of social networks among local people and its structure, and management method for building social capital for regional biodiversity restoration is an issue for the future.

It was described that local people are trying to retrieve the connection with Mt. Omine in order to obtain ecosystem services which are different from the past. To retrieve the people's interaction with *Satoyama*, recognizing the benefits from *Satoyama* for people is important. It is difficult to use *Satoyama* in the same way as in the past; therefore, adding new present values different from the past farming or life supporting ones may be needed for conserving or restoring the *Satoyama* (Washitani 2001). However, the way people maintain the *Satoyama* should not be different from the past. Therefore, the stakeholders of *Satoyama* restoration should explore its values in the present situation and learn from the past methods and wisdom in the relationship between people and *Satoyama*. Providing opportunities for re-connecting human being and nature must be paid more attention not only for protecting native species, but also to enhance human well-being (Miller 2005). The process of exploring ways to restore *Satoyama* could be opportunities for local people to realize the values of the nature around them and conserve not only for biodiversity but also culture in each region.

Acknowledgements We are grateful to every support for giving opportunities and writing this chapter. We are deeply indebted to Keitaro Ito, Mahito Kamada, Ian Ruxton, Tomomi Sudo, and Shwe Yee Lin for advising and improving this study. We would like to thank also for all students in Keitaro Ito's laboratory in Kyushu Institute of Technology and all of local people in Fukutsu city who have been involved in this project. This study was supported by the Biodiversity Conservation support project fund of the Ministry of the Environment, Japan, Fukuoka Forestation support project fund of Fukuoka prefecture and Kakenhi, Japan Society for Promotion of Science (JSPS), Grant-in-Aid for Scientific Research (B) (No. 15H02870) in 2015-2019.

References

Benedict Mark A, McMahon ET (2006) Green infrastructure: linking landscapes and communities. Island Press, Washington DC

Duraiappah AK, Nakamura K (2012) The Japan *Satoyama* Satoumi assessment: objectives focus and approach. In: Duraiappah AK, Nakamura K, Takeuchi K, Watanabe M, Nishi M (eds) Satoyama-Satoumi ecosystems and human well-being: socio-ecological production landscape of Japan. United Nation University Press, Tokyo

Fukamachi K (2017) Sustainability of terraced paddy fields in traditional Satoyama landscapes of Japan. J Environ Manag 202:543–549

Fukutsu city (2017) The second environmental master plan and regional biodiversity strategy in Fukutsu. http://city.fukutsu.lg.jp/pdf/kurashi/kankyo_01.pdf. Accessed 10 Sep 2019

Fukutsu city (2019). http://city.fukutsu.lg.jp/top.php. Accessed 10 Sep 2019

Kamada M (2017) Satoyama landscape of Japan -past, present, and future. In: Hong S-K, Nakagoshi N (eds) Landscape ecology for sustainable society. Springer, Cham

Katoh K, Sakai S, Takahashi T (2009) Factors maintaining species diversity in *satoyama*, a traditional agricultural landscape of Japan. Biol Conserv 142:1930–1936

Miller JR (2005) Biodiversity conservation and the extinction of experience. Trends Ecol Evol 20:430–434

Ministry of the Environment (2012) The national biodiversity strategy of Japan. https://www.biodic.go.jp/biodiversity/about/initiatives/files/2012-2020/01_honbun.pdf. Accessed 10 Sep 2020

Morimoto Y (2011) What is *Satoyama*? Points for discussion on its future direction. Landsc Ecol Eng 7:163–171

Nan (1999) Building a network theory of social capital. Connections 22(1):28–51

Putnam RD (2000) Bowling alone: the collapse and revival of American community. Simon and Schuster, New York

Saito O, Shibata H (2012) Satoyama and satoumi, and ecosystem services: a conceptual framework. In: Duraiappah AK, Nakamura K, Takeuchi K, Watanabe M, Nishi M (eds) Satoyama-Satoumi ecosystems and human well-being: socio-ecological production landscape of Japan. United Nation University Press, Tokyo, pp 17–59

Shinohara Y, Misumic Y, Kubota T, Nankod K (2019) Characteristics of soil erosion in a moso-bamboo forest of western Japan: comparison with a broadleaved forest and a coniferous forest. Catena 172:451–460

Someya T, Takemura S, Miyamoto S, Kamada M (2010) Predictions of bamboo forest distribution and associated environmental factors using natural environmental information GIS and digital national land information in Japan. Japan Assoc Landsc Ecol 15(2):41–54 (in Japanese with English abstract)

Suzuki S, Nakagoshi N (2011) Sustainable Management of Satoyama Bamboo Landscapes in Japan. In: Hong SK, Kim JE, Wu J, Nakagoshi N (eds) Landscape ecology in Asian cultures. Ecological research monographs. Springer, Tokyo. https://doi.org/10.1007/978-4-431-87799-8_15

Suzuki S, Kikuchi A, Nakagoshi N (2008) Structure and species composition of regenerated vegetation after clear cutting of Phyllostachys pubescens culms. Landsc Ecol Manag 12 (2):43–51 (in Japanese with English abstract)

Tomita R (2007) The nature restoration from the perspective of people and society. In: Washitani I, Kitoh S (eds) Biodiversity monitoring for nature restoration. University of Tokyo Press, Tokyo (in Japanese)

Wang J, Banzhaf E (2018) Towards a better understanding of green infrastructure: a critical review. Ecol Indic 85:758–772

Washitani I (2001) Traditional sustainable ecosystem 'SATOYAMA' and biodiversity crisis in Japan conservation ecological perspective. Glob Environ 5(2):119–133

Zerbe S, Pieretti L, Elsen S, Asanidze Z, Asanidze I, Mumladze L (2020) Forest restoration potential in a deforested mountain area: an ecosociological approach towards sustainability. For Sci 66(3):326–336

Chapter 15
Green Infrastructure as a Planning Response to Urban Warming: A Case Study of Taipei Metropolis

Wan-Yu Shih and Leslie Mabon

Abstract As concern increases over high temperatures in cities due to a combination of climate change and urban heat island effects, there is a rising interest in the cooling services that green infrastructure may provide. This article evaluates this role for green infrastructure as part of strategic land use planning to counter urban warming, through consideration of the case of Taipei Metropolis. Based on the findings from Taipei, this chapter argues that whilst strategic greenspace planning may indeed offer significant potential in cooling urban environments, this needs to be considered within a wider context of urban development patterns, meso-scale climate, and geographical features. Moreover, Taipei also illustrates the challenges of developing land use interventions—and indeed decision-making processes themselves—that ensure benefit to the most vulnerable people of the society. It is also cautioned that greening does not offer a universal solution to the problems of urban warming, and that awareness ought to be shown to the importance of balancing up different ecosystem services in planning processes.

Keywords Greenspace cooling effect · Urban heat island Green Infrastructure · Urban planning

15.1 The Risk of Urban Heat Hazard

The increase in frequency and duration of excessive heat is a major concern for public health in highly populated cities at the tropics. Urban areas are generally several degrees warmer than rural surroundings due to the replacement of natural

W.-Y. Shih (✉)
Department of Urban Planning and Disaster Management, Ming-Chuan University, Taoyuan City, Taiwan
e-mail: shih@mail.mcu.edu.tw

L. Mabon
Social Science, Scottish Association for Marine Science, The University of the Highland and Islands, Oban, Scotland, UK

K. Ito (ed.), *Urban Biodiversity and Ecological Design for Sustainable Cities*,
https://doi.org/10.1007/978-4-431-56856-8_15

surfaces with hard pavements and buildings during the process of urbanisation, which the phenomenon is well-known as the urban heat island effect (Gago et al. 2013; Yu et al. 2017). In recent years, global climate change has increased the magnitute, frequency, and duration of extreme high temperature, which further worsens thermal discomfort in cities (Revi et al. 2014). This has led to hazardous heatwave events in cities across a range of climate types (WMO and WHO 2015). Increasing mortality rates and heat related health illness are reported worldwide regardless of the development status of the country (Norton et al. 2015; WMO and WHO 2015). According to Mora et al. (2017), around 30% of the world's population is exposed to potentially lethal heat for at least 20 days a year. By 2100, this number is projected to increase to 48% even under a scenario of drastic reduction of greenhouse gas emission (Mora et al. 2017). Together with the urban heat island effect, the economic losses due to global warming could be 2.6 times higher for cities (Estrada et al. 2017). At the worst impacted cities, the cost of extreme high temperature could reach up to 10.9% of GDP by 2100 (ibid.).

15.2 The Cooling Effect of Green Infrastructure in Cities

Cities play a critical role in countering adverse impacts from climate change and building adaptive capacity in local societies (UN-Habitat 2017). The local-level governments which determine land uses and development of cities play a crucial role in heat risk reduction. There is a growing interest in urban climatological planning, of which green infrastructure strategy manifests an opportunity to reconstruct city-nature relationship by rehabilitating a natural process and regulating services in built environments (Kabisch et al. 2017; European Commission 2013; Shih et al. 2020). Green Infrastructure (GI) is a strategic spatial planning, which has been recognised as a critical nature-based solution for cities to attain climate resilience, health, social inclusion, and sustainability (e.g. IPCC AR5 and AR6). It refers to "an interconnected network of natural and semi-natural areas with other environmental features that strategically planned, designed and managed to sustain health nature process and provide multiple functions" (Benedict and McMahon 2012; EC 2013; Hansen and Pauleit 2014). Despite the disparity in grounded knowledge regarding the synergies and trade-offs of multiple functions associated with the spatial pattern of GI (Hansen and Pauleit 2014; Meerow and Newell 2017), scholars and practitioners increasingly argue integrating GI with cities' grey infrastructure to foster resilience and sustainability in urban societies and environments (Kabisch et al. 2017; Mell 2016; Shih et al. 2020).

Urban climatological planning aiming at optimising regulating services from green infrastructure may take place at different forms and spatial scales. In Stuttgart (Germany), a comprehensive GI strategy from a city-region level to a site level is used to reduce urban heat island effects and to improve air quality (VRS 2008). At the city-region level, the forested hills around Stuttgart is protected by law as a conservation area to generate cool and clean air. At the city level, urban greenspaces are strategically allocated through re-zoning and and the change of building codes (VRS 2008). This land use scheme enables the creation of wind corridors, which facilitate air producing in the surrounding hills to flow across the city. At the site scale, large trees having diameter above one metre at the breast height, are protected for cooling services (City of Stuttgart 2017). In Melbourne (Australia), the city government has initiated an urban forest strategy and aims to increase canopy cover from 22% to 40% by 2040 (City of Melbourne 2011). A similar strategy has been undertaken in Chicago (USA), where the municipal government aimed to add one million new trees to parks, parkways, and private yards by 2020 for mitigating urban heat (City of Chicago n.d.).

Indeed, vegetation removing latent heat from the surroundings via evapotranspiration and reducing incoming solar radiation on land surfaces through shading is an important mechinism to regulate city's temperature (Shashua-Bar et al. 2009). Empirical studies found that green infrastructure is generally several degrees cooler than impervious urban surfaces (Shih 2017a, b; Bowler et al. 2010). In the tropics, Giridharan and Emmanuel (2018) indicated that the urban-rural temperature difference in warm climates has a closer magnitude to the intra-urban temperature difference, which is mainly attributable to the distribution of vegetation. This emphasises the vital role of green infrastructure in mitigating the urban heat island effect in warm climates (Shih et al. 2020). Together with wind, the cooling effect from GI can be delivered to built-up areas and lower temperature several metres away from a green space (e.g. Shih 2017a, b; Shih et al. 2020; Narita et al. 2004). As this cooling mechanism provides better thermal comfort for urban dwellers, it also holds a great potential to reduce energy demand from buildings (Ca et al. 1998; Oliveira et al. 2011).

Increasing natural coverage in cities is fundamental for mitigating the urban heat island effects, but it is particularly challenging for densely developed cities where lands for greenery is scarce. However, previous studies suggested that urban temperature varies not only by the proportion of greenspaces (Chen et al. 2012; Ren et al. 2013; Tan and Li 2013; Chang et al. 2015), but also by their configuration (Shih 2017b; Xu et al. 2017; Zhou et al. 2011). In other words, increasing green volume is not the only solution—better cooling effect may be achieved by modifying urban morphology. It is understood that urban temeprature are jointly influenced by various factors, including (a) meso-scale climate conditions and topographical characteristics; (b) greenspace composition and configuration (Kong et al. 2014), such as greenspace coherence (Shih 2017a, b, c; Li et al. 2012), surface area (e.g. Chang et al. 2007; Tan and Li 2013; Shashua-Bar and Hoffman 2000), and shape (e.g. Li et al. 2012; Ren et al. 2013); as well as (c) geometry of buildings in three dimensions (e.g. Shih et al. 2020; Xu et al. 2017). Greater greenspace cooling effect may thus be delivered through an integrated consideration of these factors.

This chapter elaborates opportunities and challenges for optimising cooling services from GI in cities by using Taipei Metropolis as a case study area.

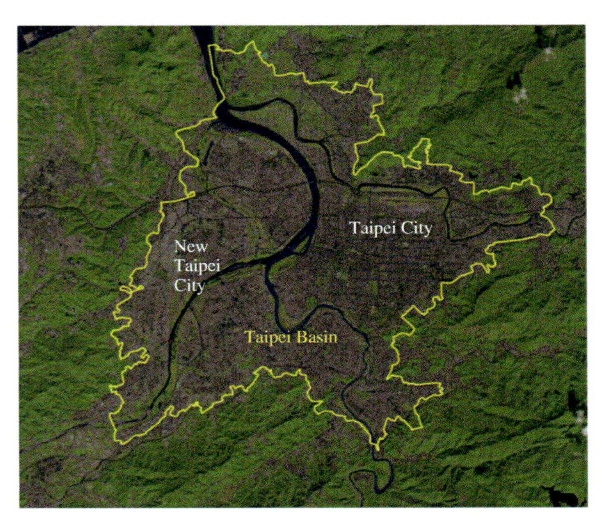

Fig. 15.1 The location of Taipei Metropolis and the studied area (Source: Google Maps, adapted from Shih 2017b)

15.3 The Warming Trend in Taipei and the Need for Adaptation

This study takes the urbanised area of Taipei Basin (25°'N, 121°'E), including parts of Taipei City and New Taipei City, as empirical study areas (Fig. 15.1). It covers approximately 2726 km^2 and has population estimated at 6.67 million by 2014. The climate in the North Taiwan is humid subtropical type with an annual average temperature about 23 °C (CWB 2014). The summer months, June, July, and August, are the warmest in a year with mean temperature at 28.8 °C (ibid). Due to global warming and rapid urbanisation in the past decades, Taipei has shown a distinct warming trend (Bai et al. 2011; Hsu et al. 2011). In the last 30 years, annual mean temperature has raised at the rate of 0.27°Cper year in Taipei City (CWB 2014) and the decadal mean number of hot days (above 35°C) has increased from 5–22 days on average to 37 days per year in the 2000s (Liu et al. 2010).

The projection from the Taiwan Climate Change Protection and Information Platform (TCCIP 2017) further indicates that the increase of summer temperatures will reach 1.125–1.25 °C between the 2021–2040 period under the IPCC RCP 8.5 scenario (i.e. 'business as usual'). Even under the RCP 2.6 scenario of radical emissions reduction, average summer temperatures are still projected to increase by between 0.625 and 0.75 °C in Taipei over the same period.

Thermal comfort in the summertime has thus become a crucial issue to be addressed within urban development. The need for heat mitigation and adaptation planning is further emphasised by Taiwan's ageing demographic characteristics. There were 14.4% and 18.07% of the population older than 65 years old in New

Taipei City and Taipei City in 2019 (DBAS-NTCG 2020; DBAS-TCG 2020). The World Population Prospect predicts that by 2050 Taiwan will hold the oldest population in the world, with a median age of 56.2 years old (UN-DESA 2015). Age is an important factor determining the sensitivity of individuals to excessive high temperature, because the physiological mechanism to regulate body temperature is declined with age (Watts et al. 2019). Excess urban heat in summer, which exacerbates air pollution, increases the spread of diseases, and triggers both physiological and mental health problems, is particularly threatening for an anging society (Watts et al. 2019; Huang et al. 2011).

15.4 Green Infrastructure and the Thermal Distribution of Taipei Metropolis

This section summarises the findings regarding thermal environments and greenspace cooling effect of Taipei basin (Shih 2017a, b, c). The principal data used for the analyses was satellite images acquired from LANDSAT 8, which periodically visited Taiwan at 10:20 am local time with a 14-days interval. Therefore, it is important to note that the findings and discussions in this article are specific to the climate conditions in the daytime. The temperatures mentioned hereafter refer to land surface temperature (LST) derived from the TIR band of LANDSAT 8 unless otherwise specified. The urbanised areas of Taipei metropolis are closely in line with the area of Taipei basin, which is surrounded by mountains and hills. Due to topographical constraints limiting development, most of the area in the mountain region remains forested. By contrast, substantial natural areas in the low-land basin areas have disappeared and been converted into buildings due to urbanisation (Shih 2010).

15.4.1 The Characteristics of Green Infrastructure

To date, the greenspace in the basin area is small, fragmented, and artificial because of radical urban development in the past decades prioritising construction and economic growth (Shih 2010). Other than parks designated in the earlier stage of the urban plan and farmlands/wetlands remaining in the Guandu areas, large greenspaces are scare in the basin areas (Fig. 15.2). In Taipei City, more than 80% of parks are smaller than 1 hectare and many of them have low coverage of vegetations (Shih 2010). Official planning policy, which has focused on the creation of small but accessible neighbourhood parks since the 1960s (Shih 2010), is one of the reasons for this green pattern. For New Taipei City, considerable amounts of agricultural lands have been developed in the last decade due to the rising housing demand for people commuting to work in Taipei City. Agricultural lands also face

Fig. 15.2 Green infrastructure (vegetated grounds) in Greater Taipei area (The map is generated by calculating the NDVI value on 25 April 2015/LANDSAT 8 imagery)

a constant encroachment of illegal factories due to the lack of effective control on land use. Although there are greenspaces persisting in some areas, most of them are facing intensive development pressure.

15.4.2 Thermal Distribution in Summer

Figure 15.3c, d shows the heterogeneity of land surface temperature across the basin areas of Taipei on 12 August 2015 and 29 July 2016. The mean air temperature recorded on each day of observation at 10 am from Taipei and Banqiao meteorological stations was 32.8 °C and 32.7 °C, respectively, in 2015, and 35 °C and 35.1 °C in 2016. The difference in land surface temperature was significant, reaching 10 °C in the urbanised basin areas. Comparing the thermal distribution on each date, a similar thermal pattern is revealed in that New Taipei City on the west of the Danshui and Xindian Rivers is constantly warmer than Taipei City on the east.

Clearly, the thermal pattern is closely related to the types of land cover. Buildings and hard pavements have higher temperature than vegetated grounds and water bodies (Fig. 15.3). On average, water has the strongest cooling intensity, followed by trees and grasses. In New Taipei City, the hot spots are associated with the location of manufacturing and densely developed areas, where vegetation is particularly scarce. This association may be influenced by the kinds of building materials

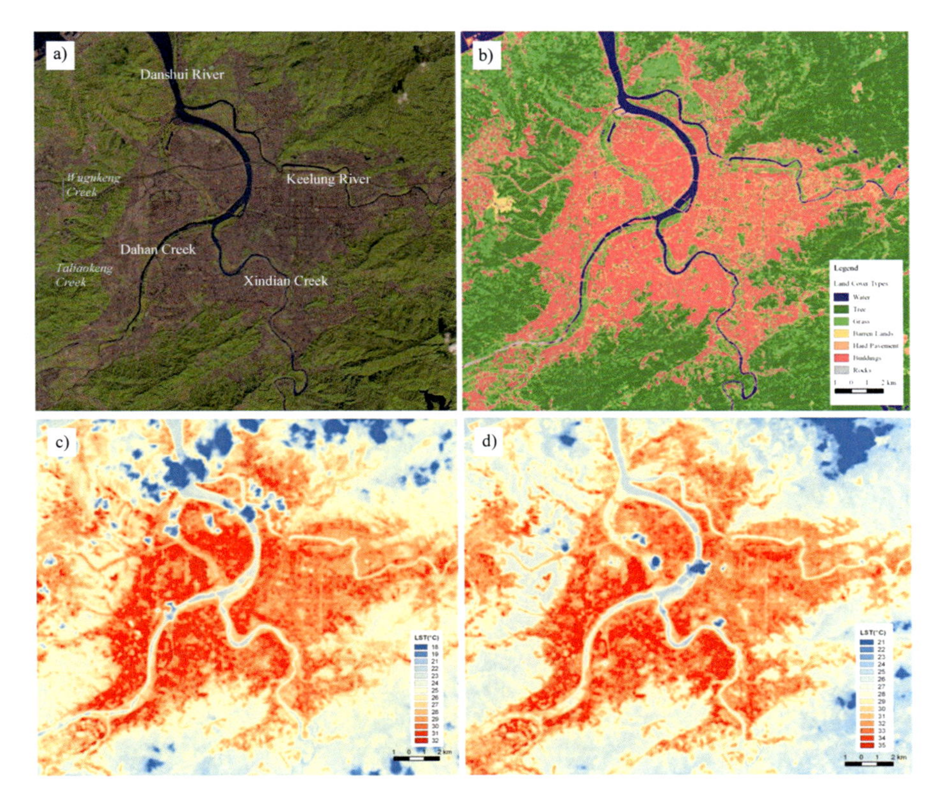

Fig. 15.3 Land cover and thermal distribution in Greater Taipei area on (**a**) natural colour of LANDSAT image; (**b**) land cover types; (**c**) LST on 12 August, 2015, (**d**) LST on 29 July 2016

used, as most factories in New Taipei City are low-rise steel buildings, which tend to absorb and store large amounts of heat (Shih 2017c). In Taipei City, hot spots are distributed in the areas of Song-san Airport, Wan-hua District, Da-tong District, and Wu-xing Street. This thermal pattern may be attributable to large and continuous hard pavement, higher building density, and the lack of greenery.

On the contrary, the coolest areas within the basin are associated with the location of major rivers, such as Danshui, Keelung, Dahan, and Xindian. However, it is noteworthy that little or insignificant cooling effect is observed alongside small creeks and ditches, particularly those surrounded by intensive development. For example, the waterfront areas alongside the Wugukeng Creek and Taliaokeng Creek are mostly steel-built factories, hence the cooling intensity from the creeks can barely be detected (Fig. 15.3). Other than rivers, the locations of cool islands are associated with open-air water bodies as well as extensive and dominant green areas, such as riverside greenspaces, wetlands, farmlands, derelict lands, large urban parks, and forested hills (Shih 2017b).

Amongst densely developed areas, lower temperatures tend to be found in large parks, gardens, and institutional grounds. Examples of such lower temperature areas include Daan Forest Park, Youth Park, Songshan Cultural and Creative Park, and Taipei Botanical Garden, as well as universities such as National Taiwan University of Arts, National Taiwan University, and Fu-Jen University. In New Taipei City, however, many cool areas are influenced by remaining agricultural lands and temporary greenspaces, which are designated for further development in the urban plan. It is expected that future development of these areas will worsen the thermal conditions in New Taipei City. In addition to large greenspaces, linear street greenery could also deliver cooling services to the city. Although street greenery is common in Taipei Metropolis, it is important to note that some street greenery demonstrates better cooling performance than the others. As showed in Fig. 15.3c, d, the cooling performance of Ren-ai and Dun-hua South boulevards is more constant and distinct. The difference of cooling magnitude of street greenery is likely related to types of street trees, width of greenery, and the width street canyons (Li et al. 2011).

15.4.3 Wind Path

River corridors and mountain valleys are important to natural ventilation of cities, particularly for those located in basins (Hebbert and Webb 2012). For Taipei Basin, land-sea breeze has a significant impact on local circulation, whilst it is further complicated by terrain roughness. During daytime summer, sea-breeze enters the basin mostly through the major river valleys of Danshui, Keelung, and Dahan; whereas at night time, land-breeze developed on the surrounding mountains flows along with river valleys and enter the city and then the sea (Lin et al. 2008; Kagiya and Ashie 2009). Whilst there has to date lacked direct observation, this circulation mechanism, together with unique topography and green infrastructure, is likely to play a critical role in regulating temperature at the urbanised area of Taipei basin.

However, urban development has been undertaken without understanding and considering this natural mechanism. It is likely that this meso-scale natural ventilation system has been weakened or even destroyed due to construction blocking the way of wind. For example, sea-land breeze should theoretically be able to reach the basin area of New Taipei City not only through the Danshui River, but also through mountain valleys on the west of Taipei Basin. Yet the temperature of these mountain valleys has increased significantly in the past decades to the extent that little cooling pattern can be observed both above and on its extended surroundings. For example, the valleys of Wugukeng Creek and Taliaokeng Creek are intensively built and some high-rise buildings are oriented against the direction of the creeks and prevailing wind. This development pattern will likely heat up cool air flowing between the mountains and the basin and also reduces wind volume and speed.

Another type of construction that might undermine the cooling function from the rivers is the existence of flooding walls, and intensive development at the waterfront areas. Although water surface within the watercourse of large rivers is significantly cooler, the cooling distance tends to be limited from the bank (Shih and Mabon 2018; Shih 2017b). Areas close to the rivers are not necessarily cooler, whereas in areas where open spaces or greenspaces are adjacent to rivers, the cooling patterns show a better extension from the watercourse despite the existence of flooding walls (Shih 2017b). Similar findings have emerged from previous studies, both in Sheffield, UK (Hathway and Sharples 2012) and in Taipei (Chen et al. 2014). This suggests that increasing greenery and openness at water front areas might improve the levels of cooling perceived from the river bank. However, due to a lack of local data regarding how urban form at the water front might influence wind and the cooling extension from the rivers, further research is needed to clarify whether and how the cooling is contributed by wind against the change of land use and land cover.

15.4.4 Greenspace Cooling Effect on Surrounding Built Environments

The temperature of greenspaces is on average cooler than built environments. The cooling intensity from greenspaces can be particularly significant in summer compared to Spring and Autumn (Shih 2017c). Yet not every kind of vegetation has a parallel cooling effect. In general, areas covered with trees are cooler than areas covered with grass, because trees have higher evapotranspiration rates that absorb heat from their surroundings, and provide shade to reduce the incoming solar radiation.

Greenspaces not only lower temperature from above, but also moderate heat from their surroundings. The cooling extension from greenspaces is more likely to be felt within 100 m when wind is relatively calm (Shih 2017a, b). The temperature of built environments is highly affected by nearby greenspaces and vice versa due to heat exchange between vegetated and not vegetated areas. A reverse heat effect from built environments might occur when the cooling magnitude of greenspaces is weak and the heat from outside the greenspaces is strong. Greenspaces below 2 hectares tend to have higher temperature fluctuation, as heat intrusion from the surroundings can penetrate further (Shih 2015, 2017a).

In this sense, temperature reduction from larger greenspaces can be more stable (Shih 2017a). Large greenspaces possess greater distance from the centre to the surrounding built environments (sources of heat), so cool spots within larger greenspaces might be better preserved if the shape of greenspaces is relatively compact. It has been found that the minimum temperature both within and around larger greenspaces tends to be lower (Shih 2017a). Although a threshold size for greenspace cooling has not yet been identified, more significant temperature reductions are found with a size interval of four hectares (Shih 2017b). In addition, a visual analysis suggested that the coolest spot within a greenspace is mostly associated with the location of water bodies (Shih and Mabon 2018). Increasing ponds, fountains, or creeks together with greenspaces is likely to amplify cooling magnitude.

Although greenspaces and water are cool islands in the city, the cooling effect extending from them to surrounding built environments is even more critical to mitigating the urban heat island effects and enhancing thermal comfort for urban dwellers. However, factors contributing to cooler greenspaces, such as larger vegetated areas, greater shape compactness, and higher degree of greenery within greenspaces, do not necessarily have an explicit effect on temperature reduction in the surrounding built environment (Shih 2017b). To extend cooling effects beyond cool islands, planning strategies should consider both greenspace features and the development characteristics of their surroundings (Shih et al. 2020; Shih and Mabon 2018; Shih 2017a, b). Lower development intensity at the adjacent areas of greenspaces can be an effective approach. This includes preserving large undeveloped lands, enhancing greenery at the immediate surrounding of greenspaces, allocating greenspaces with higher coherence rather than dispersal form, and increasing water elements in the city (Shih 2017a, b).

15.4.5 Thermal Inequity and Socio-Ecological Heterogeneity

Just as thermal environments and green cover vary across a city, differences may also exist in the extent of heat risk to which the population is exposed, and in the presence and/or accessibility of greenspace in their surroundings (Shih and Mabon 2021). These differences are not 'natural'—rather they are the result of social processes which mean that some people may have higher adaptive capacity than others, or live in a 'greener' environment. Those with lower incomes, lower education levels, and less access to mitigating technologies like air conditioning may face higher exposure to heat if they end up living in denser, lower-quality accommodation with limited surrounding greenery (e.g. Harlan et al. 2006; Byrne et al. 2016). When it comes to greenspace provision too, processes such as the 'luxury effect' (Hope et al. 2003) or 'ecological gentrification' (Dooling 2009) may result in greenery accruing to wealthier areas as cities develop. This 'double inequality'—where already marginalised and vulnerable groups have the least access to greenspace (Apparicio et al. 2016)—risks not only reducing the availability of cooling service for groups who may already be more vulnerable, but may also reduce the potential for social capital-building activities (Dinnie et al. 2013) and also physical (WHO 2016) and psychological (Fuller et al. 2007) well-being that accessible greenspace provides.

There are indications that such differences in the distribution of vulnerable populations hold true for Taipei as well. Figure 15.4 shows the distribution of greenspaces against heat intensity of Taipei Basin by neighbourhoods. Even on visual inspection, the heterogeneity of heat exposure in association with the location of greenspaces across space is clear. An initial assessment of this spatial pattern with socioeconomic factors of Taipei neighourhoods found that richer neighourhoods

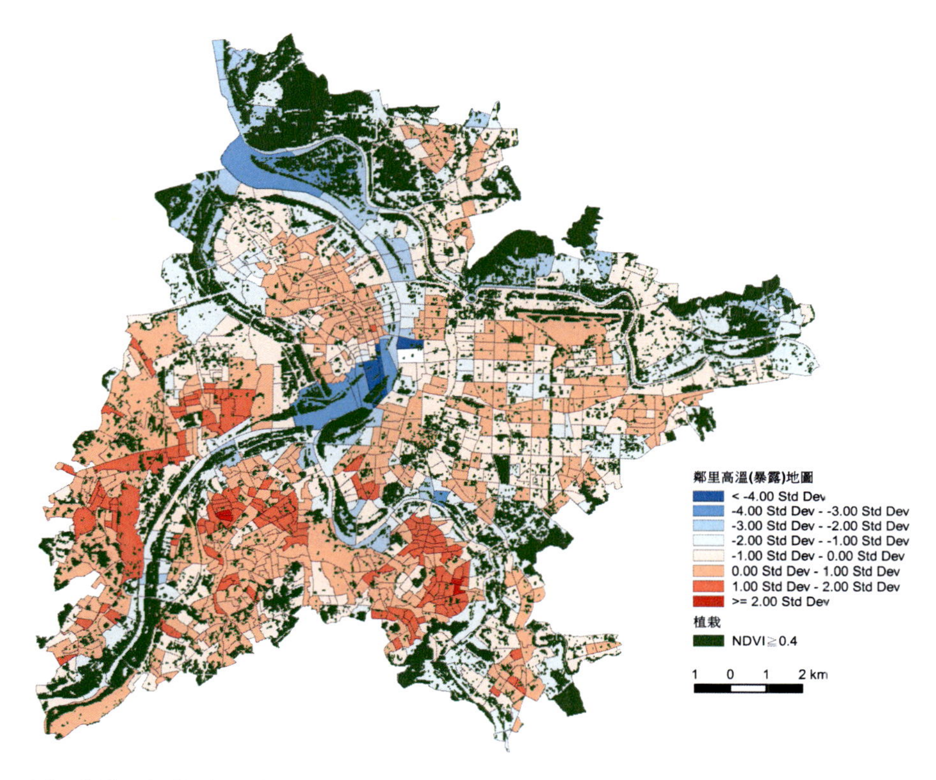

Fig. 15.4 Distribution of ageing population within Taipei Metropolis

tend to have lower heat intensity and greater greenery (Shih and Ahmed 2018). Because of the urban-rural dynamic in the past decades, lower income neighbourhoods were mostly located in New Taipei City, whilst more aging population was found in Taipei City. This may make it challenging to know which areas to prioritise for planning interventions aimed specifically at reducing heat risk. Moreover, it is also vital to remember that heat vulnerability is a product of social and cultural driver which is closely related to urban development decisions. Further research thus needs to evaluate the extent to which factors such as community participation, inequality and social cohesion (e.g. Chang et al. 2015; Cutter et al. 2003; Klinenberg 2002) affect neighbourhood vulnerability in the Taipei context specifically.

It is also vital to consider not only the *distribution* of vulnerable populations, but also which sections of society benefit from the outcomes of planning *processes*. Over the last decade, some greenspace-related planning initiatives in Taipei have been criticised by environmental groups for prioritising short-term developer profit over longer-term environmental sustainability (e.g. Taipei Times 2011a). When equity and justice are raised, these tend to be in the context of their absence within greenspace discussions in Taipei (e.g. Taipei Times 2010; Taipei Times 2011b). Such criticism of the prioritisation of economic gain over community well-being in urban development has likewise emerged in other Taipei-focused research (e.g. Raco et al. 2011; Jou et al. 2016). There is hence an indication in Taipei that the benefits of urban greening may accrue to those who are already more privileged or empowered.

More in-depth research into issues of spatial inequity in heat risk and greenspace provision is required for Taipei to verify the points made above. But there is evidence from both socio-economic data and recent greenspace planning controversies in the city to suggest there is a need in Taipei to develop processes and safeguards—perhaps through planning policy—to ensure that climate change adaptation via land use and green infrastructure benefits the most vulnerable people.

15.5 Planning Implications and Conclusions

Green infrastructure plays an important role in delivering cooling services in cities. For extending these cooling benefits to urban dwellers, planning policy however ought to consider strategies beyond greenspaces (Shih et al. 2020; Prieur-Richard et al. 2019). This study intents to draw attention to the complicated interrelationship of topography, wind circulation, and urban development patterns including green-blue spaces and built-up areas from a meso-scale to a site scale. In Taipei Metropolis, the thermal environment is significantly influenced by the unique basin topography, which forms a natural barrier to reduce sea-land breeze but at the same time enriches the area with mountain winds from the surroundings. Rapid urban development without prior awareness of this natural ventilation mechanism may have destroyed or damaged areas where clean and cool air are developed, and channels that facilitate air exchange.

Although it could be challenging to alter existing land use in such densely built areas, some modification on green infrastructure and built-up areas might be feasible in a long term to optimise cooling services and bring better thermal comfort. At a meso-scale, natural ventilation can be gradually reclaimed through the protection of mountain valleys from intensive development; cataloguing and modifying high-rise buildings which block the way of wind at mountain valleys and waterfront areas (Fig. 15.5); and preserving greenspaces along with wind path, particularly those located in the wider river corridors. In addition to air circulation, a baseline should be established to protect major cool islands, such as farmlands, wetlands, and ponds, from further fragmentation due to either uncontrolled urban sprawl or a poor consideration in official development plans. This can be achieved by strengthening the control of land zoning ordinance. Rezoning might also be used in strategic locations to protect existing cool islands that are at present undeveloped but marked for future development. In this case, transfer for development rights might be applied.

At a neighbourhood to site scale, optimising urban green infrastructure for enhancing cooling benefits requires comprehensive consideration of both greenspaces and surrounding development conditions (Prieur-Richard et al. 2019; Shih et al. 2020; Shih 2017a). As creating large greenspaces is less feasible for a

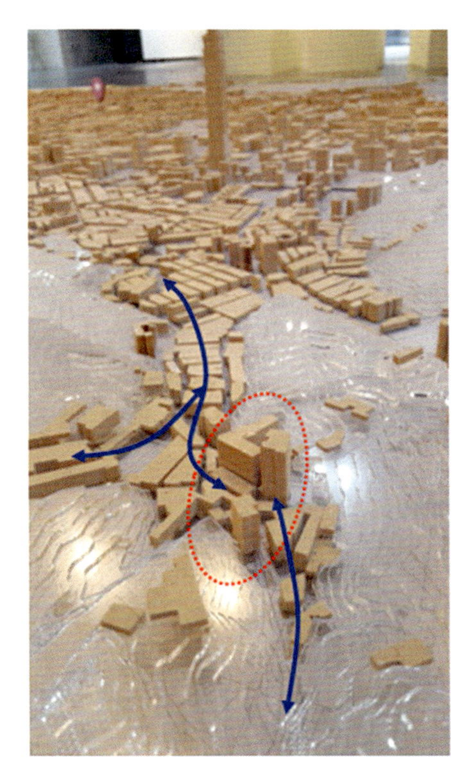

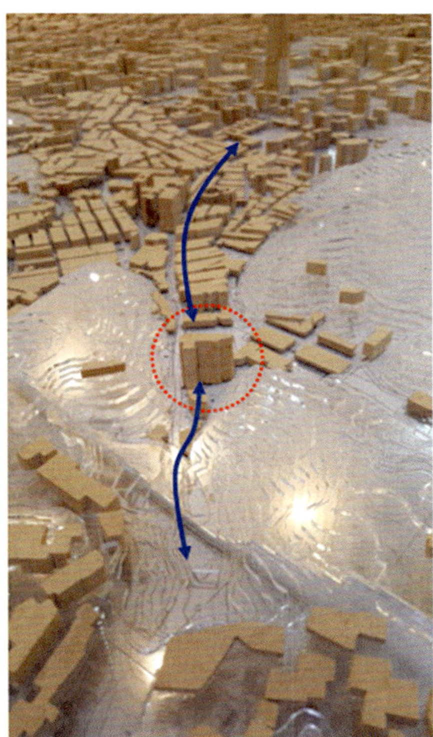

Fig. 15.5 Model showing buildings located in mountain valleys that may block wind (Model from the Taipei Vision Plan Studio)

densely developed city, mitigation strategies might focus on extending cooling services from existing cool islands and reducing heat at hot spots (Shih and Mabon 2018). In this sense, placing small greenery around large greenspaces or water bodies can help with the extension of cooling services into a wider area. Also, even though increasing small greenery at hot spots might not form a stable cool island (Shih and Mabon 2018), it can offset the accumulation of heat. This modification can be attained by greening marginal areas within different land use types, such as school grounds, institutional grounds, and streets; planting derelict lands; and/or implementing vertical greenery on roofs and walls of buildings. To this end, local governments might provide formal guidelines or incentives through building codes and urban design principles.

In addition to increasing vegetation volume and density, reclaiming water bodies can be another strategy to lower temperature in cities. In Taipei, the 'green campus' movement over the past decades has increased ecological ponds in schools for the enhancement of urban biodiversity. This is expected to have brought cooling co-benefits to the area and should be further encouraged. Also, recent public debates around and interest in recovering streams and ditches that have been covered for houses, roads, and parking lots may help to initiate action at the community level towards increasing running water within the city.

15.6 Conclusions

Urban climate is the complicated and dynamic result of the interaction between weather conditions, topography, development patterns, building geometry, and anthropogenic heat across multiple scales. Although green infrastructure, including greenspaces and water, plays an important role in mitigating urban heat island effects, an effective cooling strategy relies on planning grounded in understanding of climatological processes within the city. It is important to not just focus on greenspace itself, but to consider a much wider range of factors in an integrated manner. These include meso-scale topography, wind environments (e.g. sea-land breeze or mountain valley breeze), and development geometry in three dimensions; and also neighbourhood- and/or site-scale consideration of joint effects from not only greenspaces but also development patterns (e.g. building types, height, density). It is also vital that such 'evidence-based' planning encompasses not only the physical principles of urban climate, but also robust scholarly consideration of societal vulnerability and spatial justice in order to ensure that strategies and interventions can be developed in a way that benefits those that require them the most.

It is also important to bear in mind that greenspaces are living entities, hence there is a limitation to the cooling service they can provide. Greening is not a 'silver bullet' to the problem of excess urban heat. Greenspace has multiple functions which can be changed due to planning, and care ought to be exercised to ensure that planning greenspace to maximise cooling services does not weaken other functions. For example, a specific greenspace structure might facilitate the delivery of cooling services, but it might not provide a favourable structure for protecting urban biodiversity. Planning actions thus ought to take various ecosystem services into account and show awareness of the possible trade-offs between functions. Finally, in humid subtropical and tropical climates, increasing vegetation could enhance humidity in the air, which may have the effect of reducing thermal comfort. Effective planning actions must therefore pay cognisance not only to the cooling potential of green infrastructure, but also to its potential complexities and limitations.

References

Apparicio P, Pham TTH, Séguin AM, Dubé J (2016) Spatial distribution of vegetation in and around city blocks on the Island of Montreal: a double environmental inequity. Appl Geogr 76:128–136

Bai Y, Juang J-Y, Kondoh A (2011) Urban warming and urban heat islands in Taipei, Taiwan. In: Taniguchi M (ed) Groundwater and subsurface environments: human impacts in Asian coastal cities. Springer, Tokyo, pp 231–246
Benedict MA, McMahon ET (2012) Green infrastructure: linking landscapes and communities. Island Press, Washington, DC
Bowler D, Buyung-Ali L, Knight T, Pullin A (2010) Urban greening to cool towns and cities: a systematic review of the empirical evidence. Landsc Urban Plan 97(3):147–155
Byrne J, Ambrey C, Portanger C, Lo A, Matthews T, Baker D, Davison A (2016) Could urban greening mitigate suburban thermal inequity?: the role of residents' dispositions and household practices. Environ Res Lett 11(9):095014. https://doi.org/10.1088/1748-9326/11/9/095014
Ca V, Asaeda T, Abu E (1998) Reductions in air conditioning energy caused by a nearby park. Energ Buildings 29(1):83–92
Central Weather Bureau (CWB) (2014) The analysis of global and Taiwan temperature trend [report]. http://www.cwb.gov.tw/V7/climate/climate_info/monitoring/monitoring_7.html
Chang C-R, Li M-H, Chang S-D (2007) A preliminary study on the local cool-island intensity of Taipei city parks. Landsc Urban Plan 80(4):386–395
Chang SE, Yip JZK, van Zijll de Jong SL, Chaster R, Lowcock A (2015) Using vulnerability indicators to develop resilience networks: a similarity approach. Nat Hazards 78:1827–1841. https://doi.org/10.1007/s11069-015-1803-x
Chen X, Su Y, Li D, Huang G, Chen W, Chen S (2012) Study on the cooling effects of urban parks on surrounding environments using Landsat TM data: a case study in Guangzhou, southern China. Int J Remote Sens 33(18):5889–5914
Chen YC, Tan CH, Wei C, Su ZW (2014) Cooling effect of rivers on metropolitan Taipei using remote sensing. Int J Environ Res Public Health 11(2):1195–1210
City of Chicago (n.d.) Strategy 5: adaptation, Chicago climate action plan, City of Chicago. http://www.chicagoclimateaction.org/pages/adaptation/11.php
City of Melbourne (2011) Urban Forest Strategy, Making a great city greener 2012–2032
City of Stuttgart (2017) Urban Climate Stuttgart, section of urban climatology, office of environmental protection, City of Stuttgart. https://www.stadtklima-stuttgart.de/index.php?start_e
Cutter SL, Boruff BJ, Shirley WL (2003) Social vulnerability to environmental hazards. Soc Sci Q 84(2):242–261
Department of Budget, Accounting and Statistics, New Taipei City Government (DBAS-NTCG) (2020) New Taipei City statistical yearbook 2019. Taipei City Government
Department of Budget, Accounting and Statistics, Taipei City Government (DBAS-TCG) (2020) Taipei City statistical yearbook 2019. Taipei City Government
Dinnie E, Brown KM, Morris S (2013) Community, cooperation and conflict: negotiating the social Well-being benefits of urban greenspace experiences. Landsc Urban Plan 112:1–9
Dooling S (2009) Ecological gentrification: a research agenda exploring justice in the city. Int J Urban Reg Res 33(3):621–639
Estrada F, Botzen WW, Tol RS (2017) A global economic assessment of city policies to reduce climate change impacts. Nat Clim Chang 7(6):403–406. https://doi.org/10.1038/nclimate3301
European Commission (2013) Communication from the Commission to the European Parliament, the Council, the European Economic and Social Committee and the Committee of the Regions: Green Infrastructure (GI) - Enhancing Europe's Natural Capital. No. 52013DC0249
Fuller RA, Irvine KN, Devine-Wright P, Warren PH, Gaston KJ (2007) Psychological benefits of greenspace increase with biodiversity. Biol Lett 3:390–394
Gago EJ, Roldan J, Pacheco-Torres R, Ordoñez J (2013) The city and urban heat islands: a review of strategies to mitigate adverse effects. Renew Sust Energ Rev 25:749–758
Giridharan R, Emmanuel R (2018) The impact of urban compactness, comfort strategies and energy consumption on tropical urban heat island intensity: a review. Sustain Cities Soc 40:677–687
Hansen R, Pauleit S (2014) From multifunctionality to multiple ecosystem services? A conceptual framework for multifunctionality in green infrastructure planning for urban areas. Ambio 43 (4):516–529

Harlan SL, Brazel AJ, Prashad L, Stefanov WL, Larsen L (2006) Neighborhood microclimates and vulnerability to heat stress. Soc Sci Med 63(11):2847–2863
Hathway EA, Sharples S (2012) The interaction of rivers and urban form in mitigating the Urban Heat Island effect: a UK case study. Build Environ 58:14–22
Hebbert M, Webb B (2012) Towards a liveable urban climate: lessons from Stuttgart. In: Liveable cities: urbanising world. Routledge, Abingdon, pp 132–150
Hope D, Gries C, Zhu W, Fagan WF, Redman CL, Grimm NB, Nelson AL, Martin C, Kinzig A (2003) Socioeconomics drive urban plant diversity. PNAS 100(15):8788–8792
Hsu H-H, Chou C, Wu Y-C, Lu M-M, Chen C-T, Chen Y-M (2011) Climate change in Taiwan: scientific report 2011 (summary). National Science Council, Taipei
Huang G, Zhou W, Cadenasso ML (2011) Is everyone hot in the city? Spatial pattern of land surface temperatures, land cover and neighborhood socioeconomic characteristics in Baltimore, MD. J Environ Manag 92(7):1753–1759
Jou S-C, Clark E, Chen H-W (2016) Gentrification and revanchist urbanism in Taipei? Urban Stud 53(3):560–576
Kabisch N, Korn H, Stadler J, Bonn A (2017) Nature-based solutions to climate change adaptation in urban areas, Theory and practice of urban sustainability transitions. Springer, Berlin
Kagiya K, Ashie Y (2009) National research project on Kaze-no-michi for city planning: creation of ventilation paths of cool sea breeze in Tokyo. In: Second international conference on countermeasures to Urban Heat Islands. Heat Island Group, Berkeley
Klinenberg E (2002) Heat wave: a social autopsy of disaster in Chicago. University of Chicago Press, Chicago
Kong F, Yin H, James P, Hutyra LR, Hong HS (2014) Effects of spatial pattern of greenspace on urban cooling in a large metropolitan area of eastern China. Landsc Urban Plan 128:35–47
Li J, Song C, Cao L, Zhu F, Meng X, Wu J (2011) Impacts of landscape structure on surface urban heat islands: a case study of Shanghai, China. Remote Sens Environ 115(12):3249–3263
Li X, Zhou W, Ouyang Z, Xu W, Zheng H (2012) Spatial pattern of greenspace affects land surface temperature: evidence from the heavily urbanized Beijing metropolitan area, China. Landsc Ecol 27(6):887–898
Lin CY, Chen F, Huang JC, Chen WC, Liou YA, Chen WN, Liu SC (2008) Urban heat island effect and its impact on boundary layer development and land–sea circulation over northern Taiwan. Atmos Environ 42(22):5635–5649
Liu CM, Lin SH, Schneider SH, Root TL, Lee KT, Lu HJ, Lee PF, Ko CY, Chiou CR, Lin HJ, Dai CF, Shao KT, Huang WC, Lur HS, Shen Y, King CC (2010) Climate change impact assessment in Taiwan. Global Change Research Center, National Taiwan University, Taipei
Meerow S, Newell JP (2017) Spatial planning for multifunctional green infrastructure: growing resilience in Detroit. Landsc Urban Plan 159:62–75
Mell I (2016) Global green infrastructure: lessons for successful policy-making, investment and management. Routledge, London
Mora C, Dousset B, Caldwell IR, Powell FE, Geronimo RC, Bielecki CR, Counsell CW, Dietrich BS, Johnston ET, Louis LV, Lucas MP (2017) Global risk of deadly heat. Nat Clim Chang 7 (7):501–506
Narita K, Mikami T, Sugawara H, Honjo T, Kimura T, Kuwata K (2004) Cool-island and cold air-seeping phenomena in an Urban Park, Shinjuku Gyoen, Tokyo. Geogr Rev 77:403–420 (in Japanese)
Norton BA, Coutts AM, Livesley SJ, Harris RJ, Hunter AM, Williams NS (2015) Planning for cooler cities: a framework to prioritise green infrastructure to mitigate high temperatures in urban landscapes. Landsc Urban Plan 134:127–138
Oliveira S, Andrade H, Vaz T (2011) The cooling effect of green spaces as a contribution to the mitigation of urban heat: a case study in Lisbon. Build Environ 46(11):2186–2194
Prieur-Richard AH, Walsh B, Craig M, Melamed ML, Colbert ML, Pathak M, Connors S, Bai X, Barau AS, Bulkeley H, Cleugh H (2019) Global research and action agenda on cities and climate change science

Raco M, Imrie R, Lin W-I (2011) Community governance, critical cosmopolitanism and urban change: observations from Taipei, Taiwan. Int J Urban Reg Res 35(2):274–294

Ren Z, He X, Zheng H, Zhang D, Yu X, Shen G, Guo R (2013) Estimation of the relationship between urban park characteristics and park cool island intensity by remote sensing data and field measurement. Forests 4(4):868–886

Revi A, Satterthwaite DE, Aragón-Durand F, Corfee-Morlot J, Kiunsi RBR, Pelling M, Roberts DC, Solecki W (2014) Urban areas. In: Field CB et al (eds) Climate change 2014: impacts, adaptation, and vulnerability part a: global and sectoral aspects. Contribution of working group II to the fifth assessment report of the intergovernmental panel on climate change. Cambridge University Press, Cambridge, pp 535–612

Shashua-Bar L, Hoffman ME (2000) Vegetation as a climatic component in the design of an urban street: an empirical model for predicting the cooling effect of urban green areas with trees. Energ Buildings 31(3):221–235

Shashua-Bar L, Pearlmutter D, Erell E (2009) The cooling efficiency of urban landscape strategies in a hot dry climate. Landsc Urban Plan 92(3):179–186

Shih WY (2010) Optimising urban green networks in Taipei City: linking ecological and social functions in urban green space systems. Doctoral dissertation of Planning and Landscape, University of Manchester, United Kingdom

Shih WY (2015) The cooling effect of urban green infrastructure: does greenspace size and shape matter? The 51st ISOCARP congress: cities save the world: let's reinvent planning. Rotterdam, Netherlands, Oct 2015

Shih WY (2017a) The cooling effect of green infrastructure on surrounding built environments in a sub-tropical climate: a case study in Taipei metropolis. Landsc Res 42(5):558–573

Shih WY (2017b) Greenspace patterns and the mitigation of land surface temperature in Taipei metropolis. Habitat Int 60:69–80

Shih WY (2017c) The impact of urban development patterns on thermal distribution in Taipei. In: Urban remote sensing event (JURSE), 2017 joint. IEEE, pp 1–5

Shih W-Y, Ahmad S (2018) Spatial inequality? The influence of greenspace infrastructure on electricity consumption in Taipei's Urban neighborhoods, Global Land Programme 2018 Asia Conference, Taipei, Taiwan. Sept 2018

Shih WY, Ahmad S, Chen YC, Lin TP, Mabon L (2020) Spatial relationship between land development pattern and intra-urban thermal variations in Taipei. Sustain Cities Soc 62:102415

Shih W-Y, Mabon L (2018) Thermal environments of Taipei Basin and influence from urban green infrastructure. City Plan V45(4)

Shih W-Y, Mabon L (2021) Ways of creating usable, multipurpose greenspace in impoverished settlements in cities of the Global South: urban growth, green infrastructure loss and spatial inequity under climate change in Hanoi, Vietnam. In: Anderson P, Douglas I, Goode D, Houck M, Maddox D, Nagendra H, Tan PY (eds) Routledge handbook of urban ecology: second edition. Routledge, London

Taipei Times (2010) Quantifying the non-quantifiable. Taipei Times, 3 September 2010. http://www.taipeitimes.com/News/editorials/archives/2010/09/03/2003481966/1

Taipei Times (2011a) Parks may disappear with 'Taipei Beautiful' program. Taipei Times, 28 April 2011. http://www.taipeitimes.com/News/taiwan/archives/2011/04/28/2003501887

Taipei Times (2011b) Make tobacco factory into a park: protesters. Taipei Times, 31 October 2011. http://www.taipeitimes.com/News/taiwan/archives/2011/10/31/2003517124

Taiwan Climate Change Projection and Information Platform (2017) Projection@TCCIP. https://tccip.ncdr.nat.gov.tw/v2/future_map_en.aspx. Accessed 21 Apr 2017

Tan M, Li X (2013) Integrated assessment of the cool island intensity of green spaces in the mega city of Beijing. Int J Remote Sens 34:3028–3043

UN-DESA (2015) World population prospects: the 2015 revision. Population Division, Department of Economic and Social Affairs, United Nations, United States, New York

UN-Habitat (2017) New urban agenda: habitat III. UN Habitat, United Nations, Quito

VRS (Verband Region Stuttgart) (2008) Klimaa-tlas region Stuttgart. Verband Region Stuttgart, Stuttgart

Watts N, Amann M, Arnell N, Ayeb-Karlsson S, Belesova K, Boykoff M, Byass P, Cai W, Campbell-Lendrum D, Capstick S, Chambers J, Dalin C, Daly M, Dasandi N, Davies M, Drummond P, Dubrow R, Ebi KL, Eckelman M, Ekins P, Escobar LE, Montoya LF, Georgeson L, Graham H, Haggar P, Hamilton I, Hartinger S, Hess J, Kelman I, Kiesewetter G, Kjellstrom T, Kniveton D, Lemke B, Liu Y, Lott M, Lowe R, Sewe MO, Martinez-Urtaza J, Maslin M, McAllister L, McGushin A, Mikhaylov SJ, Milner J, Moradi-Lakeh M, Morrissey K, Murray K, Munzert S, Nilsson M, Neville T, Oreszczyn T, Owfi F, Pearman O, Pencheon D, Phung D, Pye S, Quinn R, Rabbaniha M, Robinson E, Rocklöv J, Semenza JC, Sherman J, Shumake-Guillemot J, Tabatabaei M, Taylor J, Trinanes J, Wilkinson P, Costello A, Gong P, Montgomery H (2019) The 2019 report of The Lancet Countdown on health and climate change: ensuring that the health of a child born today is not defined by a changing climate. Lancet 394(10211):1836–1878

World Health Organization (2016) Urban green spaces and health - a review of evidence. WHO, Copenhagen

World Meteorological Organization, World Health Organization (2015) Heat waves and health: guidance on warning-system development. World Meteorological Organization and World Health Organization. http://www.who.int/globalchange/publications/heatwaves-health-guidance/en

Xu Y, Ren C, Ma P, Ho J, Wang W, Lau K, Lin H, Ng E (2017) Urban morphology detection and computation for urban climate research. Landsc Urban Plan 167:212–224

Yu Z, Guo X, Jørgensen G, Vejre H (2017) How can urban green spaces be planned for climate adaptation in subtropical cities? Ecol Indic 82:152–162

Zhou W, Huang G, Cadenasso ML (2011) Does spatial configuration matter? Understanding the effects of land cover pattern on land surface temperature in urban landscapes. Landsc Urban Plan 102(1):54–63

Chapter 16
Green Infrastructure Planning for Asian Cities: The Planning Strategies, Guidelines, and Recommendations

Sadahisa Kato

Abstract Urbanization is a global trend, particularly strong in many Asian countries, Africa, and some Latin American countries now. Designing and planning for sustainable and low-carbon cities is a complex process addressing the fundamental areas of economic, environmental, and social-equitable sustainability. This chapter focuses on the environmental aspect with theories and applications of green infrastructure (GI) to support ecological and physical processes in urban regions including: hydrology, biodiversity, and cultural/recreational activities. GI is an interconnected network of waterways, hybrid hydrological/drainage systems, wetlands, both natural and designed green spaces, working farms and other cultural landscapes, and built infrastructure that provides ecological functions. GI plans apply key principles of landscape ecology to urban regions, specifically: a multi-scale approach with explicit attention to the pattern and process relationship and an emphasis on connectivity. Although GI concept and practice are gaining popularity in North America, the UK, and Europe, its systematic application in Asian cities and urban planning policies is yet to be seen. Through the examination of innovative GI application cases in New York City and five case studies of GI-like approaches to address urban green space planning issues in Japan, important GI principles are distilled and the lessons learned from these cases are used to develop specific recommendations to facilitate further application of the GI concept in Asian cities. GI is argued to become a useful green space planning tool to protect important and fragile green spaces, mitigate the lost nature, and create new green spaces in the city. Four general design and planning guidelines of GI are proposed. Based on the lessons learned from the case studies and the preceding argument, the chapter concludes with recommendation of four areas of application of the GI concept to Asian cities.

Keywords Green infrastructure · Planning guidelines · Sustainability · Landscape ecology · Connectivity · Multi-scale approach

---

S. Kato (✉)
Institute of Global Human Resource Development, Okayama University, Kita-ku, Okayama, Japan

K. Ito (ed.), *Urban Biodiversity and Ecological Design for Sustainable Cities*,
https://doi.org/10.1007/978-4-431-56856-8_16

16.1 Introduction

Developing biologically diverse, culturally rich, and just cities is a valid societal goal as more people now live in urban areas than rural areas and sustainable development is a recognized international goal (Millennium Ecosystem Assessment 2005). In landscape planning, the concept of green infrastructure (GI) has emerged as a way to provide multiple benefits to urban residents by an integrated, connected network of open spaces toward the goal of making cities more sustainable (Benedict and McMahon 2002; Gill et al. 2007; Mell 2008; Tzoulas et al. 2007). Although GI planning is gaining popularity in North America, the UK, and Europe, it is arguably lacking in Asian cities and urban planning policies. Among the three axes of sustainability (i.e., economy, environment, and social equity), this chapter focuses on the environmental aspect of sustainability with theories and applications of GI to support ecological and physical processes in broad, urban regions.

The aim of this chapter is to explore the potential of GI application in Asian cities. Although there have already been individual cases of the application of the GI concept, as defined below, in Asian cities, there is a lack of systematic application of the GI concept despite its benefits. Based on the literature review of Asian urban planning, and Japanese green space planning in particular, I will argue that GI has a large potential to become a useful green space planning tool in Asian cities as well to develop more sustainable cities. First, the chapter defines GI and describes its benefits. Second, Asian urban planning problems and those specific to green and open spaces in Japanese cities are stated. Third, three key landscape ecology principles that are useful for GI planning are discussed. Fourth, the chapter reviews and discusses innovative ecological restoration and stormwater management pilot projects in New York City and five Japanese green space planning cases to summarize important GI principles and learn from them to facilitate a more systematic GI application in Asian cities. Fifth, four general planning and design guidelines for GI are recommended. Finally, based on the case studies and the preceding argument, more specific recommendations for GI application in Asian cities are proposed to help solve the green space planning issues.

16.2 Definition, Characteristics, and Benefits of Green Infrastructure

GI is defined here as an interconnected network of open and green spaces and waterways, both natural and designed, that can provide multiple functions and services such as clean air and water, esthetics, recreation, environmental education, habitat, and increase in property values (Benedict and McMahon 2006; Gill et al. 2007; Mell 2008; Tzoulas et al. 2007). Lately, Gill et al. (2007) and Mell (2008) argue for the inclusion of open space for climate change mitigation and adaptation in the definition. GI has its precedents in the Parks, Parkways, and Boulevard System and linking conservation areas to counter habitat fragmentation (Benedict and McMahon 2006; Ishikawa 2011; Mell 2008). Following its historical precedents, GI has emerged as a popular planning concept among landscape researchers and practitioners in the UK, Europe, and North America since the late 1990s (Mell 2008).

GI exists at various scales (e.g., site, city/town, and urban region) and functions across jurisdictional boundaries (Benedict and McMahon 2006; Kambites and Owen 2006; Mell 2008; Tzoulas et al. 2007). Therefore, GI is not limited to urban greening but GI planning should be considered at multiple scales and in various planning contexts such as urban, regional, and rural planning. GI is to be distinguished from conventional built infrastructure such as roads, sewers, utility lines, hospitals, schools, and prisons (Benedict and McMahon 2002). Connectivity is a key planning concept for GI since many of the benefits of GI can be truly realized by an interconnected network of its constituting elements (Kato 2012).

Examples of GI benefits to sustainability include: enriched habitat and biodiversity, maintenance of natural landscape processes, cleaner air and water, increased recreational opportunities, improved health, and better connection to nature and sense of place (Tzoulas et al. 2007). Green space also increases property values and can decrease the costs of public infrastructure and services such as flood control, water treatment systems, and stormwater management (McMahon 2000). A caveat here is that these benefits include the effects of individual greening and/or open spaces as well as those that are characteristic to GI, a green space network. Unique roles of GI are attributable to its spatial configuration: an interconnected network of open and green spaces (Kato 2012). For example, GI is suitable to providing flood storage, improving stormwater infiltration, reducing runoff, and improving water quality (Abunnasr and Hamin 2012; Endo 2011; Gill et al. 2007; Inoue et al. 2011).

16.3 Urban Planning Problems in Asian Cities

Urban planning problems and issues in Japanese, Korean, and Chinese cities include: overpopulation of certain megacities, uncontrolled urban sprawl, natural disaster planning, environmental pollution, lack of open space, and social structural problems such as income gaps (Kyushu Chapter of the City Planning Institute of Japan 1999). Japanese cities' nature conservation issues are: loss of nature by urban sprawl, green space restoration and management in urban areas, and nature degradation due to the decreased use of Satoyama near cities (Kawakami 2008). It is argued that GI can help protect important and fragile green spaces, mitigate the lost nature, and create new green spaces in the city.

16.4 Landscape Ecology Principles for Green Infrastructure

Landscape ecology provides a theoretical perspective and the analytical tools to understand how complex and diverse landscapes, including urban areas, function with respect to specific ecological processes (Turner et al. 2001). Key ideas from landscape ecology that are relevant to GI for sustainable landscapes include: (1) a multi-scale approach, (2) the pattern and process relationship, and (3) connectivity. A multi-scale approach is based on the hierarchy theory (Allen and Starr 1982; O'Neill et al. 1986) that addresses the structure and behavior of systems that function simultaneously at multiple scales. Holling and his colleagues have developed the concept of adaptive cycle and of panarchy—adaptive cycles linked in a nested hierarchy—to study the feedbacks and processes operating across scales (Gunderson and Holling 2002). These are important concepts to develop resilient landscapes that can go through change but still maintain essential structure and feedback loops after disturbance and "surprise" (Light et al. 1995). Therefore, GI designed with multiple scales (e.g., neighborhood, region, multiple regions) in mind is one way to develop a resilient landscape.

The pattern and process dynamic is arguably the fundamental axiom of landscape ecology because the spatial composition and configuration of landscape elements directly determine how landscapes function, particularly in terms of species movement, nutrient, and water flows (Forman 1995; Turner et al. 2001). For example, using green spaces for climate change adaptation in urban areas, McMahon (2000) points out different functions which green spaces play according to their classification of corridor, patch, and matrix.

Connectivity is a property of landscapes that illustrates the relationship between landscape structure and function (Turner et al. 2001). In general, connectivity refers to the degree to which a landscape facilitates or impedes the flow of energy, materials, nutrients, species, and people across a landscape. Since GI is an interconnected network of open spaces, functional connectivity (i.e., connectivity that can support a particular function) is a prerequisite for the provision of its functions. In other words, GI is a planning method which can protect, restore, and create connectivity to protect important natural and cultural resources and to assure the services/functions which they can provide. Considering the concept of connectivity for target ecological and social/cultural functions helps landscape planners decide, for example, how to best place green spaces in urban environments. Different levels of connectivity provide a useful conceptual framework to organize green spaces across scales and challenge landscape planners to achieve connectivity in this manner (Kato 2010).

16.5 Case Studies

16.5.1 Innovative Projects by Biohabitats, Inc. in New York City in the United States

16.5.1.1 Biohabitats, Inc.

Founded by Keith Bowers in 1982, Biohabitats, Inc. is a design and consulting firm specializing in conservation planning, ecological restoration, and regenerative design. The firm features an interdisciplinary team of scientists, engineers, landscapes architects, and natural resource planners. The firm's representative projects include: the development of conservation plans to mitigate the impacts of habitat fragmentation and preserve and restore biodiversity; habitat restoration to save endangered plant species; watershed management and river restoration; and ecologically sustainable and regenerative master planning strategies for residential and commercial development, parks, campuses, and greenways. Biohabitats has involved in more than a thousand projects throughout the United States (US) and abroad that transcend the traditional discipline of landscape architecture, seeking the interplay and mutual learning among restoration ecology, conservation biology, and landscape ecology (CityCraft Ventures 2014).

16.5.1.2 Jamaica Bay Watershed Protection Plan

An estuary is an often important GI, providing many ecological, cultural, and recreational functions to surrounding areas. Biohabitats assisted the New York City Department of Environmental Protection (NYCEP) in creating a watershed protection plan for Jamaica Bay. Jamaica Bay is an estuary, encompassing the Jamaica Bay Wildlife Refuge and a unit of the Gateway National Recreation Area, and has been important to the cultural and economic development of New York City (Biohabitats 2014a). The 142-square-mile watershed contains one of the most densely populated urban areas in the US, the Boroughs of Brooklyn and Queens. Jamaica Bay has been faced with issues of effluent loading from numerous water pollution control plants, combined sewer overflows, and stormwater runoff, leading to severe water quality degradation. The salt marshes are quickly eroding and the marine estuary complex has been reduced by 50% due to landfilling and dredging operations (Biohabitats 2014a).

Biohabitats assisted the city with the technical components of the watershed protection plan and built a consensus among stakeholders to implement multifaceted protection and restoration initiatives. More specifically, Biohabitats took charge of researching the estuary ecological systems of the Bay, quantifying impacts, and making recommendations for sustainable ecological restoration and management. The firm helped organize and lead workshops aimed at addressing both technical issues and cultural and regulatory impediments to implementing a fully fledged restoration program, which will return Jamaica Bay to an ecologically rich, diverse, and resilient estuary.

16.5.1.3 Ecological Pilot Projects

Within the watershed protection plan for Jamaica Bay, Biohabitats with joint venture partners, Hydroqual/HDR and Hazen and Sawyer, is executing a range of ecological restoration and stormwater management pilot projects aimed at cleaning the water of the Bay while reestablishing previously lost ecosystems. The pilot projects summarized in Table 16.1 include: Oysters, Eelgrass, Algal Turf Scrubber®, Wave Attenuator, Macroalgae, and Mussels (Biohabitats 2014b).

Algal turf scrubbers are truly unique wastewater treatment devices that mimic a stream ecosystem in a constructed environment designed to promote algae growth. Harnessing the natural abilities of algae, bacteria, and phytoplankton to remove pollutants from water, the system filters nutrients from a small portion of the effluent from the Rockaway Wastewater Treatment Plant. The microalgae from the algal turf scrubbers are periodically harvested and can be used as a source of biofuel, along with the macroalgae harvested from Jamaica Bay (Table 16.1), creating a sustainable, "green" technology. A scaled-up operation is considered for both ATS™ and macroalgae harvesting for biofuel production; if proven feasible, the system could potentially fuel city service vehicles in the future (Biohabitats 2014b).

Also, there are ongoing efforts to reestablish eelgrass and oysters within the Bay. To date, approximately 1000 eelgrass plants, an oyster bed, and oyster reef balls

Table 16.1 Several ecological restoration pilot projects within Jamaica Bay

Name	Description
Oysters	Placing an oyster bed and oyster reef balls in two locations in the Bay to monitor the oyster growth and health and water quality over a three-year period
Eelgrass	Similar to oysters, eelgrass is planted experimentally in various places within the Bay to determine if eelgrass habitat can be restored given existing water quality
Algal Turf Scrubber®	Innovation and creation by designing and constructing an Algal Turf Scrubber® (ATS™) at the Rockaway Wastewater Treatment Plant, harvesting algae, monitoring nutrient removal over a three-year period, and converting harvested algae to biofuel
Wave attenuator	Experimentally, designing and installing a floating island wave attenuator offshore of an eroding salt marsh to determine if it can slow marsh erosion and aid in sediment accretion onshore
Macroalgae	Investigating most efficient methods for harvesting macroalgae (sea lettuce) from the Bay and microalgae from the ATS™
Mussels	Creating a wall array of mussels within a combined sewer overflow-tributary of Jamaica Bay and monitoring its biofiltration effect on water quality around the structure

have been installed in the Bay (Biohabitats 2014c). Biohabitats created first oyster bed to be restored in New York City waters (Business Wire 2014).

The floating island wave attenuator pilot projects have been carried out to protect shorelines and salt marshes from wave erosion at Brant Point since the spring of 2014. The wave attenuator design uses a series of buoyant mats planted with the cord grass *Spartina alterniflora*, whose roots would grow from the surface to the sub-aquatic environment, providing habitat for the subaqueous community (Biohabitats 2014d). The innovative technology is also expected to promote the accretion of sediments along shorelines. By introducing biological components to the wave attenuator structure, the designers expect harvesting some ecological benefits such as the removal of pollutants from water by root structure and the porous nature of the submerged matrix of the floating island (Biohabitats 2014d).

16.5.1.4 Stormwater Management

Along with ecological restoration pilot projects, Biohabitats, Hydroqual/HDR, and Hazen and Sawyer are also implementing a variety of stormwater best management practices throughout New York City. These pilot projects include: for example, bioretention ponds, stormwater swales and porous pavement integrated in parking lots, stormwater planters (raised) with underground storage, and blue roof pilot projects (Biohabitats 2014d, e). The firms capture retrofitting opportunities (e.g., buildings, paved areas, and parking lots) to test and demonstrate the effectiveness of GI for slowing and reducing stormwater runoff that enters the combined sewer system. The critical thing is that the pilot sites are monitored to document their effectiveness; data are collected including rainfall, surface inflows, surface water storage, subsurface water storage, evapotranspiration, and subsurface outflows. Biohabitats has installed, maintained, and repaired all monitoring equipment over a two-year period. It is often a problem that projects lack budget for monitoring after their completion, which was not the case for these stormwater best management pilot projects. Monitoring data is being used to develop effective GI designs and implementation strategies that can provide multiple benefits such as stormwater benefits, more pervious surfaces, improved urban wildlife habitat and neighborhood esthetics, and reduced urban heat island effects. (Biohabitats 2014d, e). The data can also be used in comparing costs associated with conventional gray infrastructure versus GI.

16.5.2 Some Cases of GI Application in Japan

Literature review of Japanese green space planning shows that there have already been some examples of GI application in Japan since the 1960s. However, in most cases, the full potential of the GI concept has not been realized and there still is a lack of systematic application of its concept in Japanese urban planning. Here I review and discuss some of the cases, regardless of the use of the term, GI. The results are

Table 16.2 A plus sign (+) denotes a specific acknowledgment of the landscape ecology principle. A solid triangle (▲) denotes a weak hint of using the principle although it is not strongly mentioned in the literature. A blank cell means that the specific landscape ecology principle was not mentioned in the literature. No cases have applied or specifically mentioned all the three key landscape ecology principles in the development of GI

Cases	Multi-scale approach	Pattern-process relationship	Connectivity
Tokyo metro regional planning	+		▲
Kohoku New Town in Yokohama		▲	+
Green parking space examples in Nagoya metro region		▲	+
Inochi-no-mori in Kyoto		▲	
Ecological networks		+	+

summarized in Table 16.2 based on the degree to which these green space planning cases use the key ideas from landscape ecology that are relevant to GI: (1) a multi-scale approach, (2) the pattern and process relationship, and (3) connectivity. The lessons learned from these cases will be used to develop recommendations (Sect. 16.7) to facilitate further application of the GI concept in Asian cities.

16.5.2.1 Tokyo Metro Regional Planning

With regard to Tokyo metro regional planning, in the 1960s there was a good amount of total green space but the concept of systematic connection among green spaces was non-existent (Takeuchi 2010). However, currently Tokyo Metropolitan Planning Division (1) recognizes the importance of planning hierarchy from national to broad regional, prefectural, and to ward, city, town, and village levels, (2) attempts to coordinate green space planning among these different levels, (3) plans the protection of riparian vegetation and corridor, and (4) incorporates the concept of adaptive planning (Kato and Ahern 2008) in its new privately owned green space protection policy (Kato and Takeuchi 2010).

16.5.2.2 Kohoku New Town in Yokohama

Kohoku New Town in Yokohama City, planned in the late 1960s, is an example of one of the first applications of the GI concept as defined in this chapter. It realized a connected, integrated network of open spaces, including diverse open spaces, pedestrian paths, and water systems (Miyagi 2010). It developed a linked system of parks as a backbone (coarser landscape element), further connecting to other green spaces in the new town (Miyagi 2010). Also, it created a network of pedestrian paths (finer landscape elements), connecting to the backbone, so that the benefits of the parks can be shared by and accessible to the residents (Miyagi 2010). Both the protective and opportunistic strategies (Ahern 1995) were used to develop the backbone by recognizing the distribution of green spaces to be protected in the valley slopes and intentionally planning most open spaces in the upper slopes (Miyagi 2010). Kohoku New Town's characteristics are that: (1) it valued open spaces equal to conventional infrastructure; (2) it used GI to shape development and conservation; and (3) it developed a layered, not hierarchical, system of linked parks and pedestrian paths (Miyagi 2010).

16.5.2.3 Green Parking Space Examples in Nagoya Metro Region

Six cases of green parking space in Nagoya metro region also incorporate the concept of adaptive planning by testing various greening technologies on the sites (Ito 2011). The significance of green parking space is not only the increase in the percentage of green cover in the metro region but also the creation of a "wind path" (Ito 2011). Nagoya City plans to make use of Hori River as a wind path to bring in cool breeze from the ocean into the city. The green parking spaces in the nearby area are expected to bring in the wind even further into the city by connecting the river and green spaces (Ito 2011). The effect of each green parking space for microclimate mitigation and other functions may be small but its cumulative effect can be significant (Ito 2011).

16.5.2.4 Inochi-no-Mori in Kyoto

Morimoto and Tabata (2010) document a case of small (0.6 ha) biotope creation in the middle of Kyoto City. The target vegetation and species composition were those of floodplain forests and wetlands, formally existed in the area. Monitoring of the site has continued for 15 years since its development. Adaptive management has been applied to the management of the biotope since its development. The managing body sets the management goal, keeps monitoring, and adaptively changes the management policies based on the monitoring result. The authors suggest that even a small created green space in the middle of a city can greatly contribute to increasing biodiversity and supporting ecosystem services by proper design, planning, and management. Urban green spaces need to be considered along with various economic activities toward the goal of increasing biodiversity and providing ecosystem services (Morimoto and Tabata 2010).

16.5.2.5 Ecological Networks

Ichinose (2010) recommends ecological network planning for biodiversity conservation in the city. The premise is that green spaces in the city can be spatially configured to develop ecological networks that function as habitat for plants and animals although we lack empirical studies to strongly link the spatial configuration of green spaces to the movement of organisms and increase in local biodiversity of certain species. Moreover, there is no established method to develop ecological networks (Ichinose 2010). While we wait for more empirical studies, we must start planning ecological networks based on what we know right now and "learn by doing" by using adaptive planning (Ichinose 2010; Kato and Ahern 2008).

Ichinose (2010) also points out the importance of scattered small green spaces in the city for improving the habitat quality of the urban matrix, which plays an important role in biodiversity conservation in the city. Although it is difficult to newly develop large green spaces or wide corridors in the city, it is possible to increase vegetation cover and improve the habitat quality of existing green spaces in the urban matrix (Ichinose 2010). For example, biotope development in city parks and schoolyards and even green parking spaces (Ito 2011) can contribute to increasing green cover in the city and improving the overall habitat quality of the urban matrix (Ichinose 2010).

16.6 Guidelines for Planning and Designing Green Infrastructure

The planning and design guidelines presented below are not "how to" design formula but "big ideas," for each landscape plan or design is unique: it is for a specific place and a particular set of issues and landscape changes. If planned and designed based on the following guidelines, GI is more likely to provide its multitude of ecological, economic, and social benefits as previously discussed.

16.6.1 Articulation of a Spatial Concept

Many Dutch planners argue for spatial concepts to interpret and apply basic spatial solutions to real places. Spatial concepts convey the essence of a plan or strategy in simple terms. Spatial concepts are often used in the framework of developing a landscape plan to express its overall goal or vision in the form of conceptual metaphors (van Lier 1998; Ahern 1999; Leitão et al. 2006). For example, "Green Heart" denotes a central protected green space (formerly a peatland) around which major infrastructure lines and urban settlements lie in the Randstad, The Netherlands (Schrijnen 2000). When implemented in landscape plans, spatial concepts can test landscape ecological theories and generate new knowledge (Ahern 1999). Since there are many possible spatial configurations to realize a spatial concept, these different spatial arrangements, each with its own hypothesis, can be compared and contrasted to select the best plan.

Another important spatial concept in sustainable planning is the "Casco" or Framework Concept in which parts of a landscape are designated as "high dynamic" and "low dynamic" areas (van Buuren and Kerkstra 1993). "High dynamic" areas undergo rapid changes or allow faster changes (e.g., urban development, intensive agriculture, active recreational uses) and thus, they are meant to be modified and accommodate the changing demands of people. Land modifying changes (e.g., force of water and wind operating on the landscape) occur slowly—thus, the naming—in "low dynamic" areas. "Low dynamic" areas include environmentally fragile areas (e.g., water recharge and discharge areas, flood plains, steep slopes) in need of protection. The significance of the concept is that "low dynamic" areas define a durable and persistent framework that may endure changes while acknowledging that the surrounding landscape will (and should) change—a combination of a durable frame with a dynamic context.

16.6.2 Strategic Thinking

Ahern (1995) put forward four planning strategies based on the existing landscape conditions on the trajectory of change, based on the assumption that landscapes keep changing. The four planning strategies are protective, defensive, offensive, and opportunistic strategies. When the existing landscape supports sustainable processes and patterns, a protective strategy may be employed (Ahern 1995). Essentially, this strategy defines an eventual or optimal landscape pattern that is proactively protected from change while the landscape around it may be allowed to change. When the existing landscape is already fragmented and core areas are already limited in area and isolated, a defensive strategy can be applied (Ahern 1995). This strategy seeks to arrest/control the negative processes of fragmentation or urbanization. An offensive strategy is based on a vision or a possible landscape configuration that is articulated, understood, and accepted as a goal. The offensive strategy differs from protective and defensive strategies in that it employs restoration or reconstruction to rebuild landscape elements in previously disturbed or fragmented landscapes. The opportunistic strategy is conceptually aligned with the concept of GI by seeking new or innovative "opportunities" to provide abiotic–biotic–cultural functions (Ndubisi 2002) in association with urban infrastructure. The strength of the typology of planning strategies is that it is flexible enough to be adaptable to a range of landscapes in various landscape contexts (Ahern 2002). The four principle strategies can work either individually or in combination (Ahern 1995).

16.6.3 Greening of Conventional Built Infrastructure

To achieve sustainability in urban landscapes, conventional (gray) infrastructure must be conceived of and understood as a genuinely possible means to improve and contribute to sustainability. For example, streets can incorporate street trees for air purification and microclimate remediation (Jim and Chen 2003) and open drainage to retain and purify water on site. If one only thinks about avoiding or minimizing impact related to infrastructure development, the possibility to innovate is greatly diminished (Ahern 2010). Existing and future infrastructure needs to be reconceived as opportunities to (re)greening the urban environment.

16.6.4 Adaptive Planning and Learning by Doing

Adaptive management has a tremendous potential to be applied in landscape planning. Adaptive management is a "management approach to embrace uncertainty and manage adaptively" (Light et al. 1995, p. 154). Although adaptive management has been widely practiced in natural resource and ecosystem management since the late 1970s (Walters and Holling 1990), it has not yet been widely integrated into or applied to landscape planning (Kato and Ahern 2008).

Every time a new plan is developed, planners face a unique situation. The inherent uniqueness in any "real-world" planning project lowers the likelihood that adequate data exists to support a scientifically defensible decision. This is the common circumstance that defines planning and that planners must face routinely.

Also, planning is a time-sensitive activity. Landscape planners often do not have the luxury to wait for all the scientific data to accumulate to support planning decisions. Landscape planning addresses heterogeneous and dynamic landscapes—a moving target—by definition. Therefore, landscape planning tries to place itself ahead of these processes and to "steer" or influence them in a proactive, anticipatory way. It can be said that landscape planning is inherently prescriptive, while science is often more descriptive. The imperative to act to meet political expectations and deadlines has hindered planning that takes a long time to initiate and implement and that requires monitoring the status/results long after the plan is complete. This is arguably the opposite of what adaptive planning requires.

In adaptive planning, each plan or design can be treated as experiments (Felson and Pickett 2005) and a large plan can be divided into several small plans that can safely fail (Lister 2007) to learn by doing, employing precautions, monitoring, and best practice (Kato and Ahern 2008). "Learning by doing" presupposes that something "uncertain" needs to be learned. Based on the evaluation of monitoring results, new and existing plans can be adapted to the lessons learned. The adaptive approach is promising for GI because the knowledge to plan and implement these systems is evolving. If experimental applications can be practiced routinely, the potential to build empirical knowledge while exploring sustainability is quite profound.

16.7 Specific Recommendations for GI Application in Asian Cities

GI is a useful green space planning tool to address nature conservation problems in Asian cities. GI can protect important green spaces, restore nature, and mitigate the lost nature by creating new green spaces in the city. Based on the lessons we learned from the earlier case studies, I propose four key ideas for the systematic application of GI concept toward solving the problems of Asian urban planning and those specific to green and open spaces. The recommendations are summarized in the conceptual representation (Fig. 16.1).

16.7.1 Waters' Edge and Watershed Planning

Since many Asian cities are located at the waters' edge, the principle of connection between the waters' edge, coastal vegetation, riparian vegetation, and valley and ridgelines as part of an interconnected system of GI is important and applicable to Asian cities. Given the expected sea level rise by climate change, future design and planning of cities at the waters' edge would need to incorporate green space for climate change adaptation and need to plan for increased resilience to natural disasters such as floods, typhoons, and tsunami.

Watershed can become a basic unit of designing and planning GI. Regional land-use planning, for example, the former Tokyo metro regional green space plan, is

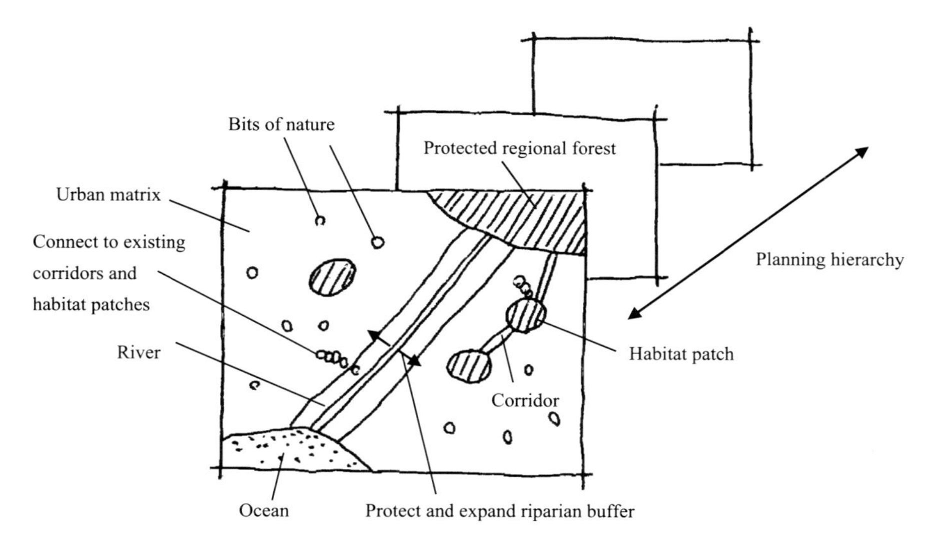

Fig. 16.1 Simplified graphic representation of the specific recommendations for GI application in Asian cities

based on watersheds (Ishikawa 2009). Watershed is inherently multi-scale: from sub-watersheds to cities, regions, and to the nation (Ishikawa 2009). The watershed scale, not administrative boundaries, is aligned with particular ecological processes (i.e., water flow and cycle). Therefore, the developed plan (i.e., the pattern on a map) has a closer connection to ecological processes than if the plan were developed bounded by administrative boundaries that are usually irrelevant to ecological processes.

16.7.2 Bits of Nature in the City

Representative urban green space planning issues in Asian cities include dense urban population and consequent small green (open) space per capita. Urban heat island effect is also an increasing environmental problem in the city. The problem is how to effectively green cities where open spaces are often scattered, small, and limited. Rooftop gardens, green walls, street trees, constructed biotopes, and green parking spaces are effective ways to increase greenery in the city, making use of limited surface area. These scattered green spaces, "bits of nature," even if they are not connected, can increase the overall habitat quality of the urban matrix (Ichinose 2010). Moreover, although the effect of each green space may be small for remediating the urban heat island effect and for providing other social and ecological services (e.g., purifying air and water, providing some habitat for plants and animals, and increasing esthetics), the cumulative effect of all the increased greenery can be significant.

16.7.3 Connect Habitat Patches and Bits of Nature to Existing Corridors

Green spaces should be connected to access paths (roads) so that urban residents can receive their benefits. This is the issue of accessibility. When species conservation is the main planning goal, green spaces as habitat patches can be connected by corridors to form ecological networks (Ichinose 2010). Ecological networks are included in the concept of GI. Both emphasize connectivity among green spaces. Ecological networks focus on creating a connected network of habitat for flora and fauna. GI is a broader concept, considering all types of open and green spaces and waterways, both natural and designed, and the benefits are more inclusive and multifaceted.

The "bits of nature" described above can be connected to river corridors as "wind path" (Ito 2011) and riparian vegetation. By this connection, both benefits can be shared. As mentioned, the cumulative effect of small green spaces can be significant. When they are planned to connect to existing corridors such as rivers and ridgelines, the effect of the corridors can be brought further into the city. Also, bits of nature, if carefully spatially planned and created, can act as "stepping stones" to facilitate the movement of certain organisms. In this way, bits of nature can become a part of functional ecological network.

16.7.4 Multi-Scale Approach

As illustrated in the important landscape ecology principles, a multi-scale approach is key to developing GI in Asian cities. An interconnected network of green spaces needs to be created across scales from neighborhoods to cities and to regions. For example, rooftop gardens, rain barrels, street bioswales, city and regional parks, constructed wetlands, and preserved regional forests can form a drainage network. Cross-scale networks of green spaces increase resilience with increased response and functional diversity (Kato 2010).

16.7.5 Summary and Conclusion

More specific recommendations to facilitate further GI application in Asian cities have been provided. To develop a systematic and strategic connection of green spaces (i.e., GI), we need to recognize hierarchical planning levels and coordinate green space planning across these different levels. Watersheds, which are inherently hierarchical, are recommended as a basic planning unit for GI and regional planning. GI can shape urban form and provide a framework for growth. If a GI is proactively planned, developed, and maintained, it has the potential to guide urban development by providing a framework for economic growth and nature conservation (Schrijnen 2000). Adaptive planning is a suited planning tool to test various GI techniques and further develop its concept.

Green spaces created as part of GI serve to increase the percentage of green cover in the city. Created and restored green spaces (e.g., biotope development in city parks and schoolyards, green parking spaces, and rooftop gardens), even if they are small and scattered, can contribute to improving the overall habitat quality of the urban matrix. Even though the effect of each green space may be small, the cumulative effect of each green space can be significant to provide social and ecological services. These scattered green spaces, when strategically and proactively planned to connect to each other and to existing corridors such as rivers and ridgelines, can reinforce their functions. For example, green spaces connected to a river corridor can bring cooler air further into the city to remediate the urban heat island effect. Connected habitat patches by corridors can facilitate the movement and dispersal of organisms and contribute to biodiversity conservation in the city. Finally, although I have focused on the ecological functions of GI in this chapter,

GI has the potential to include light infrastructure such as pedestrian paths and light rails in the network to provide more social and economic functions.

16.8 Conclusions

Although GI planning is gaining popularity in North America, the UK, and Europe, and some individual cases of application have been observed in Asian cities, GI is arguably lacking in Asian cities and urban planning policies despite its multitude of benefits. To facilitate further application of the GI concept in Asian cities, this chapter has succinctly reviewed the concept of GI and the functions it can provide, stated green space planning problems in Japanese cities, laid out key principles of landscape ecology for GI, reviewed innovative GI application in New York City, US and five green space planning and design cases in Japan, suggested broad guidelines for planning and designing GI, and made more specific recommendations for GI application in Asian cities. GI benefits are valued more in urban and suburban areas where green space is limited and natural environment is highly altered. In cities, GI can become a part of the means to control climate change along with the sustainable design of housing and larger scale infrastructure development. GI plans apply key principles of landscape ecology to urban environments, specifically: a multi-scale approach with explicit attention to pattern-process relationships, and an emphasis on connectivity.

GI needs to be strategically planned and designed with broad guidelines and landscape ecology principles in mind. A systems approach to planning GI as an integrated whole is needed (McMahon 2000; Mell 2008). A long-term thinking is also necessary to include GI as part of planning sustainable landscapes (Kambites and Owen 2006; Mell 2008). A strategic systems approach is suggested to ensure the functions of GI to be properly understood (Gill et al. 2007; Mell 2008).

Moreover, an adaptive management approach (Gunderson et al. 2008) could be tested in a planning process for GI. In an adaptive approach to planning, plans are made with the best knowledge available, but with explicit acknowledgment of uncertainty, followed by monitoring and re-evaluation of plans in order to close the loop and to "learn by doing" (Kato and Ahern 2008; Light et al. 1995). Adaptive planning is appropriate for testing an emerging landscape planning concept such as GI.

GI, based on its precedents, links parks and natural areas for human benefits and counters habitat fragmentation (Benedict and McMahon 2002). GI provides multiple functions for human benefits. GI engages key partners and involves diverse stakeholders (Benedict and McMahon 2002; Kambites and Owen 2006). Therefore, a process of developing GI is a perfect testing ground for a transdisciplinary approach. GI provides a framework for both nature conservation and urban development (Austin 2014). Future GI research needs to address the concept and planning of multifunctional GI to promote ecosystem services in the city. Also, more research is needed on quantifying the economic value of GI's benefits. A systematic application of the GI concept to Asian cities is in its beginning and we need to accumulate applied pilot cases to document their benefits.

Acknowledgement The content of the chapter is a modification of the following peer-reviewed publication: Kato, S. (2011) Green Infrastructure for Asian Cities: The Spatial Concepts and Planning Strategies, *Journal of the 2011 International Symposium on City Planning*: 161–170. The author would like to acknowledge academic guidance by Jack Ahern, University of Massachusetts Amherst, USA, and stimulus discussion with: Takanori Fukuoka, Kobe University, Japan; Isaac Brown, Coastal Management Resources, USA; and the GI Study Group organized by Yuki Iwasa, the Ministry of Land, Infrastructure, Transport and Tourism, Japan.

References

Abunnasr Y, Hamin EM (2012) The green infrastructure transect: an organizational framework for mainstreaming adaptation planning policies. In: Resilient cities 2 (K. Otto-Zimmermann, ed.), local sustainability 2. Springer, New York, pp 205–217

Ahern J (1995) Greenways as a planning strategy. Landsc Urban Plan 33(1–3):131–155

Ahern J (1999) Spatial concepts, planning strategies, and future scenarios: a framework method for integrating landscape ecology and landscape planning. In: Klopatek JM, Gardner RH (eds) Landscape ecological analysis: issues and applications. Springer, New York, pp 175–201

Ahern JF (2002) Greenways as strategic landscape planning: theory and application. Ph.D. Dissertation, Wageningen University, The Netherlands

Ahern J (2010) Planning and design for sustainable and resilient cities: theories, strategies, and best practices for green infrastructure. In: Novotny V, Ahern J, Brown P (eds) Water-centric sustainable communities. Wiley, Hoboken, pp 135–176

Allen TFH, Starr TB (1982) Hierarchy: perspectives for ecological complexity. University of Chicago Press, Chicago

Austin G (2014) Green infrastructure for landscape planning: integrating human and natural systems. Routledge, Abingdon, Oxon

Benedict MA, McMahon ET (2002) Green infrastructure: smart conservation for the 21st century. Renew Resour J 20(3):12–17

Benedict MA, McMahon ET (2006) Green infrastructure: linking landscape and communities. Island Press, Washington, D.C.

Biohabitats (2014a) New York projects—Jamaica bay watershed protection plan. http://www.biohabitats.com/wp-content/uploads/JamaicaBayWatershedProtectionPlan.pdf. Accessed 18 Oct 2014

Biohabitats (2014b) New York City CSO-PlaNYC green infrastructure initiatives—BMP and ecological pilot projects. http://www.biohabitats.com/wp-content/uploads/CSO_PLANYC_2pages1.pdf. Accessed 18 Oct 2014

Biohabitats (2014c) New York City CSO-PlaNYC green infrastructure initiatives—ecological pilot projects. http://www.biohabitats.com/projects/jamaica-bay-ecosystem-restoration-pilots/. Accessed 24 Oct 2014

Biohabitats (2014d) New York City CSO-PlaNYC green infrastructure initiatives—ecological pilot projects. http://www.biohabitats.com/wp-content/uploads/JamaicaBayEcosystemRestorationPilots_expanded_four8.5x113.pdf. Accessed 24 Oct 2014

Biohabitats (2014e) New York City CSO-PlaNYC green infrastructure initiatives—BMP pilots. http://www.biohabitats.com/wp-content/uploads/NYCStormwaterPilots2.pdf. Accessed 18 Oct 2014

Business Wire (2014) Good news for the earth: biohabitats acquires natural systems international. http://www.businesswire.com/news/home/20110216005203/en/Good-News-Earth-Biohabitats-Acquires-Natural-Systems#.VDyLaBY0_Lk. Accessed 14 Oct 2014
CityCraft Ventures (2014) Keith bowers. http://www.citycraftventures.com/the-team/keith-bowers/. Accessed 14 Oct 2014
Endo A (2011) A study of the green infrastructure planning as stormwater management policy in American cities: in case of the long term stormwater control plan in City of Philadelphia, PA. J City Plan Inst Jpn 46(3):649–654 (in Japanese)
Felson AJ, Pickett STA (2005) Designed experiments: new approaches to studying urban ecosystems. Front Ecol Environ 3:549–556
Forman RTT (1995) Land mosaics: the ecology of landscapes and regions. Cambridge University Press, Cambridge
Gill SE, Handley JF, Ennos AR, Pauleit S (2007) Adapting cities for climate change: the role of the green infrastructure. Built Environ 33(1):115–133
Gunderson LH, Holling CS (eds) (2002) Panarchy: understanding transformations in human and natural systems. Island Press, Washington, D.C.
Gunderson L, Peterson G, Holling CS (2008) Practicing adaptive management in complex social-ecological systems. In: Norberg J, Cumming GS (eds) Complexity theory for a sustainable future. Columbia University Press, New York, pp 223–245
Ichinose T (2010) Ecological network planning in Japanese cities. City Plan Rev 59(5):38–41 (in Japanese)
Inoue K, Sugimoto M, Shimizu H, Onishi A, Murayama A, Otsuki A (2011) Effect of perviousness oriented streets design in Nagoya city: applying the concept of Green Infrastructure, Transactions of AIJ. J Archit Plan 76(660):335–340 (in Japanese)
Ishikawa M (2009) Creating new horizon of city and environmental planning through watershed planning. City Plan Rev 58(3):17–20 (in Japanese)
Ishikawa M (2011) Green infra-structure as social common capital. Civ Eng 66(10):10–15 (in Japanese)
Ito T (2011) Urban Environmental Design for Green Parking. City Plan Rev 60(1):53–56 (in Japanese)
Jim CY, Chen SS (2003) Comprehensive greenspace planning based on landscape ecology principles in compact Nanjing city, China. Landsc Urban Plan 65(3):95–116
Kambites C, Owen S (2006) Renewed prospects for green infrastructure planning in the UK. Plan Pract Res 21(4):483–496
Kato S (2010) Greenspace conservation planning framework for urban regions based on a forest bird-habitat relationship study and the resilience thinking. Ph.D. Dissertation, University of Massachusetts Amherst, Massachusetts. http://scholarworks.umass.edu/open_access_dissertations/212
Kato S (2011) Green infrastructure for Asian cities: the spatial concepts and planning strategies. In: Journal of the 2011 international symposium on city planning. Korea Planners Association, Seoul, pp 161–170
Kato S (2012) An overview of green infrastructure's contribution to climate change adaptation. In: Proceedings of the 13th international symposium of landscape architecture, Korea, China, and Japan. The Korean Institute of Landscape Architecture, Seoul, pp 224–228
Kato S, Ahern J (2008) 'Learning by doing': adaptive planning as a strategy to address uncertainty in planning. J Environ Plan Manag 51(4):543–559
Kato N, Takeuchi T (2010) Midori kakuho no sougouteki na houshinn ni tsuite. City Plan Rev 59 (3):78–79 (in Japanese)
Kawakami M (2008) Toshikeikaku. Morikita Shuppan, Tokyo, 156 pp
Kyushu Chapter of the City Planning Institute of Japan (1999) Ajia no toshikeikaku. Kyushu University Press, Fukuoka, 165 pp
Leitão AB, Miller J, Ahern J, McGarigal K (2006) Measuring landscapes: a planner's handbook. Island Press, Washington, DC

Light SS, Gunderson LH, Holling CS (1995) The everglades: evolution of management in a turbulent ecosystem. In: Gunderson LH, Holling CS, Light SS (eds) Barriers and bridges to the renewal of ecosystems and institutions. Columbia University Press, New York, pp 103–168

Lister N-M (2007) Sustainable large parks: ecological design or designer ecology? In: Czerniak J, Hargreaves G (eds) Large parks. Princeton Architectural Press, New York, pp 35–57

McMahon ET (2000) Green infrastructure. Planning Commissioners Journal 37:4–7

Mell IC (2008) Green infrastructure: concepts and planning. FORUM Ejournal 8:69–80

Millennium Ecosystem Assessment (2005) Ecosystems and human well-being: synthesis. Island Press, Washington, D.C.

Miyagi S (2010) Invisible structure of natural environment underlies the suburban new towns envisioned and developed through 1960s. City Plan Rev 59(2):30–33 (in Japanese)

Morimoto Y, Tabata K (2010) Biodiversity oriented urban greenery and its management. City Plan Rev 59(5):13–17 (in Japanese)

Ndubisi F (2002) Ecological planning: a historical and comparative synthesis. The Johns Hopkins University Press, Baltimore

O'Neill RV, DeAngelis D, Waide J, Allen TFH (1986) A hierarchical concept of ecosystems. Princeton University Press, Princeton, NJ

Schrijnen PM (2000) Infrastructure networks and red-green patterns in city regions. Landsc Urban Plan 48(3–4):191–204

Takeuchi T (2010) The legacy of the green belt policy in Tokyo. City Plan Rev 59(2):58–62 (in Japanese)

Turner MG, Gardner RH, O'Neill RV (2001) Landscape in theory and practice: pattern and process. Springer, New York

Tzoulas K, Korpela K, Venn S, Yli-Pelkonen V, Kaźmierczak A, Niemela J, James P (2007) Promoting ecosystem and human health in urban areas using green infrastructure: a literature review. Landsc Urban Plan 81(3):167–178

van Buuren M, Kerkstra K (1993) The framework concept and the hydrological landscape structure: a new perspective in the design of multifunctional landscapes. In: Vos CC, Opdam P (eds) Landscape ecology of a stressed environment. Chapman & Hall, London, pp 219–243

van Lier HN (1998) The role of land use planning in sustainable rural systems. Landsc Urban Plan 41(2):83–91

Walters CJ, Holling CS (1990) Large-scale management experiments and learning by doing. Ecology 71(6):2060–2068

Index

K. Ito (ed.), *Urban Biodiversity and Ecological Design for Sustainable Cities*,
https://doi.org/10.1007/978-4-431-56856-8

Printed in the United States
by Baker & Taylor Publisher Services